AF566954

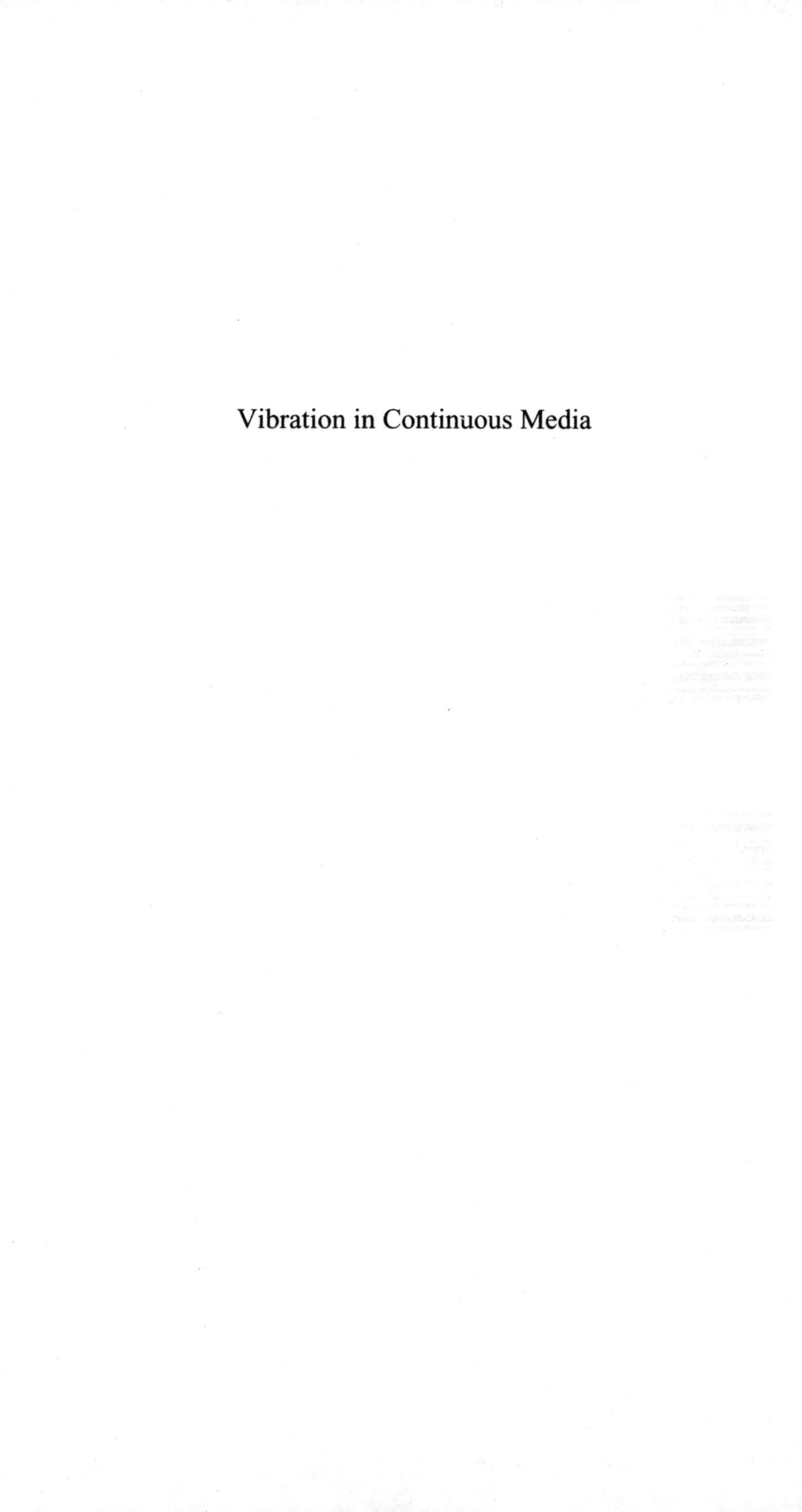

Vibration in Continuous Media

Vibration in Continuous Media

Jean-Louis Guyader

Series Editors
Société Française d'Acoustique

First published in France by Hermes Science/Lavoisier entitled "Vibrations des milieux continus"
First published in Great Britain and the United States in 2006 by ISTE Ltd

ISTE Ltd
6 Fitzroy Square
London W1T 5DX
UK
www.iste.co.uk

ISTE USA
4308 Patrice Road
Newport Beach, CA 92663
USA

First South Asian Edition 2007

Library of Congress Cataloging-in-Publication Data

Guyader, Jean-Louis.
[Vibrations des milieux continus, English]
Vibration in continuous media/Jean-Louis Guyader.
p. cm.
Includes bibliographical references and index.
ISBN 13: 978-1-905209-27-9
ISBN 10: 1-905209-27-4
1. Vibration. 2. Continuum mechanics. I. Title.
QA935.G8813 2006

2006015557

British Library Cataloguing-in-Publication Data
A CIP record for this book is available from the British Library
ISBN 10: 1-905209-27-4
ISBN 13: 978-1-905209-27-9

Printed in Brijbasi Art Press Ltd., I-72, Sector-9, Noida, U.P. India.

Table of Contents

Preface

This book, which deals with vibration in continuous media, originated from the material of lectures given to engineering students of the National Institute of Applied Sciences in Lyon and to students preparing for their Master's degree in acoustics.

The book is addressed to students of mechanical and acoustic formations (engineering students or academics), PhD students and engineers wanting to specialize in the area of dynamic vibrations and, more specifically, towards medium and high frequency problems that are of interest in structural acoustics. Thus, the modal expansion technique used for solving medium frequency problems and the wave decomposition approach that provide solutions at high frequency are presented.

The aim of this work is to facilitate the comprehension of the physical phenomena and prediction methods; moreover, it offers a synthesis of the reference results on the vibrations of beams and plates. We are going to develop three aspects: the derivation of simplified models like beams and plates, the description of the phenomena and the calculation methods for solving vibration problems. An important aim of the book is to help the reader understand the limits hidden behind every simplified model. In order to do that, we propose simple examples comparing different simplified models of the same physical problem (for example, in the study of the transverse vibrations of beams).

The first few chapters are devoted to the general presentation of continuous media vibration and energy method for building simplified models. The vibrations of continuous three-dimensional media are presented in Chapter 1 and the equations which describe their behavior are established thanks to the conservation laws which govern the mechanical media. Chapter 2 presents the problem in terms of variational formulation. This approach is fundamental in order to obtain, in a systematic way, the equations of the simplified models (also called condensed media), such as beams, plates or shells. These simplified continuous media are often sufficient models to describe the vibrational behavior of the objects encountered in practice However, their importance is also linked to the richness of the information which is accessible thanks to the analytical solutions of the equations which characterize them. Nevertheless, since these models are obtained through *a priori* restriction of possible three-dimensional movements and stresses, it is necessary to master the underlying hypothesis well, in order to use them advisedly. The aim of Chapters 3 and 4 is to provide these hypotheses in the case of beams and plates. The derivation of equations is done thanks to the variational formulations based on Reissner and Hamilton's functionals. The latter is the one which is traditionally used, but we have largely employed the former, as the limits of the simplified models obtained in this way are established more easily. This approach is given comprehensive coverage in this book, unlike others books on vibrations, which dedicate very little space to the establishment of simplified models of elastic solids.

Chapters 5, 6 and 7 deal with the different aspects of the behavior of beams and plates in free vibrations. The vibrations modes and the modal decomposition of the response to initial conditions are described, together with the wave approach and the definition of image source linked to the reflections on the limits. We must also insist on the influence of the "secondary effects", such as shearing, in the problems of bending plates. From a general point of view, the discussion of the phenomena is done on two levels: that of the mechanic in terms of modes and that of the acoustician in terms of wave's propagation. The notions of phase speed and group velocity will also be exposed.

We will provide the main analytical results of the vibrations modes of the beams and rectangular or circular plates. For the rectangular plates, even quite simple boundary conditions often do not allow analytical calculations. In this case, we will describe the edge effect method which gives a good approximation for high order modes.

Chapter 8 is dedicated to the introduction of damping. We are going to show that the localized source of damping results in the notion of complex modes and in a difficulty of resolution which is much greater than the one encountered in the case of distributed damping, where the traditional notion of vibrations modes still remains.

The calculation of the forced vibratory response is at the center of two chapters. We will start by discussing the modal decomposition of the response (Chapter 9), where we are going to introduce the classical notions of generalized mass, stiffness and force. Then we will continue with the decomposition in forced waves (Chapter 10) which offers an alternative to the previous method and is very effective for the resolution of beam problems.

For the modal decomposition, the response calculations are conducted in the frequency domain and time domain. The same instances are treated in a manner which aims to highlight the specificities of these two calculation techniques. Finally we will study the convergence of modal series and the way to accelerate it.

In the case of forced wave decomposition, we will show how to treat the case of distributed and non-harmonic excitations, starting from the solution for a localized, harmonic excitation. This will lead us to the notion of integral equation and its key idea: using the solution of a simple case to treat a complicated one.

Chapters 11 and 12 deal with the problem of approximating the solutions of vibration problems, using the Rayleigh-Ritz method. This method employs directly the variational equations of the problems. The classical approach, based on Hamilton's functional, is used and the convergence of the solutions studied is illustrated through some examples. The Rayleigh-Ritz quotient – which stems directly from this approach – is also introduced.

A second approach is proposed, based on the Reissner's functional. This is a method which has not been at the center of accounts in books on vibrations; however, it presents certain advantages, which will be discussed in some examples.

Chapter 1

Vibrations of Continuous Elastic Solid Media

1.1. Objective of the chapter

This work is addressed to students with a certain grasp of continuous media mechanics, in particular, of the theory of elasticity. Nevertheless, it seems useful to recall in this chapter the essential points of these domains and to emphasize in particular the most interesting aspects in relation to the discussion that follows.

After a brief description of the movements of the continuous media, the laws of conservation of mass, momentum and energy are given in integral and differential form. We are thus led to the basic relations describing the movements of continuous media.

The case of small movements of continuous elastic solid media around a point of static stable equilibrium is then considered; we will obtain, by linearization, the equations of vibrations of elastic solids which will be of interest to us in the continuation of this work.

At the end of the chapter, a brief exposition of the equations of linear vibrations of viscoelastic solids is outlined. The equations in the temporal domain are given as well as those in the frequency domain, which are obtained by Fourier transformation. We then note a formal analogy of elastic solids equations with those of the viscoelastic solids, known as principle of correspondence.

Generally, the presentation of these reminders will be brief; the reader will find more detailed presentations in the references provided at the end of the book.

1.2. Equations of motion and boundary conditions of continuous media

1.2.1. *Description of the movement of continuous media*

To observe the movement of the continuous medium, we introduce a Galilean reference mark, defined by an origin O and an orthonormal base $\vec{e}_1, \vec{e}_2, \vec{e}_3$. In this reference frame, a point M, at a fixed moment T, has the co-ordinates (x_1, x_2, x_3).

The Euler description of movement is carried out on the basis of the four variables (x_1, x_2, x_3, t); the Euler unknowns are the three components of the speed $\vec{U}$ of the particle which is at the point M at the moment t.

$$\vec{U} = U_i(x_1, x_2, x_3, t) \qquad [1.1]$$

Derivation with respect to time of quantities expressed with Euler variables is particular; it must take into account the variation with time of the co-ordinates x_i of the point M.

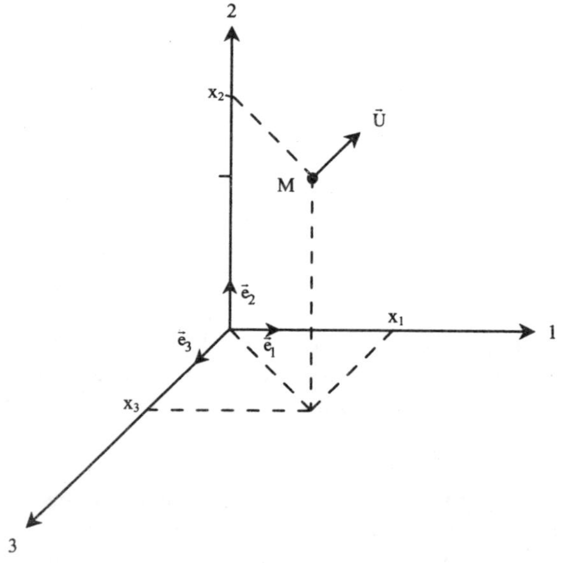

Figure 1.1. *Location of the continuous medium*

For example, for each acceleration component γ_i of the particle located at the point M, we obtain by using the chain rule of derivation:

$$\gamma_i = \frac{dU_i}{dt} = \frac{\partial U_i}{\partial t} + \sum_{j=1}^{3} \frac{\partial U_i}{\partial x_j} \frac{\partial x_j}{\partial t},$$

and noting that:

$$U_j = \frac{\partial x_j}{\partial t},$$

we obtain the expression of the acceleration as the total derivative of the velocity:

$$\gamma_i = \frac{dU_i}{dt} = \frac{\partial U_i}{\partial t} + \sum_{j=1}^{3} \frac{\partial U_i}{\partial x_j} U_j ;$$

or in index notation:

$$\gamma_i = \frac{dU_i}{dt} = \frac{\partial U_i}{\partial t} + U_{i,j} U_j . \qquad [1.2]$$

In the continuation of this work we shall make constant use of the index notation, which provides the results in a compact form. We shall briefly point out the equivalences in the traditional notation:

– partial derivation is noted by a comma:

$$\frac{\partial U_i}{\partial x_j} = U_{i,j} ;$$

– an index repeated in a monomial indicates a summation:

$$\sum_{j=1}^{3} U_{i,j} U_j = U_{i,j} U_j .$$

The Lagrangian description is an alternative to the Euler description of the movement of continuous media. It consists of introducing Lagrange variables

(a_1, a_2, a_3, t), where (a_1, a_2, a_3) are the co-ordinates of the point where the particle is located at the moment of reference t_0. The Lagrange unknowns are the co-ordinates x_i of the point M where the particle is located at the moment t:

$$x_i = \phi_i (a_1, a_2, a_3, t). \quad [1.3]$$

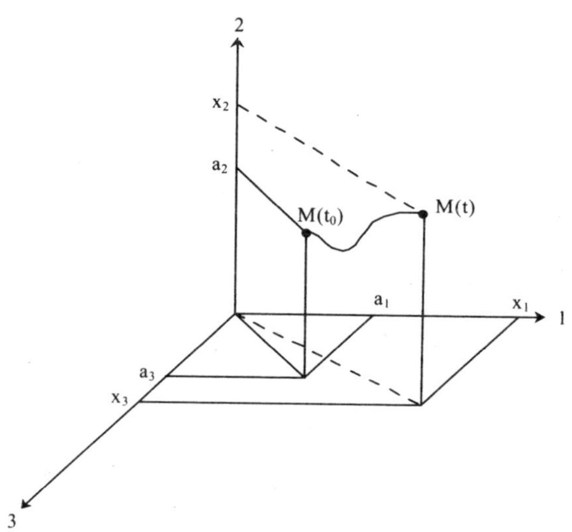

Figure 1.2. *Initial a_i and instantaneous x_i co-ordinates*

a_j being independent of time, the speed or the acceleration of the particle M with co-ordinates x_i is deduced from it by partial derivation:

$$U_i (a_j, t) = \frac{\partial \phi_i}{\partial t}(a_j, t) \qquad \gamma_i (a_j, t) = \frac{\partial^2}{\partial t^2}\phi_i (a_j, t). \quad [1.4]$$

The Lagrangian description is direct: it identifies the particle; the Euler description is indirect: it uses variables with instantaneous significance, which eventually proves to be interesting for the motion study of continuous media; it is the reason for the frequent use of Euler's description. The two descriptions are, of course, equivalent; the demonstration thereof can be found in the titles on the mechanics of continuous media provided in the references section.

1.2.2. *Law of conservation*

Laws that govern the evolution of continuous media over time are the laws of conservation: conservation of mass, conservation of momentum and conservation of energy. These laws can be expressed in an integral form [1.5] or in a differential form [1.6] with the boundary condition [1.7].

The general form of the conservation law is provided in this section; it will be detailed in the next sections with the conservation of mass, momentum and energy.

Let us consider a part D of the continuous medium whose movement is being observed. Let us also introduce its boundary $\overline{D}$ and n_j the direction cosines of the exterior normal $\vec{n}$, which is supposed to exist in all the points of $\overline{D}$. V is the volume of the continuous medium and $\overline{V}$ is the surface delimiting it. These quantities are defined in Figure 1.3.

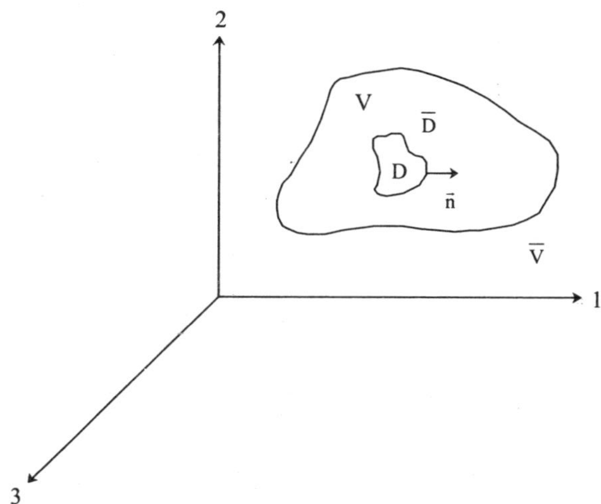

Figure 1.3. *Continuous medium* V *with boundary* $\overline{V}$ *and part* D *with boundary* $\overline{D}$

The integral form of a conservation equation, in a very general case, is given by the following equation:

$$\frac{d}{dt}\int_D A_i + \int_{\overline{D}} \alpha_{ij}\, n_j = \int_D B_i \, . \qquad [1.5]$$

$\frac{d}{dt}$ indicates the total derivative, i.e. the derivative with respect to time when the derived quantity is followed in its movement. A_i and B_i are vector quantities, in the general case of dimension 3, but may also be scalar values, in the particular case of dimension 1.

From a physical point of view:

$\frac{d}{dt}\int_D A_i$ represents the fluctuation over time of a physical value, attached to the part D of the continuous medium, whose movement is being followed.

$\int_D \alpha_{ij} n_j$ represents the action of the exterior surface on D.

$\int_D B_i$ represents the action of the exterior volume on D.

The law of conservation [1.5] thus translates the fact that the fluctuation over time of a quantity attached to the part D, followed in its movement, results from the actions of surface and volume affecting the part D of the considered continuous medium from the outside.

We may associate a differential form to the integral form of the conservation equation.

The differential form of the conservation law:

$$\frac{\partial A_i}{\partial t} + (A_i U_j + \alpha_{ij})_{,j} = B_i \quad \text{in } V, \tag{1.6}$$

$$\alpha_{ij}\, n_j = C_i \quad \text{on } \bar{V}. \tag{1.7}$$

Equation [1.6] supposes that A_i, α_{ij}, B_i and C_i are continuously derivable in any point of V. This assumption, which we adopt, excludes the existence of discontinuity surfaces in volume V. For a detailed account of discontinuity surfaces we refer the reader to specialized works on continuous media mechanics.

The boundary condition [1.7] translates the equality of the projection of the tensor α_{ij} following the external normal to an external action of surface contact C_i. This action of contact will generally be a given in a problem; we shall see, however, that sometimes it will be preferable to modify the boundary condition, in order to more easily introduce the action of the exterior upon the continuous medium.

1.2.3. *Conservation of mass*

This law of conservation postulates that the mass of a part D of the continuous medium, whose movement is followed, remains constant over time.

To give the integral form of this conservation law, let us intróduce the density $\rho(M, t)$; under these conditions the law of conservation of mass is written:

$$\frac{d}{dt}\int_D \rho\,(M,t) = 0\,. \qquad [1.8]$$

Equation [1.8] is a particular case of the general form [1.5]. The associated differential form is deduced from it:

$$\frac{d}{dt}\rho + (\rho\, U_j)_{,j} = 0\,. \qquad [1.9]$$

Equation [1.9] is called continuity relation.

1.2.4. *Conservation of momentum*

A fundamental law of mechanics is introduced. To apply this law to every part D of the continuous medium, it is necessary to define the external efforts applied to D. These are of two kinds:

– efforts exerted on D by systems external to the continuous medium, which are remote actions or forces of volume written $f_i(M, t)$;

– efforts exerted on D through surface actions on $\overline{D}$; these are actions of local contact verifying the two following conditions:

a) at each point M of the boundary $\overline{D}$ and at every moment t, these efforts are represented by a density of force T_i ,

b) the vector T_i at the moment t depends only on the point M and the unitary vector normal to $\overline{D}$ in M.

Let us state [1.10], where σ_{ij} is a second-order tensor, called a stress tensor:

$$T_i = \sigma_{ij}\, n_j\,. \qquad [1.10]$$

Note: in [1.10], T_i is the ith component of the resulting stress for the vector $\vec{n}$; σ_{ij} is the ij^{th} component of the stress tensor. Somewhat abusing the language, the σ_{ij} will also be called stresses.

Let us write the fundamental law of the dynamics applied to a part D of the continuous medium. Equality of the dynamic torque and the torque of the external efforts applied to D led to the two relations [1.11] and [1.12]; O is a point related to the point of reference, which we take as the origin without restricting the general case:

$$\frac{d}{dt}\int_D \rho U_i = \int_D \sigma_{ij}\, n_j + \int_D f_i\,, \qquad [1.11]$$

$$\frac{d}{dt}\int_D (x_l\,\rho U_k - x_k\,\rho U_l) = \int_D (x_l\,\sigma_{kj} - x_k\,\sigma_{lj})\; n_j + \int_D (x_l\, f_k - x_k\, f_l) \qquad [1.12]$$

with $(1,k) = \{(1,2), (2,3), (3,1)\}$.

Relations [1.11] and [1.12] express the conservation of momentum. Their expressions can also be given in vectorial notation:

$$\frac{d}{dt}\int_D \rho\vec{U} = \int_D \vec{T} + \int_D \vec{f}\,,$$

$$\frac{d}{dt}\int_D \overrightarrow{OM}\wedge\rho\vec{U} = \int_D \overrightarrow{OM}\wedge\vec{T} + \int_D \overrightarrow{OM}\wedge\vec{f}.$$

The associated partial derivative equation [1.11] is:

$$\frac{d}{dt}(\rho U_i) + (\rho U_j U_i)_{,j} = \sigma_{ij,j} + f_i \qquad \text{in } V. \qquad [1.13]$$

By using the continuity equation [1.9] in [1.13] and after appropriate grouping, we obtain:

$$\rho\left(\frac{d}{dt}U_i + U_j\, U_{i,j}\right) = \sigma_{ij,j} + f_i \qquad \text{in } V. \qquad [1.14]$$

The first member of [1.14] represents $\rho\gamma_i$ where γ_i is the acceleration of the particle located at the point M, which we calculated in [1.2]. Equation [1.14] thus appears as a generalization of the point mechanics. It bears the name of the equation of motion.

Let us now exploit the law of conservation [1.12], by writing the associated partial derivative equation:

$$\frac{d}{dt}(x_l\,\rho U_k - x_k\,\rho U_l) + \left[\,(x_l\,\rho U_k - x_k\,\rho U_l)\,U_j - (x_l\,\sigma_{kj} - x_k\,\sigma_{lj})\right]_{,j} = x_l\,f_k - x_k\,f_l \qquad [1.15]$$

with (l,k) = {(1,2), (2,3), (3,1)}.

Let us take the example of the couple (l,k) = (1,2) and develop the derivations. After rearranging the terms we obtain:

$$x_1\left(\frac{d}{dt}(\rho U_2) + (\rho U_2 U_j)_{,j} - \sigma_{2j,j} - f_2\right) - x_2\left(\frac{d}{dt}(\rho U_1) + (\rho U_1 U_j)_{,j} - \sigma_{1j,j} - f_1\right) = \sigma_{21} - \sigma_{12}.$$

Taking into account the relation [1.13] the first member is nil; it is thus noted that:

$$\sigma_{12} = \sigma_{21}.$$

Proceeding in an identical manner for couples (2,3) and (3,1), we obtain the general relation of reciprocity of stresses:

$$\sigma_{ij} = \sigma_{ji}. \qquad [1.16]$$

The conservation of momentum involves the symmetry of the stress tensor.

1.2.5. *Conservation of energy*

At every moment the total derivative of the energy E (D) of a part D of the continuous medium is the sum of the power of the external efforts exerted on D and the rate of heat received by D.

Energy E (D) is the sum of kinetic and potential energy, i.e.:

$$E(D) = \int_D \rho\left(e + \frac{1}{2}U_i^2\right) \qquad [1.17]$$

with e as the specific potential energy.

The integral form of the law of conservation of energy is given by [1.18], where q_j is the heat flow vector. The minus sign is related to taking into account the external normal, thus $q_j n_j$ represents the heat flow emitted by the continuous medium.

$$\frac{d}{dt}\int_D \rho\left(e + \frac{1}{2}U_i^2\right) = \int_D \sigma_{ij} n_j U_i - q_j n_j + \int_D f_i U_i \,. \qquad [1.18]$$

The differential form of the law of conservation of energy results from [1.18]; we obtain all the calculations done:

$$\frac{d}{dt}\left(\rho\left(e + \frac{1}{2}U_i^2\right)\right) + \left(\rho U_j\left(e + \frac{1}{2}U_i^2\right) - U_i\sigma_{ij} + q_i\right)_{,j} = f_i U_i \quad \text{in V}. \qquad [1.19]$$

It follows from transforming [1.19] using relations [1.9] and [1.14]:

$$\rho\left(\frac{\partial}{\partial t}e + U_i\, e_{,i}\right) = \sigma_{ij}\, U_{i,j} - q_{j,j} \qquad \text{in V}. \qquad [1.20]$$

This partial derivative equation has a simple physical interpretation, since the total derivative of specific potential energy appears in the term between the brackets (on the left-hand side of the equation). Thus the variation of specific potential energy results from the power of interior efforts ($\sigma_{ij}\, U_{i,j}$) and from a contribution of heat ($-q_{j,j}$).

1.2.6. *Boundary conditions*

The boundary conditions represent the natural prolongation of the conservation equations, over the surface $\overline{V}$ of the continuous medium. They are obtained through the relation [1.7] given in the general case of a conservation law, which will have to be further specified by the conservation of mass, momentum and energy.

Let us note first of all that the conservation of mass [1.8] does not involve a boundary condition because the term α_{ij} does not appear in [1.8].

Equation [1.11] of the conservation of momentum involves the boundary condition:

$$\sigma_{ij}\, n_j = F_i \qquad \text{on } \bar{V}. \qquad [1.21]$$

F_i represent the components of the external surface forces applied to the continuous medium.

Equation [1.12] of the conservation of momentum involves the boundary condition:

$$x_l\, \sigma_{kj}\, n_j - x_k\, \sigma_{lj}\, n_j = x_l\, F_k - x_k\, F_l \qquad \text{on } \bar{V}, \qquad [1.22]$$

with (l,k) = {(1,2), (2,3), (3,1)}.

The second member represents the moment of external surface forces applied to V. The verification of the boundary condition [1.21] involves the verification of [1.22] which, therefore, does not bring any additional information.

The conservation of energy involves the boundary condition:

$$q_i\, n_i + \sigma_{ij}\, n_j U_i = \Pi + F_i U_i \qquad \text{on } \bar{V}. \qquad [1.23]$$

Π is the amount of heat introduced into the continuous medium, by action of contact at its boundary surface. $F_i U_i$ is the power introduced by the surface forces applied to $\bar{V}$.

By using the relation [1.21] in [1.23], we obtain:

$$q_i\, n_i = \Pi \qquad \text{on } \bar{V}. \qquad [1.24]$$

The formulation of a problem of continuous media mechanics is summarized to finding the density $\rho(M, t)$, speed $U_i(M,t)$, stress $\sigma_{ij}(M,t)$ and the specific energy density $e(M, t)$, knowing the forces exiting the volume $f_i(M,t)$ and the surface $F_i(M,t)$ as well as the quantity of heat input Π (M, t). All these quantities are related by the 4 partial derivative equations [1.9], [1.14], [1.16], [1.20] to be verified in the volume V and the two boundary conditions [1.21], [1.24] to be verified over the surface $\bar{V}$.

1.3. Study of the vibrations: small movements around a position of static, stable equilibrium

1.3.1. *Linearization around a configuration of reference*

Linearized equations that we are going to establish only reflect a physical reality if the continuous medium keeps the positions close to those, which it occupies in the configuration of reference, during its movement. We choose a Lagrange position of reference, and the displacement of the particle M is expressed by the formula:

$$x_i = a_i + W_i(a_j, t). \qquad [1.25]$$

x_i is the ith co-ordinate of particle M whose movement is being followed (Euler's variable). a_i is the ith co-ordinate of particle M in the configuration of reference (Lagrange's variable). $W_i(a_j, t)$ is the ith co-ordinate of the displacement of point M around its position in the situation of reference. We suppose that this displacement as well as its derivatives are small:

$$\left|\frac{dW_i}{dt}\right| << 1 \text{ and } \left|\frac{dW_i}{dx_j}\right| << 1. \qquad [1.26]$$

We will examine the consequences of the assumption [1.26]:

a) Let us at first consider a regular function $f(x_i, t)$, and let us express its value in the vicinity of the position of reference. The components x_i of the position of the point M are close to the co-ordinates a_i, of the same point M that had occupied it in the position of reference; consequently, a first approximation of the value of the function may be obtained by considering the first terms of its development in a Taylor series in the vicinity of a_i:

$$f(x_i, t) = f(a_i, t) + \sum_{j=1}^{3} (x_j - a_j) \frac{\partial f}{\partial x_j}(a_i, t),$$

that is, taking into account the decomposition of movement [1.25]:

$$f(x_i, t) = f(a_i, t) + \sum_{j=1}^{3} W_j(a_i, t) \frac{\partial f}{\partial x_j}(a_i, t). \qquad [1.27]$$

Taking into account the regularity of $f(x_i, t)$, the partial derivative $\dfrac{\partial f}{\partial x_j}(a_i, t)$ is bounded. From [1.26] and [1.27] we deduce that in the first approximation:

$$f(x_i, t) = f(a_i, t). \tag{1.28}$$

b) Let us now take the derivative $\dfrac{\partial f}{\partial a_j}$; by using the chain derivation formula it follows:

$$\frac{\partial f}{\partial a_j} = \sum_{i=1}^{3} \frac{\partial f}{\partial x_i} \frac{\partial x_i}{\partial a_j}.$$

Introducing the form [1.25] of the movement x_i, we shall obtain:

$$\frac{\partial f}{\partial a_j} = \frac{\partial f}{\partial x_j} + \sum_{i=1}^{3} \frac{\partial f}{\partial x_i} \frac{\partial W_i}{\partial a_j}.$$

The second term of the right-hand side member being infinitely small, it can be deduced that in the first approximation:

$$\frac{\partial f}{\partial a_j} = \frac{\partial f}{\partial x_j}. \tag{1.29}$$

c) Let us calculate the total derivative of a regular function $G(x_i, t)$:

$$\frac{dG}{dt} = \frac{\partial G}{\partial t}(x_i, t) + \sum_{j=1}^{3} \frac{\partial G}{\partial x_j}(x_i, t)\; U_j(x_i, t),$$

that is, taking into account the decomposition of movement [1.25]:

$$\frac{dG}{dt} = \frac{\partial G}{\partial t}(x_i, t) + \sum_{j=1}^{3} \frac{\partial G}{\partial x_j}(x_i, t)\; \frac{\partial W_j}{\partial t}(a_i, t).$$

The function $G(x_i, t)$ being regular, $\frac{\partial G}{\partial x_j}(x_i, t)$ is bounded, the second term of the second member is infinitely small; we thus have at first approximation:

$$\frac{dG}{dt} = \frac{\partial G}{\partial t}(x_i, t),$$

i.e. also taking into account [1.28]:

$$\frac{dG}{dt} = \frac{\partial G}{\partial t}(a_i, t).$$

To sum up, for small movements:

$$\frac{dG}{dt} = \frac{\partial G}{\partial t}(x_i, t) + \sum_{j=1}^{3} \frac{\partial G}{\partial x_j}(x_i, t)\, U_j(x_i, t) = \frac{\partial G}{\partial t}(a_i, t). \qquad [1.30]$$

The distinction between the Euler and Lagrangian descriptions is no longer necessary: on the one hand the initial and current co-ordinates a_i and x_i can be assimilated and the particulate derivative can be replaced by the partial derivative with respect to time. This is true for regular functions, i.e. not for discontinuity surfaces.

Let us examine the effects of [1.28], [1.29] and [1.30] on the equations describing the behavior of the continuous medium.

The equation of conservation of mass [1.9] becomes:

$$\frac{\partial \rho}{\partial t}(a_i, t) = 0 \qquad \text{in } V,$$

that is:

$$\rho(a_i, t) = \rho(a_i) \qquad \text{in } V. \qquad [1.31]$$

During small movements, the density of the continuous medium does not vary over time. This property is valid only at first approximation; at a higher degree of accuracy, there is an additional small term, which fluctuates with time. In linear acoustics, this small disturbance must be preserved in calculations as it intervenes in the ideal gas law of the acoustic medium. In the case of elastic solids considered here, the constant term is sufficient to describe the conservation of mass.

Equations [1.14] and [1.16], translating the conservation of momentum, become:

$$\rho\,(a_i)\frac{\partial^2 W_i}{\partial t^2}(a_i\ ,t)=\sum_{j=1}^{3}\frac{\partial \sigma_{ij}}{\partial a_j}\ (a_i\ ,t)+f(a_i\ ,t)\qquad \text{in V,} \tag{1.32}$$

$$\sigma_{ij}\ (a_i\ ,t)=\sigma_{ji}\ (a_i\ ,t)\qquad \text{in V.} \tag{1.33}$$

Equation [1.20], characterizing the conservation of energy, becomes:

$$\rho\,(a_i)\frac{\partial e}{\partial t}(a_i\ ,t)=\sum_{i=1}^{3}\sum_{j=1}^{3}\sigma_{ij}\ (a_i\ ,t)\frac{\partial^2 W_i}{\partial t\,\partial a_j}\ (a_i\ ,t)-\sum_{j=1}^{3}\frac{\partial q_j}{\partial a_j}\ (a_i\ ,t)\qquad \text{in V.} \tag{1.34}$$

Boundary conditions:

$$\sum_{j=1}^{3}\sigma_{ij}(a_i\ ,t)\ n_j(a_i)=F_i(a_i\ ,t)\qquad \text{on } \bar{V}, \tag{1.35}$$

$$\sum_{j=1}^{3}-q_j(a_i\ ,t)\ n_j(a_i)=\Pi(a_i\ ,t)\qquad \text{on } \bar{V}. \tag{1.36}$$

Equations [1.31] to [1.36] constitute the linearized model of general equations within the framework of small movements, around a configuration of reference, defined by the relations [1.25] and [1.26].

All quantities appearing in the linearized equations [1.31] to [1.36] are variables of the pair (a_i, t); thus, for the study of small movements, the equations and the boundary conditions are inscribed directly on the configuration of reference.

In the continuation of the course, we will often consider the case of adiabatic movements. This assumption involves $q_i(a_i , t) = 0$; there follows a modification of the equation of energy [1.34] and boundary condition [1.36] which become:

$$\rho \frac{\partial e}{\partial t} = \sigma_{ij} \frac{\partial W_{i,j}}{\partial t} \quad \text{in V ,} \tag{1.37}$$

$$\Pi = 0 \quad \text{on } \bar{V} . \tag{1.38}$$

The boundary condition [1.38] translates the impossibility for the adiabatic medium to exchange heat.

The equation of energy [1.37] shows that the variation of specific potential energy is due only to the power of interior efforts.

We have used the index notation in [1.37], and from now we will make constant use of it.

1.3.2. *Elastic solid continuous media*

The unknowns of a problem of vibration of an elastic solid are: W_i , σ_{ij} and e. The calculation reveals 10 independent quantities (taking into account the symmetry of the stress tensor). However, the equations of continuity, movement and energy provide only 5 relations at each point. Thus, information is missing to determine the solution of the problem; that is the stress-strain relation of the continuous medium.

The stress-strain relation is characteristic of material; it connects the stress tensor to that of the strain of the continuous medium. In the case of small movements, considered here, the behavior of the continuous medium is well represented by the law of elastic behavior. The stress-strain relation is of the type:

$$\sigma_{ij}(a_i ,t) = C_{ijkl}(a_i)\ \varepsilon_{kl}(a_i ,t) \quad \text{on V .} \tag{1.39}$$

The quantity $\varepsilon_{kl}(a_i , t)$ is a symmetrical second-order tensor; it is the strain tensor defined by the relation:

$$\varepsilon_{kl}(a_i , t) = \frac{1}{2}\left(\frac{\partial W_k}{\partial a_l} + \frac{\partial W_l}{\partial a_k} \right),$$

or in the index notation:

$$\varepsilon_{kl}(a_i , t) = \frac{1}{2}(W_{k,l} + W_{l,k}).$$

The tensor of the 4th order $C_{ijkl}(a_i)$ characterizes the elastic properties of the continuous medium. In the references provided at the end of the chapter, a detailed presentation may be found. Let us note here that this tensor has the properties of symmetry [1.40]:

$$C_{ijkl} = C_{jikl} = C_{ijlk} = C_{klij}. \qquad [1.40]$$

Taking into account the properties [1.40] of the stress-strain relation, we obtain a second expression equivalent to [1.39]:

$$\sigma_{ij}(a_i , t) = C_{ijkl}(a_i)\ W_{kl}(a_i , t). \qquad [1.39']$$

1.3.3. *Summary of the problem of small movements of an elastic continuous medium in adiabatic mode*

The problem consists of finding W_i, σ_{ij} and e, knowing f_i, F_i, C_{ijkl} and ρ, verifying:

$$\rho\frac{\partial^2 W_i}{\partial t^2} = \sigma_{ij,\,j} + f_i \qquad \text{in V}, \qquad [1.41a]$$

$$\rho\frac{\partial e}{\partial t} = \sigma_{ij}\left(\frac{\partial W_i}{\partial t}\right)_{,j} \qquad \text{in V}, \qquad [1.41b]$$

$$\sigma_{ij} = C_{ijkl}\ \varepsilon_{kl} = C_{ijkl} W_{k,l} \qquad \text{in V}. \qquad [1.41c]$$

Boundary conditions:

$$\sigma_{ij}\ n_j = F_i \qquad \text{on } \overline{V}. \qquad [1.41d]$$

The use of [1.4c] in [1.41b] makes it possible to integrate the equation of energy, which becomes [1.41e]:

$$\rho e = \frac{1}{2}\,\varepsilon_{ij}\,C_{ijkl}\,\varepsilon_{kl} = \frac{1}{2}\,\varepsilon_{ij}\,\sigma_{ij} \qquad \text{in V}. \qquad [1.41e]$$

The knowledge of σ_{ij} and ε_{ij} implies that of e; there are thus only two unknowns in the present problem: σ_{ij} and ε_{ij}, in order to determine which equations [1.41a], [1.4c] and [1.41] need to be integrated.

These equations are well adapted to the description of vibrations of solids whose displacements remains close to the static position of equilibrium, which is taken as a configuration of reference.

1.3.4. *Position of static equilibrium of an elastic solid medium*

The vibrations of continuous media occur around a position of stable static equilibrium. Consequently, the first stage of the study of the vibrations consists of determining this position of static equilibrium.

Let us consider the position of the continuous medium at rest, when no force is applied to it, as a configuration of reference and suppose that the position of static equilibrium is close to this position of reference.

In these conditions it is possible to use equations [1.41], obtained with the assumption of small movements, to describe the state of static equilibrium. In fact, the task is to find $W_j^S(a_i)$ and $\sigma_{ij}^S(a_i)$ verifying equations [1.41] when the forces $F_j^S(a_i)$ and $f_j^S(a_i)$ are applied. The reader will note that all the quantities appearing in the static problem are independent of time; it follows that derivations of these quantities with respect to time are nil and the partial derivative equations [1.41a, c, d] are reduced to:

$$\sigma_{ij,j}^S + f_i^S = 0 \qquad \text{in V}, \qquad [1.42]$$

$$\sigma_{ij}^S = C_{ijkl}\,\varepsilon_{kl}^S \qquad \text{in V}, \qquad [1.43]$$

$$\sigma_{ij}^S\,n_j = F_i^S \qquad \text{in } \overline{V}. \qquad [1.44]$$

The equations of continuity and energy are automatically verified since density and specific internal energy are constant over time, and equal to ρ^s and e^s.

1.3.5. *Vibrations of elastic solid media*

Vibrations of elastic solids are small movements around a position of static equilibrium, generated by dynamic forces $f_i^D(a_j, t)$ and $F_i^D(a_j, t)$ superimposed with static forces, so that:

$$f_i(a_j t) = f_i^S(a_j) + f_i^D(a_j, t) \text{ and } F_i(a_j t) = F_i^S(a_j) + F_i^D(a_j, t). \qquad [1.45]$$

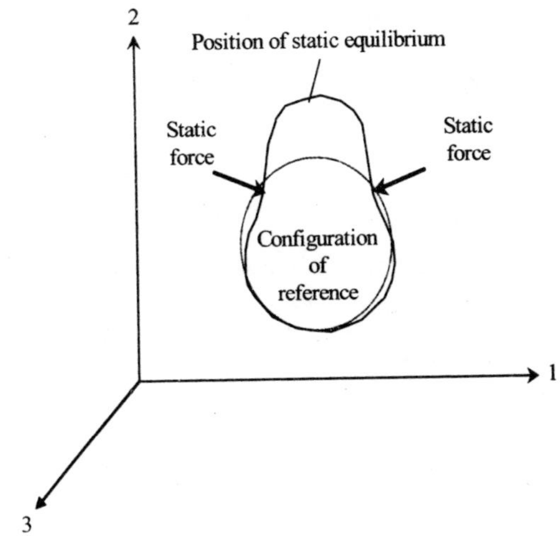

Figure 1.4. *Configuration of reference and position of static equilibrium*

We will make the assumption that the application of dynamic forces introduces only small movements, i.e. dynamic stresses are sufficiently weak, and that static equilibrium is stable. The movement is described by linearized equations [1.41], in which we reveal the division of quantities into static values characteristic of the state of equilibrium (exponent S) and dynamic values characterizing the vibrations (exponent D):

$$\rho(a_i, t) = \rho^S(a_i),$$

$$W_j(a_i, t) = W_j^S(a_i) + W_j^D(a_i, t),$$

$$\sigma_{ij}(a_i, t) = \sigma_{ij}^S(a_i) + \sigma_{ij}^D(a_i, t),$$

$$e(a_i, t) = e^S(a_i) + e^D(a_i, t). \qquad [1.46]$$

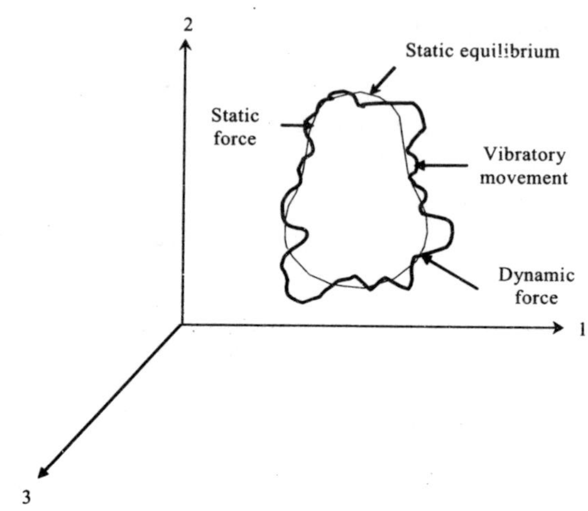

Figure 1.5. *Position of static equilibrium and vibratory movement*

Let us introduce the decompositions [1.46] into equations [1.41a, c, d]. After a rather simple calculation, we find:

$$\rho^S \frac{\partial^2 W_i^D}{\partial t^2} = \sigma_{ij,j}^S + \sigma_{ij,j}^D + f_i^S + f_i^D \qquad \text{in V}, \tag{1.47}$$

$$\sigma_{ij}^S + \sigma_{ij}^D = C_{ijkl}\left\{\varepsilon_{kl}^S + \varepsilon_{kl}^D\right\} \qquad \text{in V}, \tag{1.48}$$

$$\sigma_{ij}^S\, n_j + \sigma_{ij}^D\, n_j = F_i^S + F_i^D \qquad \text{on } \bar{V}. \tag{1.49}$$

These equations can be simplified by taking account of the static equilibrium conditions [1.42], [1.43], [1.44] to become:

$$\rho^S \frac{\partial^2 W_i^D}{\partial t^2} = \sigma_{ij,j}^D + f_i^D \qquad \text{in V}, \tag{1.50}$$

$$\sigma_{ij}^D = C_{ijkl}\, \varepsilon_{kl}^D \qquad \text{in V}. \tag{1.51}$$

Boundary conditions:

$$\sigma_{ij}^{D}\, n_j = F_i^{D} \qquad \text{on } \overline{V}. \tag{1.52}$$

A remarkable property of equations [1.50] to [1.52] is their quasi-independence of the state of static equilibrium around which the system vibrates. In fact, the only influence of static equilibrium is related to the density of the medium. Under normal conditions, ρ^S varies very little and the vibrations of a continuous medium are not affected by a modification of the static position. The variations of the field of gravity in particular do not modify appreciably the vibratory state of the continuous medium.

1.3.6. *Boundary conditions*

The boundary condition [1.52] translates the equality of the normal dynamic stresses tensor projection with external surface forces applied to the elastic solid. The external forces are supposed to be given in the problem, which presents a difficulty in practice. Indeed, they result from actions of contact with other mechanical media that are generally unknown. To overcome this difficulty, the two following simplified configurations are generally introduced:

Free surface $\overline{V}_L$: this situation is to be considered when the external actions on the surface are sufficiently weak to be regarded as nil. We would then write:

$$\sigma_{ij}^{D}\, n_j = 0 \qquad \text{on } \overline{V}_L. \tag{1.53}$$

Constrained surface $\overline{V}_E$: this situation occurs when external actions on the surface are very strong and tend to impose a given displacement on the surface considered. The external force applied under these conditions strongly depends on the response of the continuous medium. It is preferable to model the boundary condition on an embedded surface by consequence of application of the external force, i.e. an imposed displacement D_i. We would then write:

$$W_i^{D} = D_i \qquad \text{on } \overline{V}_E. \tag{1.54}$$

The two models [1.53] and [1.54] are extreme cases, and one can consider types of intermediate boundary conditions having a certain flexibility.

We will see in the chapters related to the vibrations of beams or plates how to introduce this type of boundary conditions. However, let us specify, at this general level, that taking into account sophisticated boundary conditions involves the need for measuring quantities describing the behavior of the boundaries, which poses large experimental problems. For mechanical problems of vibrations, it is often at the level of boundary conditions that the uncertainty of modeling is the strongest. In the continuation, we will often suppose that the continuous medium is either constrained, or free, or subjected to known external forces, on the surface limiting the elastic solid. Constrained surfaces $\overline{V}_E$, free surfaces $\overline{V}_L$, and those where external forces $\overline{V}_F$ are given, are disjointed; consequently:

$$\overline{V_E} \cap \overline{V}_L = \varnothing ,$$

$$\overline{V_L} \cap \overline{V}_F = \varnothing ,$$

$$\overline{V_E} \cap \overline{V}_F = \varnothing ,$$

$$\overline{V_E} \cup \overline{V_L} \cup \overline{V_F} = \overline{V} .$$

1.3.7. *Vibrations equations*

The problem of vibrations of elastic solids is stated as follows: to find the fields of stress σ_{ij} and of displacement W_i verifying the equations:

– Equations of motion:

$$\rho \frac{\partial^2 W_i}{\partial t^2} = \sigma_{ij,j} + f_i \qquad \text{in } V . \qquad [1.55]$$

– Stress-strain relation:

$$\sigma_{ij} = C_{ijkl} W_{k,l} \qquad \text{in } V . \qquad [1.56]$$

– Boundary conditions:

$$\sigma_{ij}\, n_j = 0 \qquad \text{on } \overline{V}_L , \qquad [1.57]$$

$$W_i = D_i \qquad \text{on } \overline{V}_E , \qquad [1.58]$$

$$\sigma_{ij}\, n_j = F_i \qquad \text{on } \bar{V}_F . \tag{1.59}$$

Note: in order to be concise, we give up the exponents S and D in the notation for static and dynamic states. Let us also recall that all the quantities appearing in the equations of vibrations [1.55] to [1.59] are of a dynamic nature, except density which is characteristic of the static position.

1.3.8. *Notes on the initial conditions of the problem of vibrations*

As long as the continuous medium has not been subjected to dynamic excitation (volume forces, surface forces or displacements imposed at the limits), it is in a static equilibrium. It follows that for an application of the vibratory state at the moment t_0, we will take the following initial conditions:

$$W_i(a_i\,, t_0) = 0\,, \tag{1.60}$$

$$\frac{\partial W_i}{\partial t}(a_i\,, t_0) = 0\,. \tag{1.61}$$

The partial derivative equation in time being of the second-order, the two initial conditions are sufficient.

It is sometimes interesting to describe the vibrations of a continuous medium starting at a moment t_1, posterior to the application of forces. At this moment, the system is no longer in a static equilibrium, but is in a given vibratory state (displacement $X_i(a_i)$ and speed $V_i(a_i)$) so that the following initial conditions would have to be taken:

$$W_i\left(a_i\,, t_1\right) = X\left(a_i\right), \tag{1.62}$$

$$\frac{\partial W_i}{\partial t}\left(a_i\,, t_1\right) = V_i\left(a_i\right). \tag{1.63}$$

Let us note to close this point that in many vibratory problems, the interest lies in the forced "movement" which is independent of the initial conditions, the latter then not being specified.

1.3.9. *Formulation in displacement*

The vibration problem defined in section 1.3.7 has displacement and stress fields as unknowns. It is a mixed formulation in stress and displacement. It is often interesting to reduce the number of unknown functions, and therefore the number of equations, to thus have a more compact formulation. This reduction is carried out by substitution of the stress field by its expression as a function of displacements [1.56] in equations [1.55], [1.57] and [1.59]. We then obtain a formulation, which now only depends on vibratory displacements.

To find the displacement field W_i verifying:

$$\rho \frac{\partial^2 W_i}{\partial t^2} = (C_{ijkl} W_{k,l})_{,j} + f_i \qquad \text{in } V, \qquad [1.64]$$

$$W_i = D_i \qquad \text{on } \overline{V}_E, \qquad [1.65]$$

$$C_{ijkl}\, W_{k,l}\, n_j = 0 \qquad \text{on } \overline{V}_L, \qquad [1.66]$$

$$C_{ijkl}\, W_{k,l}\, n_j = F_i \qquad \text{on } \overline{V}_F. \qquad [1.67]$$

This formulation in displacement has the clear advantage of decreasing the number of unknowns, since at any point of the continuous medium, it is sufficient to establish the displacements. This reduction of the number of unknowns is, however, made at the expense of the simplicity of resolution of equations, which see their order of spatial derivation increasing.

1.3.10. *Vibration of viscoelastic solid media*

We will see in the following chapters that vibrations of continuous media are characterized by the presence of resonances, for which the damping of the vibrating system plays a capital role. The elastic systems that we have considered are not dissipative and consequently will not be representative of the vibratory answer to resonances. To take account of the dissipation parameter, it is necessary to consider a behavior relation more complex than that of linear elasticity: linear viscoelasticity.

Viscoelastic materials have some rigidity but dissipate more energy by internal friction. Contrary to elasticity, where the stress changes instantaneously with strain, viscoelasticity introduces a memory effect: the stress at a certain moment depends on all the former strain. It is thus necessary to utilize time in the stress-strain relation of a viscoelastic material; numerous models have been elaborated according to the type of dependence on time. Within the narrow framework of these reminders, we will limit ourselves to the following model:

$$\sigma_{ij}(a_i\ ,t) = \lambda_{ijkl} \otimes \frac{\partial W_{k,l}}{\partial t}\ . \qquad [1.68]$$

This stress-strain relation characterized by a product of temporal convolution, noted as $\otimes$, shows that the stress field at the moment t depends on the former strain of the continuous medium.

To summarize, vibrations of the viscoelastic continuous medium are described by the fields of stress $\sigma_{ij}(a_i\ , t)$ and of displacements $W_i(a_i\ , t)$ verifying:

$$\rho \frac{\partial^2 W_i}{\partial t^2} = \sigma_{ij,j} + f_i \qquad \text{in V}, \qquad [1.69]$$

$$\sigma_{ij}(a_i\ ,t) = \lambda_{ijkl} \otimes \frac{\partial W_{k,l}}{\partial t} \qquad \text{in V}, \qquad [1.70]$$

$$\sigma_{ij}\ n_j = 0 \qquad \text{on } \bar{V}_L, \qquad [1.71]$$

$$W_i = D_i \qquad \text{on } \bar{V}_E, \qquad [1.72]$$

$$\sigma_{ij}\ n_j = F_i \qquad \text{on } \bar{V}_F. \qquad [1.73]$$

Equations [1.69] to [1.73] may be brought to equations of the type of those obtained for the elastic medium by introducing Fourier transforms $\tilde{W}$, $\tilde{\sigma}$ and $\tilde{C}$ of the values W, σ and C:

$$\tilde{W}_i(a_i\ , \omega) = \int_{-\infty}^{+\infty} W_i(a_i\ , t)\ e^{-j\omega t}\, dt\ , \qquad [1.74]$$

$$\tilde{\sigma}_{ij}(a_i\ , \omega) = \int_{-\infty}^{+\infty} \sigma_{ij}(a_i\ , t)\ e^{-j\omega t}\, dt\ , \qquad [1.75]$$

$$\widetilde{C}_{ijkl}(a_i, \omega) = \int_{-\infty}^{+\infty} C_{ijkl}(a_i, t)\, e^{-j\omega t}\, dt\,, \qquad [1.76]$$

where ω is the angular frequency.

Taking the Fourier transform of equations [1.69] – [1.73], it follows:

$$-\rho\omega^2 \widetilde{W}_i = \widetilde{\sigma}_{ij,j} + \widetilde{f}_i\,, \qquad [1.77]$$

$$\widetilde{\sigma}_{ij} = \widetilde{C}_{ijkl} \widetilde{W}_{k,l}\,, \qquad [1.78]$$

$$\tilde{\sigma}_{ij}\, n_j = 0 \qquad \text{on } \overline{V}_L\,, \qquad [1.79]$$

$$\tilde{W}_i = \tilde{D}_i \qquad \text{on } \overline{V}_E\,, \qquad [1.80]$$

$$\tilde{\sigma}_{ij}\, n_j = \tilde{F}_i \qquad \text{on } \overline{V}_F\,. \qquad [1.81]$$

The Fourier transformation makes it possible to replace the convolution product characteristic of the stress-strain relation in time domain by a simple product in frequency domain.

Let us now take the Fourier transform of the vibrations equations of the elastic system [1.55] – [1.59]; it follows:

$$-\rho\omega^2 \widetilde{W}_i = \widetilde{\sigma}_{ij,j} + \widetilde{f}_i\,, \qquad [1.82]$$

$$\widetilde{\sigma}_{ij} = C_{ijkl} \widetilde{W}_{k,l}\,, \qquad [1.83]$$

$$\tilde{\sigma}_{ij}\, n_j = 0 \qquad \text{on } \overline{V}_L\,, \qquad [1.84]$$

$$\tilde{W}_i = \tilde{D}_i \qquad \text{on } \overline{V}_E\,, \qquad [1.85]$$

$$\tilde{\sigma}_{ij}\, n_j = \tilde{F}_i \qquad \text{on } \overline{V}_F\,. \qquad [1.86]$$

A formal analogy known as the principle of correspondence is noted between the two systems of equations [1.77] – [1.81] and [1.82] – [1.86].

In fact, all the equations of elastic and viscoelastic systems are identical except for those related to the stress-strain relation. For the elastic medium, the tensor of elasticity moduli is independent of time and, thus, remains unchanged after Fourier transformation of the stress-strain relation (i.e. real and independent of the angular frequency ω). For the viscoelastic medium, the Fourier transform of the viscoelastic modules which are variable with time appears in [1.78]; there follow two significant consequences:

– on the one hand $\widetilde{C}_{ijkl}$ is variable with the angular frequency ω;

– on the other hand $\widetilde{C}_{ijkl}$ is a complex number. The real part represents the elastic effect, the imaginary part that of dissipation. These coefficients are called complex modules.

The complex module translates a phase shift between stress and strain; this vision, characteristic of the representation in frequency domain, is the consequence of the delay between stress and displacements characterizing the viscoelastic medium in time domain.

Loss factors η_{ijkl} are often introduced:

$$\widetilde{C}_{ijkl} = \mathrm{Re}\left\{\widetilde{C}_{ijkl}\right\} \; (1 + j\eta_{ijkl}) .$$

In short, in frequency domain, the elastic or viscoelastic continuous media have the same equations. This leads, due to a preoccupation with simplicity, to the undertaking of studies of the elastic medium in time domain, and the introduction of viscoelasticity afterwards, by complex modules in frequency domain.

The resolution of the equations in frequency domain yields

$$\tilde{W}_i(a_i, \omega) \text{ and } \tilde{\sigma}_{ij}(a_i, \omega) .$$

Temporal solutions should be expressed thereafter; this is of course achieved by inverse Fourier transformation:

$$W_i(a_i, t) = \frac{1}{2\pi} \int_{-\infty}^{+\infty} \widetilde{W}_i(a_i, \omega) \; e^{+j\omega t} \, d\omega ,$$

$$\sigma_{ij}(a_i, t) = \frac{1}{2\pi} \int_{-\infty}^{+\infty} \widetilde{\sigma}(a_i, \omega) \; e^{+j\omega t} \, d\omega .$$

Let us note that in general the inverse transformation is not carried out because the physical interpretation of the results is in fact easier in frequency domain.

Note: the viscoelastic medium is not conservative and the adiabatic assumption of behavior is no longer realistic, the equation of energy which has to be considered is not [1.37] anymore, but the complete equation [1.34].

1.4. Conclusion

This chapter constitutes an introduction to the governing equations of the vibrations of elastic solid continuous media. Its essential goal is to present the assumptions underlying the equations, which we will come back to use in the continuation. These reminders would not in any way be capable of replacing a thorough study of the works specialized in this field, but constitute instead the minimum knowledge necessary for the good understanding of what follows.

A basic comment that we will make in conclusion of the chapter relates to the great complexity of the equations obtained. Indeed, the partial derivative equations which describe the vibrations, although already simplified compared to the general case, do not have known analytical solutions. We are thus confronted with the alternative of an approximate resolution, which can be only numerical, or with the introduction of additional assumptions, simplifying the problem sufficiently to lead to analytical solutions. It is the second option that has generally been exploited; it has led to the mono and bi-dimensional continuous media that engineers have called beams, plates and shells. The methodology of passage of the tri-dimensional medium into simplified media is thus of capital importance. In the following chapters we provide the methodology by using the variational approach.

Chapter 2

Variational Formulation for Vibrations of Elastic Continuous Media

2.1. Objective of the chapter

The equations describing vibrations of elastic solid media in 3 dimensions have been provided in Chapter 1. We will demonstrate that it is possible to obtain them by calculating the extrema values of energy functionals. Moreover, this approach lays the theoretical foundation which enables the construction of models of condensed elastic continuous media. This would allow passing from a 3D to a 2D or 1D problem.

Variational formulation uses directional derivation and can thus appear to be a more cumbersome version of the traditional formulation. It is, in fact, at the level of searching for approximate solutions that variational formulation assumes its full importance, since it suffices to restrict functional spaces where the extremalization is carried out. It is this step which will enable us in the following chapters to systematically obtain the equations of beams and plates without any difficulty other than the choice of assumptions restricting the field of movement and stress according to the geometry of the continuous medium.

The object of this chapter is double. On the one hand, we present the basic idea of variational formulation which consists of passing from a local aspect of forces equilibriums to a global energy aspect. In addition, we obtain results, which will be useful in the continuation, i.e. the functionals of Reissner and Hamilton, as well as Euler equations associated with the extremalization of various types of functionals. For uniformity of presentation, we will suppose that all the functions that appear in the equations are sufficiently regular so that integrals exist.

2.2. Concept of the functional, bases of the variational method

2.2.1. *The problem*

The equations of the vibrations of a continuous medium that we have determined in Chapter 1 (equations [1.55] to [1.59]) are:

$$\rho \frac{\partial^2 W_i}{\partial t^2} = \sigma_{ij,j} + f_i \quad \text{in } V \times \left] t_0, t_1 \right[, \qquad [2.1]$$

$$\sigma_{ij} = C_{ijkl} \varepsilon_{kl} \quad \text{in } V \times \left] t_0, t_1 \right[, \qquad [2.2]$$

$$\sigma_{ij} n_j = 0 \quad \text{over } \overline{V}_L \times \left] t_0, t_1 \right[, \qquad [2.3]$$

$$W_i = D_i \quad \text{over } \overline{V}_E \times \left] t_0, t_1 \right[, \qquad [2.4]$$

$$\sigma_{ij} n_j = F_i \quad \text{over } \overline{V}_F \times \left] t_0, t_1 \right[. \qquad [2.5]$$

Moments t_0 and t_1 are arbitrary.

The basis of the variational method consists of posing the problem in a different, global form leading to the implicit respect of equations [2.1] to [2.5].

2.2.2. *Fundamental lemma*

The result that we will point out is fundamental in the sense that it contains the basic idea of the variational method, which consists of passing from a local to a global presentation.

Let us recall first of all that a family F of open sets D of a volume V is dense in V if for any point M $\in$ V and for all neighborhood of the point M, there exists at least one set D of the family which lies inside the neighborhood. To solidify the ideas, let us give without demonstration three families dense in V:

– all of the open balls inside of V;

– all of the open cubes whose edges are parallel to the axes of co-ordinates;

– all of the open sets of V.

The fundamental lemma is stated as follows: let f(M) be a function defined and continuous in V and F a family of sets D dense in V. If for any set D belonging to the family the integral [2.6] is nil,

$$\int_D f(M)\,dv = 0\,, \qquad [2.6]$$

then the function f(M) is identically nil in V.

To show the lemma, let us presume that in a certain point $M_0 \in V$, $f(M_0)$ is not nil. Let us take for example the case $f(M_0) > 0$. It is possible, taking into account the continuity of f(M), to find a neighborhood of M_0 such that in this neighborhood $f(M) > (1/2)\, f(M_0)$. Let us consider then a set D of the family F interior to the neighborhood. We have:

$$\int_D f(M)\,dv > \frac{1}{2} f(M_0) \int_D dv > 0\,,$$

which contradicts the hypothesis [2.6].

2.2.3. *Basis of variational formulation*

Let us take the example of the equation of motion [2.1]. The use of the fundamental lemma makes it possible to write it in the form of:

$$\int_D \left(\rho \frac{\partial^2 W_i}{\partial t^2} - \sigma_{ij,j} - f_i \right) dv = 0. \qquad [2.7]$$

D belongs to a family of dense open sets in $V \times \left] t_0, t_1 \right[$.

However, formula [2.7] is not very practical and we would rather write:

$$\int_{t_0}^{t_1}\int_V \left(\rho \frac{\partial^2 W_i}{\partial t^2} - \sigma_{ij,j} - f_i \right) W_i^* dv dt = 0 \quad \forall W_i^* \in \Omega^*\left(V \times \left] t_0, t_1 \right[\right). \qquad [2.8]$$

The space $\Omega^*(V \times]t_0,t_1[)$ is that of functions with real values, indefinitely derivable, definite on $V \times]t_0,t_1[$.

The functions $W_i^*(x_1, x_2, x_3, t)$ can be interpreted as vibratory movements, which are, however, not compelled to verify all the conditions imposed on real movements of the continuous medium (in particular, boundary conditions); in this sense, they are merely virtual movements.

To demonstrate the equivalence of [2.8] and [2.1], it is enough to consider, initially, the field of movement $W_i^*(x_1, x_2, x_3, t)$ provided below:

$$W_i^* = \begin{cases} W_1^\varepsilon(x_1, x_2, x_3, t) \\ 0 \\ 0 \end{cases} \qquad [2.9]$$

Let $D_\varepsilon(X_1, X_2, X_3, T)$ be the open ball, of band ε, centered in (X_1, X_2, X_3, T), i.e.:

$$D_\varepsilon(X_1, X_2, X_3, T) = \left\{(x_1, x_2, x_3, t) \middle| (x_1 - X_1)^2 + (x_2 - X_2)^2 + (x_3 - X_3)^2 + (t - T)^2 < \varepsilon^2 \right\},$$

the functions $W_1^\varepsilon(x_1, x_2, x_3, t)$ are strictly positive inside $D_\varepsilon(X_1, X_2, X_3, T)$ of $V \times \left]t_0, t_1\right[$ and nil outside of it.

Such functions exist and can be construed in the following fashion:

If $(x_1, x_2, x_3, t) \in D_\varepsilon(X_1, X_2, X_3, T)$

$$W_1^\varepsilon(x_1, x_2, x_3, t) = \exp\left(\frac{1}{(x_1 - X_1)^2 + (x_2 - X_2)^2 + (x_3 - X_3)^2 + (t - T)^2 - \varepsilon^2}\right).$$

If $(x_1, x_2, x_3, t) \notin D_\varepsilon(X_1, X_2, X_3, T)$

$$W_1^\varepsilon(x_1, x_2, x_3, t) = 0.$$

For the field of virtual movement [2.9], the condition [2.8] is translated by:

$$\int_{t_0}^{t_1}\int_V \left(\rho\frac{\partial^2 W_1}{\partial t^2} - \sigma_{1j,j} - f_1\right) W_1^\varepsilon \, dvdt = 0 \quad \forall W_1^\varepsilon. \qquad [2.10]$$

Let us suppose that there exists a point (X_1, X_2, X_3, T) where equation [2.11] is verified (the case where the value would be negative would be treated in an identical manner):

$$\left(\rho\frac{\partial^2 W_1}{\partial t^2} - \sigma_{1j,j} - f_1\right)(X_1, X_2, X_3, T) > 0. \qquad [2.11]$$

We then can, taking continuity into account, find a neighborhood of (X_1, X_2, X_3, T) such that for any point (x_1, x_2, x_3, t) in this neighborhood:

$$\left(\rho\frac{\partial^2 W_1}{\partial t^2} - \sigma_{1j,j} - f_1\right)(x_1, x_2, x_3, t) > \frac{1}{2}\left(\rho\frac{\partial^2 W_1}{\partial t^2} - \sigma_{1j,j} - f_1\right)(X_1, X_2, X_3, T). \quad [2.12]$$

Let us consider an open ball $D_\varepsilon(X_1, X_2, X_3, T)$ inside the neighborhood. Due tc [2.11] and owing to the fact that the function W_1^ε is strictly positive in the open bal and nil outside of it, we have:

$$\int_{t_0}^{t_1}\int_V \left(\rho\frac{\partial^2 W_1}{\partial t^2} - \sigma_{1j,j} - f_1\right) W_1^\varepsilon \, dvdt > \frac{1}{2}\left(\rho\frac{\partial^2 W_1}{\partial t^2} - \sigma_{1j,j} - f\right)(X_1, X_2, X_3, T)\int_{t_0}^{t_1}\int_{D_\varepsilon(X_1, X_2, X_3, T)} dvdt \; ;$$

i.e. since the volume of the ball is strictly positive:

$$\int_{t_0}^{t_1}\int_V \left(\rho\frac{\partial^2 W_1}{\partial t^2} - \sigma_{1j,j} - f_1\right) W_1^\varepsilon \, dvdt > 0. \quad [2.1$$

This inequality contradicts [2.10]. Consequently, it is deduced from it that tl condition [2.8] involves:

$$\rho\frac{\partial^2 W_1}{\partial t^2} - \sigma_{1j,j} - f_1 = 0 \quad \forall (x_1, x_2, x_3, t) \in V \times \left]t_0, t_1\right[.$$

Similarly, considering the fields of virtual movements $\begin{pmatrix} 0 \\ w_2^\varepsilon \\ 0 \end{pmatrix}$ and $\begin{pmatrix} 0 \\ 0 \\ w_3^\varepsilon \end{pmatrix}$, could be shown that condition [2.8] involves:

$$\rho\frac{\partial^2 W_2}{\partial t^2} - \sigma_{2j,j} - f_2 = 0 \quad \forall (x_1, x_2, x_3, t) \in V \times \left]t_0, t_1\right[,$$

$$\rho\frac{\partial^2 W_3}{\partial t^2} - \sigma_{3j,j} - f_3 = 0 \quad \forall (x_1, x_2, x_3, t) \in V \times \left]t_0, t_1\right[.$$

To sum up, there is an equivalence between equations [2.8] and [2.1], which appear as global and local formulations of the continuous media equation of motion.

From a physical point of view, this result shows that if the integral over time of the work of the force of volume $\rho\dfrac{\partial^2 W_i}{\partial t^2} - \sigma_{ij,j} - f_i$ is nil for any displacement $W_i^* \in \Omega^*\left(V \times \left]t_0, t_1\right[\right)$, then the force of volume is nil. The integral [2.8] defines a functional Ξ that associates a real number $\Xi(W_i, \sigma_{ij}, W_i^*)$ to each triplet $(W_i, \sigma_{ij}, W_i^*)$:

$$\begin{array}{llll} \Xi: & \Omega \times \Sigma \times \Omega^* & \longrightarrow & \mathrm{IR} \\ & (W_i, \sigma_{ij}, W_i^*) & & \Xi(W_i, \sigma_{ij}, W_i^*) \end{array}$$

with:

Ω: the set of fields of indefinitely derivable displacements, defined over $V \times \left]t_0, t_1\right[$, with real values;

Σ: the set of fields of tensors of indefinitely derivable stress, defined over $V \times \left]t_0, t_1\right[$, with real values;

Ω^*: the set of fields of indefinitely derivable virtual displacements, defined over $V \times \left]t_0, t_1\right[$, with real values.

If the pair (W_i, σ_{ij}) does not verify the equation of motion [2.1], the functional $\Xi(W_i, \sigma_{ij}, W_i^*)$ is not nil.

If the pair (W_i, σ_{ij}) verifies the equation of motion [2.1], the functional $\Xi(W_i, \sigma_{ij}, W_i^*)$ is nil.

2.2.4. *Directional derivative*

The transformation of the local presentation [2.1] to global presentation [2.8] is the basic idea of the variational method; however, a second transformation is generally carried out, which consists of displaying [2.8] as the directional derivative of a simpler functional. Let us introduce the functional Ψ:

$$\begin{array}{llll} \Psi: & \Omega \times \Sigma & \longrightarrow & \mathrm{IR} \\ & (W_i, \sigma_{ij}) & & \Psi(W_i, \sigma_{ij}) \end{array}$$

with:

$$\Psi\left(W_i, \sigma_{ij}\right) = \int_{t_0}^{t_1}\int_V \left\{\frac{1}{2}\rho\left(\frac{\partial W_i}{\partial t}\right)^2 - \frac{1}{2}\sigma_{ij}\,\varepsilon_{ij} + f_i W_i\right\} \text{dvdt}. \quad [2.14]$$

The integrand contains three terms: the first is the kinetic energy, the second is the deformation energy and the third is the potential energy of the volume efforts.

Let us break up the field of displacement W_i in the following manner:

$$W_i = \overline{W}_i + \lambda W_i^*. \quad [2.15]$$

$\overline{W}_i$ is the field of displacement verifying the equation of motion, for a fixed stress field σ_{ij} (this notation has nothing to do with that of Chapter 1 where the bar indicated the boundary of a space):

$$\rho\frac{\partial^2 \overline{W}_i}{\partial t^2} - \sigma_{ij,j} - f_i = 0 \quad \forall\,(x_1, x_2, x_3, t) \in V \times \left]t_0, t_1\right[;$$

λ is a real number, W_i^* is a field of virtual displacement.

Let us replace W in [2.14] with its expression [2.15]:

$$\Psi\left(\overline{W}_i + \lambda \overline{W}_i^*, \sigma_{ij}\right) = \int_{t_0}^{t_1}\int_V \left\{\frac{1}{2}\rho\left(\frac{\partial \overline{W}_i}{\partial t} + \lambda\frac{\partial \overline{W}_i^*}{\partial t}\right)^2 - \frac{1}{2}\sigma_{ij}\left(\overline{\varepsilon}_{ij} + \lambda \overline{\varepsilon}_{ij}^*\right) + f_i\left(\overline{W}_i + \lambda \overline{W}_i^*\right)\right\} \text{dvdt}. \quad [2.16]$$

The directional derivative of the functional [2.14] with respect to W_i at the point $(\overline{W}_i, \sigma_{ij})$ noted $\delta_W \Psi(\overline{W}_i, \sigma_{ij}, W_i^*)$ is by definition:

$$\delta_W \Psi(\overline{W}_i, \sigma_{ij}, W_i^*) = \frac{d}{d\lambda}\Psi(\overline{W}_i + \lambda W_i^*, \sigma_{ij})\Big|_{\lambda=0}; \quad [2.17]$$

that is:

$$\delta_W \Psi(\overline{W}_i, \sigma_{ij}, W_i^*) = \int_{t_0}^{t_1}\int_V \left\{\rho\frac{\partial \overline{W}_i}{\partial t}\frac{\partial W_i^*}{\partial t} - \sigma_{ij}\, W_{i,j}^* + f_i\, W_i^*\right\} \text{dvdt}. \quad [2.18]$$

Let us transform equation [2.18]. By using Green's formulae and integration by parts, it follows:

$$\delta_W \Psi(\overline{W}_i, \sigma_{ij}, W_i^*) = \int_{t_0}^{t_1}\int_V \left(-\rho \frac{\partial^2 \overline{W}_i}{\partial t^2} + \sigma_{ij,j} + f \right) W_i^* \, dv dt + \int_{t_0}^{t_1}\int_{\overline{V}} \sigma_{ij} n_j W_i^* \, d\overline{v} dt + \left[\int_V \rho \frac{\partial \overline{W}_i}{\partial t} W_i^* \, dv \right]_{t_0}^{t_1} \qquad [2.19]$$

In the first integral of the second member we recognize the expression defined in [2.8] as the global form of the equation of motion. Thus, if we suppose that, on the one hand, W_i^* is nil over $\overline{V}$ at any moment t and that, on the other hand, it is nil over V at moments t_0 and t_1, the global form of the equation of motion is obtained by writing [2.20], that is by setting equal to 0 the directional derivative of the functional Ψ for all virtual displacements satisfying the boundary condition $W_i^* = 0$ over the surface $\overline{V}$ and the initial and final conditions $W_i^*(M, t_0) = 0$, $W_i^*(M, t_1) = 0$ for any point M of the volume V:

$$\delta_W \Psi(\overline{W}_i, \sigma_{ij}, W_i^*) = 0 \quad \forall\, W_i^* . \qquad [2.20]$$

A significant point appears here: it relates to the boundary conditions and the initial and final conditions of virtual displacement. Indeed it can be observed that the directional derivative of [2.14] in addition to the global form of the equation of motion reveals two additional terms which disappear when virtual displacement verifies the conditions: $W_i^* = 0$ over the surface $\overline{V}$, $W_i^*(M, t_0) = 0$ and $W_i^*(M, t_1) = 0$, for any point M of the volume V.

Let us suppose now that virtual displacements verify $W_i^*(M, t_0) = 0$ and $W_i^*(M, t_1) = 0$, for any point M of the volume V, but are left without the boundary condition. Taking into account [2.19], the relation [2.20] then becomes:

$$\delta_W \Psi(\overline{W}_i, \sigma_{ij}, W_i^*) = \int_{t_0}^{t_1}\int_V \left(-\rho \frac{\partial^2 \overline{W}_i}{\partial t^2} + \sigma_{ij,j} + f \right) W_i^* \, dv dt + \int_{t_0}^{t_1}\int_{\overline{V}} \sigma_{ij} n_j W_i^* \, d\overline{v} dt = 0 \quad \forall\, W_i^* . \qquad [2.21]$$

All of the virtual displacements nil over $\overline{V}$ are contained in the set of virtual displacements free over $\overline{V}$; the relation [2.21] thus implies:

$$\delta_W \Psi\left(\overline{W}_i, \sigma_{ij}, W_i^*\right) = \int_{t_0}^{t_1}\int_V \left(-\rho \frac{\partial^2 \overline{W}_i}{\partial t^2} + \sigma_{ij,j} + f\right) W_i^* \, dvdt = 0$$

$$\forall\, W_i^* \left| W_i^* = 0 \text{ over } \overline{V}\right.,$$

that is:

$$-\rho \frac{\partial^2 \overline{W}_i}{\partial t^2} + \sigma_{ij,j} + f = 0 \qquad \text{in} \qquad V \times \left]t_0, t_1\right[. \qquad [2.22]$$

Taking this result into account, equation [2.21] is reduced to:

$$\int_{t_0}^{t_1}\int_{\overline{V}} \sigma_{ij}\, n_j\, W_i^*\, d\overline{v}dt = 0 \quad \forall\, W_i^*. \qquad [2.23]$$

Considering now the virtual displacements not nil over $\overline{V}$, we deduce from the fundamental lemma that:

$$\sigma_{ij}\, n_j = 0 \qquad \text{over} \qquad \overline{V} \times \left]t_0, t_1\right[. \qquad [2.23']$$

If virtuai displacements are left free over $\overline{V}$, the nullity of the directional derivative [2.21] leads to the two relations [2.22] and [2.23'], that is to the equation of motion and the limiting condition of the free boundary type.

It is necessary to note here the importance of functional spaces where the directional derivation is carried out since they lead to different equations being verified.

Let us take the case of space Ω_0 of fields of displacements nil at the edges and equal to displacement solutions at the two moments, initial and final:

$$\Omega_0 = \left\{ W_i(M,t) \left| W_i(M,t) = 0 \quad \forall M \in \overline{V},\ \forall t \in \left]t_0, t_1\right[\right.\right.$$
$$\left. \text{and } W_i(M,t_0) = \overline{W}_i(M,t_0),\, W_i(M,t_1) = \overline{W}_i(M.t_1) \quad \forall M \in V \right\}. \qquad [2.24]$$

Introducing virtual displacements $W_i^*(M,t)$ verifying the relation [2.25],

$$W_i(M,t) = \overline{W}_i(M,t) + \lambda W_i^*(M,t), \qquad [2.25]$$

we note, since $W_i(M,t)$ and $\overline{W}_i(M,t)$ are elements of the space, that $W_i^*(M,t)$ must verify the relations [2.26] and [2.27]:

$$W_i^*(M,t) = 0 \quad \forall M \in \overline{V}\ ,\ \forall t \in \left]t_0, t_1\right[, \qquad [2.26]$$

$$W_i^*(M,t_0) = W_i^*(M,t_1) = 0 \quad \forall M \in V. \qquad [2.27]$$

The relations [2.26] and [2.27] confirm that virtual displacements do not form part of the space Ω_0 of real displacements.

By using virtual displacements compatible with Ω_0 in [2.19], we obtain:

$$\delta_W \Psi(\overline{W}_i, \sigma_{ij}, W_i^*) = 0 \quad \forall W_i^* \Leftrightarrow -\rho\frac{\partial^2 \overline{W}_i}{\partial t^2} + \sigma_{ij,j} + f_i = 0$$
$$\forall (M,t) \in V \times \left]t_0, t_1\right[. \qquad [2.28]$$

Let us now take the case of the space Ω_L containing the fields of displacements free on the surface $\overline{V}$ and equal to solution displacements at the two moments t_0 initial and t_1 final:

$$\Omega_1 = \left\{ W_i(M,t) \,\middle|\, W_i(M,t_0) = \overline{W}_i(M,t_0), W_i(M,t_1) = \overline{W}_i(M,t_1) \quad \forall M \in V \right\} \qquad [2.29]$$

The decomposition [2.25] shows then that in general:

$$W_i^*(M,t) \neq 0 \quad \forall M \in \overline{V}\ ,\ \forall t \in \left]t_0, t_1\right[. \qquad [2.30]$$

Using [2.30] in [2.19] we arrive at [2.21] that leads to the results [2.22] and [2.23]. In more mathematical terms we write [2.31]:

$$\delta_W \Psi(\overline{W}_i, \sigma_{ij}, W_i^*) = 0 \quad \forall W_i^*$$

$$\Leftrightarrow \begin{cases} -\rho\dfrac{\partial^2 \overline{W}_i}{\partial t^2} + \sigma_{ij,j} + f_i = 0 & \forall (M,t) \in V \times \left]t_0, t_1\right[\\ \sigma_{ij} n_j = 0 & \forall (M,t) \in \overline{V} \times \left]t_0, t_1\right[. \end{cases} \qquad [2.31]$$

The functional space where the directional derivation of a functional is carried out leads, as we have just demonstrated, to respecting different boundary conditions:

– for Ω_0 the boundary condition $W_i = 0 \quad \forall (M, t) \in \overline{V} \times \left]t_0, t_1\right[$ is prescribed;

– for Ω_1 the boundary condition $\sigma_{ij} n_j = 0 \quad \forall (M, t) \in \overline{V} \times \left]t_0, t_1\right[$ is deduced from the calculation of extremum.

This duality of the prescribed and deduced boundary conditions depending on the functional space where the directional derivation is carried out is general. Similarly, we could also obtain deduced initial and final conditions rather than those which we have prescribed:

$$W_i(M, t_0) = \overline{W_i}(M, t_0), W_i(M, t_1) = \overline{W_i}(M, t_1) \quad \forall M \in V.$$

However, the problems seldom arise in terms of initial and final conditions. Indeed, all the conditions are deferred to the initial moment. Consequently, the use of the variational technique to determine these conditions is generally unnecessary and in the continuation we will always place ourselves in the case of prescribed initial and final conditions.

2.2.5. *Extremum of a functional calculus*

We can give a physical image of the directional derivative based on the well-known concept of extremum.

Let us consider the functional of the preceding section and more precisely the expression [2.16]. For a fixed stress field σ_{ij}, the solution displacement $\overline{W_i}(M, t)$ is also fixed. If, moreover, we consider a particular field of virtual displacements W_i^* (M,t), the functional becomes nothing more than a function of the real variable λ.

We write this function down as: $f_{W_i^*}(\lambda)$. That is:

$$f_{W_i^*}(\lambda) = \Psi(\overline{W_i} + \lambda W_i^*, \sigma_{ij}). \qquad [2.32]$$

Under these conditions, writing:

$$\delta_W \Psi(\overline{W_i}, \sigma_{ij}, W_i^*) = \frac{d}{d\lambda} \Psi(\overline{W_i} + \lambda W_i^*, \sigma_{ij}) \Big|_{\lambda=0} = 0 \qquad [2.33]$$

amounts to having:

$$\partial f_{W_i^*} / \partial \lambda \,(0) = 0 \,. \qquad [2.34]$$

The traditional results of real functions show that the function $f_{W_i^*}(\lambda)$ presents an extremum in 0, and that, taking into account [2.32], it is equal to $\Psi(\overline{W}_i, \sigma_{ij})$. If the condition [2.33] is verified for any virtual displacement $W_i^*(M,t)$, all the functions $f_{W_i^*}(\lambda)$ present an extremum in 0 equal to $\Psi(\overline{W}_i, \sigma_{ij})$.

Consequently, stating that the directional derivative $\delta_W \Psi(\overline{W}_i, \sigma_{ij}, W_i^*)$ is nil for any virtual displacement $W_i^*(M,t)$, means that the functional presents an extremum for the pair $(\overline{W}_i, \sigma_{ij})$ (i.e. is stationary):

$$\delta_W \Psi(\overline{W}_i, \sigma_{ij}, W_i^*) = 0 \quad \forall\, W_i^* \Leftrightarrow \Psi(\overline{W}_i, \sigma_{ij}) = \underset{\Omega}{\mathrm{Extr}}\, \Psi(W_i, \sigma_{ij}) \,. \qquad [2.35]$$

As we have already stressed, the functional space where the calculation of extremum is carried out must be specified. We will thus use the notation [2.35] for saying that the extremum is obtained over the space Ω.

The extremum can be a maximum or a minimum; to establish that the second derivative of $f_{W_i^*}(\lambda)$ needs to be calculated. There are two cases:

– the extremum is a minimum if: $\dfrac{d^2 f_{W_i^*}}{d\lambda^2}(0) > 0$; [2.36a]

– the extremum is a maximum if: $\dfrac{d^2 f_{W_i^*}}{d\lambda^2}(0) < 0$. [2.36b]

There is no use in establishing whether the extremum of a functional is a maximum or a minimum in terms of the equivalence of the variational formulation and traditional formulation of the elastic solids vibration problems, since the only condition of stationarity is necessary.

2.3. Reissner's functional

2.3.1. *Basic functional*

The Reissner's functional is of the mixed type, that is it depends on the two variables (W_i, σ_{ij}). It will make it possible to find the set of equations [2.1] – [2.5] describing the vibrations of a continuous medium.

Let us introduce two functional spaces $\Omega_R(V \times]t_0, t_1[)$ and $\Sigma_R(V \times]t_0, t_1[)$.

$\Omega_R(V \times]t_0, t_1[)$ is the space of the fields of kinematically acceptable displacements $W_i(M, t)$. These displacements are real, defined over $V \times]t_0, t_1[$, indefinitely derivable and verifying the imposed displacement boundary conditions [2.37a] and the initial and final conditions [2.37b] and [2.37c]:

$$W_i(M, t) = D_i(M, t) \quad \forall (M, t) \in \overline{V}_E \times]t_0, t_1[, \qquad [2.37a]$$

$$W_i(M, t_0) = \overline{W}_i(M, t_0) \quad \forall M \in V, \qquad [2.37b]$$

$$W_i(M, t_1) = \overline{W}_i(M, t_1) \quad \forall M \in V. \qquad [2.37c]$$

$\Sigma_R(V \times]t_0, t_1[)$ is the space of the real stress fields $\sigma_{ij}(M,t)$, defined over $V \times]t_0, t_1[$ and indefinitely derivable.

The Reissner's functional $R_1(W_i, \sigma_{ij})$ is defined by:

$$\begin{aligned} R_1 : \ \Omega_R \times \Sigma_R &\longrightarrow \mathbb{R} \\ (W_i, \sigma_{ij}) &\longrightarrow R_1(W_i, \sigma_{ij}) \end{aligned}$$

with:

$$R_1(W_i, \sigma_{ij}) = \int_{t_0}^{t_1} \left(\int_V \left[\frac{1}{2}\rho \left(\frac{\partial W_i}{\partial t} \right)^2 - \sigma_{ij}\, \varepsilon_{ij} + f_i W_i + \frac{1}{2} \sigma_{ij}\, S_{ijkl}\, \sigma_{kl} \right] dv + \int_{\overline{V}} F_i W_i \, d\overline{v} \right) dt, \qquad [2.38]$$

S_{ijkl} is the inverse tensor of C_{ijkl} characteristic of the elastic law of behavior [2.2]:

$$\varepsilon_{ij} = S_{ijkl}\, \sigma_{kl}.$$

The following result can be stated: the directional derivatives of Reissner's functional $\delta_W R_1$ and $\delta_\sigma R_1$ are equal to 0 for a pair $(\overline{W}_i, \overline{\sigma}_{ij}) \in \Omega_R \times \Sigma_R$ if and only if the pair $(\overline{W}_i, \overline{\sigma}_{ij})$ verifies equations [2.1] – [2.5]. It can also be said that the pair $(\overline{W}_i, \overline{\sigma}_{ij})$ render the Reissner's functional R_1 stationary over the product space $\Omega_R \times \Sigma_R$ if and only if it verifies equations [2.1] – [2.5]. We will note:

$$R_1(\overline{W}_i, \overline{\sigma}_{ij}) = \frac{\text{Extr}}{\Omega_R \times \Sigma_R} R_1(W_i, \sigma_{ij}) \Leftrightarrow (\overline{W}_i, \overline{\sigma}_{ij}) \text{ verifies equations [2.1]} - \text{[2.5]}.$$

To show the result stated previously it suffices to use the results of section 2.2. Let us calculate the directional derivatives $\delta_W R_1$ and $\delta_\sigma R_1$:

$$\delta_W R_1(\overline{W}_i, \overline{\sigma}_{ij}, W_i^*) = \frac{d}{d\lambda} R_1(\overline{W}_i + \lambda W_i^*, \overline{\sigma}_{ij}), \qquad [2.39]$$

$$\delta_\sigma R_1(\overline{W}_i, \overline{\sigma}_{ij}, \sigma_{ij}^*) = \frac{d}{d\lambda} R_1(\overline{W}_i, \overline{\sigma}_{ij} + \lambda \sigma_{ij}^*), \qquad [2.40]$$

For $\delta_W R_1$, after using the formulas of Green and of integration by parts over time we obtain [2.41]:

$$\delta_W R_1(\overline{W}_i, \overline{\sigma}_{ij}, W_i^*) = \int_{t_0}^{t_1} \left(\int_V \left(-\rho \frac{\partial^2 \overline{W}_i}{\partial t^2} + \overline{\sigma}_{ij} + f_i \right) W_i^* \, dv - \int_{V_F} (\overline{\sigma}_{ij} n_j - F_i) W_i^* \, d\overline{v} - \int_{V_L} \overline{\sigma}_{ij} n_j W_i^* \, d\overline{v} \right) dt. \qquad [2.41]$$

Similarly, for $\delta_\sigma R_1$, we obtain the relation [2.42]:

$$\delta_\sigma R_1(\overline{W}_i, \overline{\sigma}_{ij}, \sigma_{ij}^*) = \int_{t_0}^{t_1} \int_V (S_{ijkl} \overline{\sigma}_{kl} - \overline{\varepsilon}_{ij}) \sigma_{ij}^* \, dv dt. \qquad [2.42]$$

After setting the directional derivatives equal to 0, equations [2.43], [2.44] and [2.45] are obtained from $\delta_W R_1$ and equation [2.46] from $\delta_\sigma R_1$:

$$\delta_W R_1(\overline{W}_i, \overline{\sigma}_{ij}, W_i^*) = 0 \quad \forall W_i^*$$
$$\Leftrightarrow -\rho \frac{\partial^2 \overline{W}_i}{\partial t^2} + \overline{\sigma}_{ij,j} + f_i = 0 \quad \forall (M,t) \in V \times \left]t_0, t_1\right[, \qquad [2.43]$$

$$\overline{\sigma}_{ij}\, n_j = 0 \quad \forall (M,t) \in \overline{V}_L \times \left]t_0, t_1\right[, \qquad [2.44]$$

$$\overline{\sigma}_{ij}\, n_j = F_i \quad \forall (M,t) \in \overline{V}_F \times \left]t_0, t_1\right[; \qquad [2.45]$$

$$\delta_\sigma R_1(\overline{W}_i, \overline{\sigma}_{ij}, \sigma^*_{ij}) = 0 \quad \forall \sigma^*_{ij}$$
$$\Leftrightarrow \overline{\varepsilon}_{ij} - S_{ijkl}\, \overline{\sigma}_{kl} = 0 \quad \forall (M,t) \in V \times \left]t_0, t_1\right[. \qquad [2.46]$$

The four relations [2.1], [2.2], [2.3] and [2.5] are easily found; the relation [2.4] has already been prescribed taking into account the choice of functional space Ω_R which imposed it *a priori*:

$$\overline{W}_i = D_i \quad \forall (M,t) \in \overline{V}_E \times \left]t_0, t_1\right[.$$

2.3.2. *Some particular cases of boundary conditions*

a) *Constrained medium*

Consider the case where: $\overline{V}_E = \overline{V}$ and $\overline{V}_L = \overline{V}_F = \varnothing$.

The functional R_1 is reduced to [2.47]:

$$R_1(W_i, \sigma_{ij}) = \int_{t_0}^{t_1} \left(\int_V \left[\frac{1}{2} \rho \left(\frac{\partial W_i}{\partial t} \right)^2 - \sigma_{ij}\, \varepsilon_{ij} + f_i W_i + \frac{1}{2} \sigma_{ij}\, S_{ijkl}\, \sigma_{kl} \right] dv \right) dt. \qquad [2.47]$$

The functional space where extremalization is carried out is still defined by $\Omega_R \times \Sigma_R$; however, the condition [2.37a] $W_i = D_i$ must in this case be verified for the entire boundary $\overline{V}$ of the continuous medium. If the imposed displacement is nil, we will simply have $W_i = 0$ over $\overline{V} \times]t_0, t_1[$.

b) *Free medium*

Consider the case where: $\overline{V}_L = \overline{V}$ and $\overline{V}_E = \overline{V}_F = \varnothing$.

The Reissner's functional is still given by [2.47], the functional space is $\Omega_R \times \Sigma_R$; however, $\overline{V}_E$ being reduced to the empty set, the condition [2.37a] is no longer prescribed; the kinematically acceptable displacements are free over the boundary of the continuous medium.

2.3.3. *Case of boundary conditions effects of rigidity and mass*

In certain cases we are brought to introduce boundary conditions intermediate between constrained and free surface. These conditions are characterized by a rigidity K and a mass μ; mathematically we use the following model type:

$$\sigma_{ij} n_j = -KW_i - \mu \frac{\partial^2 W_i}{\partial t^2} \quad \forall (M,t) \in \overline{V}_K \times \left]t_0, t_1\right[; \qquad [2.48]$$

$\overline{V}_K$ is part of the boundary $\overline{V}$ where the boundary condition [2.48] must be verified.

This type of boundary condition is in fact the most general; the traditional conditions of constrained and of free surface are borderline cases of this condition: we obtain a free surface by setting $K = 0$ and $\mu = 0$; a constrained surface by setting $\mu = 0$ and making K tend towards infinity.

The problem of vibration of elastic solids thus consists of finding the pair $(\overline{W_i}, \overline{\sigma_{ij}})$ verifying equations [2.1] – [2.5], given at the beginning of the chapter, and equation [2.48] above.

Moreover, we will have:

$$\overline{V} = \overline{V}_L \cup \overline{V}_E \cup \overline{V}_F \cup \overline{V}_K$$

and:

$$\overline{V}_L \cap \overline{V}_E = \overline{V}_L \cap \overline{V}_F = \overline{V}_L \cap \overline{V}_K = \overline{V}_F \cap \overline{V}_E = \overline{V}_K \cap \overline{V}_E = \overline{V}_F \cap \overline{V}_K = \varnothing .$$

The variational form of this problem can be stated as follows: the couple $(\overline{W_i}, \overline{\sigma_{ij}})$ render the Reissner's functional $R_2(W_i, \sigma_{ij})$ stationary over the product space $\Omega_R \times \Sigma_R$ if and only if it verifies equations [2.1] to [2.5] and [2.48], with:

$$R_2(W_i, \sigma_{ij}) = \int_{t_0}^{t_1} \left(\int_V \left[\frac{1}{2}\rho \left(\frac{\partial W_i}{\partial t} \right)^2 - \sigma_{ij}\varepsilon_{ij} + f_i W_i + \frac{1}{2}\sigma_{ij} S_{ijkl} \sigma_{kl} \right] dv + \int_{\overline{V}_F} F_i W_i \, d\overline{v} \right) dt \qquad [2.49]$$

$$+ \int_{t_0}^{t_1} \left(\int_{\overline{V}_K} \frac{1}{2} \left(KW_i^2 + \mu \left(\frac{\partial W_i}{\partial t} \right)^2 \right) d\overline{v} \right) dt .$$

The demonstration, identical to that of section 2.3.1, is left to the reader by way of exercise.

2.4. Hamilton's functional

2.4.1. *The basic functional*

The Hamilton's functional is a functional that depends only on the field of displacements W_i. It allows finding the equations of formulation in displacement of the problems of elastic solid media vibrations. These equations have been provided in Chapter 1, equations [1.64] – [1.67]. We will remind them here:

$$\rho\frac{\partial^2 W_i}{\partial t^2} = (C_{ijkl}\,\varepsilon_{kl})_{,j} + f_i \quad \forall (M,t) \in V \times \left]t_0, t_1\right[, \qquad [2.50]$$

$$C_{ijkl}\,\varepsilon_{kl}\,n_j = 0 \quad \forall (M,t) \in \overline{V}_L \times \left]t_0, t_1\right[, \qquad [2.51]$$

$$W_i = D_i \quad \forall (M,t) \in \overline{V}_E \times \left]t_0, t_1\right[, \qquad [2.52]$$

$$C_{ijkl}\,\varepsilon_{kl}\,n_j = F_i \quad \forall (M,t) \in \overline{V}_F \times \left]t_0, t_1\right[. \qquad [2.53]$$

We have the following result: the field of displacement $\overline{W_i}$ renders the Hamilton's functional H_l stationary over the space of kinematically admissible displacements Ω_R if and only if it verifies equations [2.50] – [2.53].

$$H_l(\overline{W_i}) = \underset{\Omega_R}{\text{Extr}}\, H_l(W_i) \Leftrightarrow \overline{W_i} \text{ verifies equations [2.50] – [2.53]} \qquad [2.54]$$

with:

$$H_l(W_i) = \int_{t_0}^{t_1}\left(\int_V \left(\frac{1}{2}\rho\left(\frac{\partial W_i}{\partial t}\right)^2 - \frac{1}{2}\varepsilon_{ij}\,C_{ijkl}\,\varepsilon_{kl} + f_i W_i\right)dv + \int_{\overline{V}_F} F_i W_i\,d\overline{v}\right)dt. \qquad [2.55]$$

To demonstrate this result, let us calculate the directional derivative of $\delta_W H_l$:

$$\delta_W H_l(\overline{W_i}, W_i^*) = \int_{t_0}^{t_1}\left(\int_V \left(-\rho\frac{\partial^2 \overline{W_i}}{\partial t^2} + (C_{ijkl}\,\overline{\varepsilon}_{kl})_{,j} + f_i\right)W_i^* dv\right.$$
$$\left. - \int_{\overline{V}} C_{ijkl}\,\overline{\varepsilon}_{kl}\,n_j\,W_i^*\,d\overline{v} + \int_{\overline{V}_F} F_i W_i^*\,d\overline{v}\right)dt + \left[\int_V \rho\frac{\partial \overline{W_i}}{\partial t}W_i^*\,dv\right]_{t_0}^{t_1}. \qquad [2.56]$$

The calculation of the directional derivative over the space Ω_R defined in [2.37] involves the nullity of the last term of the left member of [2.56] and the following equality:

$$\int_{\overline{V}} C_{ijkl}\,\overline{\varepsilon}_{kl}\,n_j\,W_i^*\,d\overline{v} = \int_{\overline{V}_F \cup \overline{V}_L} C_{ijkl}\,\overline{\varepsilon}_{kl}\,n_j\,W_i^*\,d\overline{v}$$

The directional derivative [2.56] is thus reduced to:

$$\delta_W H_1(\overline{W}_i, W_i^*) = \int_{t_0}^{t_1}\left(\int_V \left(-\rho\frac{\partial^2 \overline{W}_i}{\partial t^2} + (C_{ijkl}\,\overline{\varepsilon}_{kl})_{,j} + f_i\right) W_i^*\,dv - \int_{\overline{V}_F}(C_{ijkl}\,\overline{\varepsilon}_{kl}\,n_j - F_i)\,W_i^*\,d\overline{v} - \int_{\overline{V}_L} C_{ijkl}\,\overline{\varepsilon}_{kl}\,n_j\,W_i^*\,d\overline{v}\right) dt. \qquad [2.57]$$

It follows that the nullity of the directional derivative for any virtual displacement W_i^* implies that $\overline{W}_i$ must respect the relations [2.50], [2.51] and [2.53]. In other words:

$$\delta_W H_1(\overline{W}_i, W_i^*) = 0 \quad \forall\, W_i^* \Leftrightarrow \overline{W}_i \text{ verifies [2.50], [2.51], [2.53].}$$

Moreover, as the boundary condition [2.52] has been prescribed by the choice of the functional space Ω_R, we duly obtain the result [2.54].

2.4.2. *Some particular cases of boundary conditions*

a) *Boundary conditions presenting effects of mass and spring*

We employ the same notations as in section 2.3.3. The Hamilton's functional H_2 becomes in this case:

$$H_2(W_i) = \int_{t_0}^{t_1}\left(\int_V \left(\frac{1}{2}\rho\left(\frac{\partial W_i}{\partial t}\right)^2 - \frac{1}{2}\varepsilon_{ij}\,C_{ijkl}\,\varepsilon_{kl} + f_i W_i\right) dv + \int_{\overline{V}_F} F_i W_i\,d\overline{v} + \int_{\overline{V}_K}\left(\frac{1}{2}\mu\left(\frac{\partial W_i}{\partial t}\right)^2 - \frac{1}{2}K\,(W_i)^2\right) d\overline{v}\right) dt. \qquad [2.58]$$

The displacement $\overline{W}_i$, which is the solution of the problem, must verify:

$$H_2(\overline{W}_i) = \underset{\Omega_R}{\text{Extr}}\, H_2(W_i)\,.$$

b) *Constrained medium*

In this case $\overline{V}_E = \overline{V}$. The functional H_1 is reduced to:

$$H_1(W_i) = \int_{t_0}^{t_1} \left(\int_V \left(\frac{1}{2}\rho \left(\frac{\partial W_i}{\partial t} \right)^2 - \frac{1}{2}\varepsilon_{ij}\, C_{ijkl}\, \varepsilon_{kl} + f_i W_i \right) dv \right) dt\,. \qquad [2.59]$$

We still have the result [2.54]; however, the condition [2.37a] which provides the definition of the space Ω_R must be verified over the entire boundary $\overline{V}$.

c) *Free medium*

In this case $\overline{V}_L = \overline{V}$. The Hamilton's functional is provided by [2.55] and the space Ω_R is no longer subject to the condition [2.37a], since $\overline{V}_E$ is reduced to the empty set. The kinematically admissible displacements are thus left free over the entire boundary $\overline{V}$ of the continuous medium.

2.5. Approximate solutions

The exact solutions of the problems of elastic solid media vibration are generally impossible to find and we must be satisfied with approximations. A way of obtaining these approximations consists of using the geometrical characteristics of the continuous medium to determine *a priori* simplified sets of displacements and stress. These are the hypotheses of condensation which we will explain in detail in the following chapters. Let us simply say here that spaces Ω_R and Σ_R are restricted to spaces Ω_c and Σ_c verifying: $\Omega_c \subset \Omega_R \subset \Sigma_c \subset \Sigma_R$.

The variational technique may be used directly for the study of approximate solutions. Indeed, it is enough to carry out the calculation of extremum on the subspace corresponding to the hypotheses of condensation. The approximate solution of the problem $(\widetilde{W}_i, \widetilde{\sigma}_{ij})$ can be found by writing:

$$R_1(\widetilde{W}_i, \widetilde{\sigma}_{ij}) = \underset{\Omega_C \times \Sigma_C}{\text{Extr}}\, R_1(W_i, \sigma_{ij})\,.$$

It is rather difficult to determine if the pair $(\widetilde{W}_i, \widetilde{\sigma}_{ij})$ is close to the exact solution pair $(\overline{W}_i, \overline{\sigma}_{ij})$. We can only affirm that in the product space $\Omega_C \times \Sigma_C$ the pair $(\widetilde{W}_i, \widetilde{\sigma}_{ij})$ is the best possible approximation. To the extent that the space $\Omega_C \times \Sigma_C$ has been well selected, the approximate solution will be realistic. There are, however, methods to quantify the validity of the approximation. On this subject we may address ourselves to the article of Guyader [GUY 86] which uses a residual functional to justify the assumption of thin plates.

It is the ease of obtaining the approached formulations, by simple restriction of functional spaces where we carry out the calculation of extremum that constitutes the main attraction of the variational method.

2.6. Euler equations associated to the extremum of a functional

2.6.1. *Introduction and first example*

Euler equations state the conditions that the functions, on which a functional depends to become stationary, must verify. They thus make it possible to dispense with the often long calculation of the directional derivatives. We present the method of acquiring the Euler equations with two examples: firstly, when the unknown function depends only on the variable of space (which corresponds to the problems of statics of beams) and, secondly, in the next section, when the unknown function depends on two variables of space and time (which corresponds to the problems of vibrations of plates or shells). We will finally draw up a summary of the various types of functionals and associated Euler equations, which will be useful to us thereafter.

To begin with let us consider a functional of the form [2.60]:

$$\Lambda\big(y(x)\big) = \int_0^L F\big(y(x)\big)dx . \qquad [2.60]$$

The functions y(x) are not constrained to verify any boundary condition in y(x) and x = L (afterwards we will consider the case of prescribed boundary conditions).

We can interpret each function y(x) as a path of integration upon which the value on the functional calculus depends. Calculating the extremum of the functional consists in determining the path of integration $\overline{y}(x)$ placing the value of the functional at its extremum. The calculation can be carried out thanks to the directional derivative as we have highlighted in the preceding sections.

Let us calculate the directional derivative of the functional in the neighborhood of the function $\overline{y}(x)$ that is supposed to return the extremum of the functional. We pose [2.61] and then carry out the calculation [2.62]:

$$y(x) = \overline{y}(x) + \lambda y^*(x) \qquad [2.61]$$

$$\frac{d\Lambda}{d\lambda}(\overline{y} + \lambda y^*)\bigg|_{\lambda=0} = \int_0^L \frac{dF}{d\lambda}(\overline{y} + \lambda y^*)\bigg|_{\lambda=0} dx\,. \qquad [2.62]$$

Observing the rules of composed derivations, and supposing that the function F depends on y(x) and its derivatives $\frac{d^i}{dx^i}$ for an i variable to the nth order, it follows:

$$\frac{d\Lambda}{d\lambda}(\overline{y} + \lambda y^*)\bigg|_{\lambda=0} = \int_0^L \left[\frac{\partial F}{\partial y}(\overline{y})\, y^* + \frac{\partial F}{\partial y_{,x}}(\overline{y})\, y^*_{,x} + \dots + \frac{\partial F}{\partial y_{,x^n}}(\overline{y})\, y^*_{,x^n}\right] dx\,. \qquad [2.63]$$

Let us transform [2.63] by integration by parts. After all the calculations it results in:

$$\begin{aligned}
&\frac{d\Lambda}{d\lambda}(\overline{y} + \lambda y^*)\bigg|_{\lambda=0} \\
&= \int_0^L \left(\frac{\partial F}{\partial y}(\overline{y}) - \frac{d}{dx}\frac{\partial F}{\partial y_{,x}}(\overline{y}) + \dots + (-1)^n \frac{d^n}{dx^n}\frac{\partial F}{\partial y_{,x^n}}(\overline{y})\right) y^*\, dx \\
&+ \left[y^*(x)\left(\frac{\partial F}{\partial y_{,x}} + \dots + (-1)^{n-1}\frac{d^{n-1}}{dx^{n-1}}\frac{\partial F}{\partial y_{,x^n}}\right)\right]_0^L \\
&+ \left[y^*_{,x}(x)\left(\frac{\partial F}{\partial y_{,x^2}} + \dots + (-1)^{n-2}\frac{d^{n-2}}{dx^{n-2}}\frac{\partial F}{\partial y_{,x^n}}\right)\right]_0^L \\
&+ \left[y^*_{,x^{n-1}}(x)\left(\frac{\partial F}{\partial y_{,x^n}}\right)\right]_0^L
\end{aligned} \qquad [2.64]$$

Setting the directional derivative [2.64] equal to 0 for any virtual displacement $y^*(x)$ implies verifying the following equations:

1) an equation to be verified in the domain $x \in]0, L[$:

$$\frac{\partial F}{\partial y}(\bar{y}) - \frac{d}{dx}\frac{\partial F}{\partial y_{,x}}(\bar{y}) + \dots + (-1)^n \frac{d^n}{dx^n}\frac{\partial F}{\partial y_{,x^n}}(\bar{y}) = 0; \qquad [2.65]$$

2) boundary conditions to be verified in $x = 0$ and $x = L$:

$$\frac{\partial F}{\partial y_{,x^1}}(\bar{y}) - \frac{d}{dx}\frac{\partial F}{\partial y_{,x^2}}(\bar{y}) + \dots + (-1)^{n-1} \frac{d^{n-1}}{dx^{n-1}}\frac{\partial F}{\partial y_{,x^n}}(\bar{y}) = 0, \qquad [2.66]$$

and for $1 < i \leq n$:

$$\frac{\partial F}{\partial y_{,x^i}}(\bar{y}) - \frac{d}{dx}\frac{\partial F}{\partial y_{,x^{i+1}}}(\bar{y}) + \dots + (-1)^{n-i} \frac{d^{n-i}}{dx^{n-i}}\frac{\partial F}{\partial y_{,x^n}}(\bar{y}) = 0. \qquad [2.67]$$

The number of terms to be cumulated in the expression [2.67] varies with index i: only the first $(n - i + 1)$ terms are to be considered in the sum. For example, for $i = 2$, there are $n - 1$ terms to be cumulated, while for $i = n$, there remains only one term and the limiting condition [2.67] is reduced to:

$$\frac{\partial F}{\partial y_{,x^n}}(\bar{y}) = 0$$

Let us now consider the case where the functions y(x) are constrained to verify the boundary conditions:

$$y(0) = c_0^0 \quad \text{and} \quad y(L) = c_L^0, \qquad [2.68]$$

and for $1 \leq i \leq n - 1$:

$$\frac{d^i y}{dx^i}(0) = c_0^i \quad \text{and} \quad \frac{d^i y}{dx^i}(L) = c_L^i. \qquad [2.69]$$

Taking into account the decomposition [2.61] and owing to the fact that the function $\bar{y}(x)$ is a particular y(x) function, it follows:

$$y^*(0) = 0 \quad \text{and} \quad y^*(L) = 0, \qquad [2.70]$$

and for $1 \leq i \leq n-1$:

$$\frac{d^i y^*}{dx^i}(0) = 0 \quad \text{and} \quad \frac{d^i y^*}{dx^i}(L) = 0 \qquad [2.71]$$

Setting the directional derivative equal to 0 over the space of functions y(x) that we have just defined, taking into account the cancellation of the terms with boundaries in [2.64] (consequently, of [2.70] and [2.71]), leads to the verification of solely equation [2.65]. The boundary conditions are now prescribed by [2.68] and [2.69].

In short the calculation of extremum of the functional [2.60] gives the Euler equation [2.65] to be verified in the domain]0, L[, and to the alternative choice of prescribed or deduced boundary conditions ([2.66] or [2.68], [2.67] or [2.69]) that has to be determined according to the problem considered.

As an example, let us take the case of $n = 1$.

The Euler equation to be verified $\forall x \in]0, L[$ is reduced to:

$$\frac{\partial F}{\partial y}(\overline{y}) - \frac{d}{dx}\frac{\partial F}{\partial y_{,x}}(\overline{y}) = 0. \qquad [2.72]$$

The boundary conditions to choose alternatively are:

– for $x = 0$:

either: $\overline{y}(0) = c_0^0$, [2.73]

or: $\dfrac{\partial F(\overline{y})}{\partial y_{,x}}(0) = 0$; [2.74]

– for $x = L$:

either: $\overline{y}(L) = c_L^0$, [2.75]

or: $\dfrac{\partial F(\overline{y})}{\partial y_{,x}}(L) = 0$. [2.76]

If the functional depends on several functions $y_i(x)$:

$$\Lambda\left(y_1(x), \ldots, y_n(x)\right) = \int_0^L F\left(y_1(x), \ldots, y_n(x)\right) dx. \qquad [2.77]$$

A calculation similar to the preceding developments shows that the Euler equations that we stipulated for a function y(x) must be verified for each function $\overline{y}_i(x)$ in order to render the functional stationary.

Henceforth, in order to be concise, we will not mention anymore that the equations are verified by the particular function $\overline{y}_i(x)$. We will note, for example, instead of [2.72]:

$$\frac{\partial F}{\partial y} - \frac{d}{dx}\frac{\partial F}{\partial y_{,x}} = 0 . \qquad [2.78]$$

2.6.2. *Second example: vibrations of plates*

The functional [2.60] was of the type describing the problems of statics of beams, since it depended only on the functions of the single variable of space x. In order to obtain the Euler equations for a more general case, we will consider the case of the functionals of the type describing the problems of vibrations of plates (see Chapter 4), i.e. of the type [2.79]:

$$\Lambda\left(W(x_1, x_2, t)\right) = \int_{t_0}^{t_1}\int_S F\left(W(x_1, x_2, t)\right) dx_1\, dx_2\, dt . \qquad [2.79]$$

The functions $W(x_1, x_2, t)$ that we consider hereafter are constrained to verify the initial and final conditions:

$$W(x_1, x_2, t_0) = \overline{W}(x_1, x_2, t_0)$$

$$\text{and } W(x_1, x_2, t_1) = \overline{W}(x_1, x_2, t_1) \quad \forall (x_1, x_2) \in S . \qquad [2.80]$$

The function $\overline{W}(x_1, x_2, t)$ is the solution that we seek and we break up the functions $W(x_1, x_2, t)$ in the usual manner:

$$W(x_1, x_2, t) = \overline{W}(x_1, x_2, t) + \lambda W^*(x_1, x_2, t) . \qquad [2.81]$$

The calculation of the directional derivative $\delta_W \Lambda$ using the rules of chain derivation yields:

$$\delta_W \Lambda(\overline{W}, W^*) = \int_{t_0}^{t_1}\int_S \left(\frac{\partial F}{\partial W}(\overline{W})\, W^* + \frac{\partial F}{\partial W_{,t}}(\overline{W})\, W^*_{,t} \right.$$

$$+ \frac{\partial F}{\partial W_{,1}}(\overline{W})\, W^*_{,1} + \frac{\partial F}{\partial W_{,2}}(\overline{W})\, W^*_{,2} + \frac{\partial F}{\partial W_{,11}}(\overline{W})\, W^*_{,11} \qquad [2.82]$$

$$\left. + \frac{\partial F}{\partial W_{,12}}(\overline{W})\, W^*_{,12} + \frac{\partial F}{\partial W_{,22}}(\overline{W})\, W^*_{,22} \right) dx_1\, dx_2\, dt .$$

In order to write [2.82] we have supposed that the function F depended on W as well as its partial derivative of the 1st order with respect to time and of the 2nd order with respect to the variables of space.

Hereafter, in order to avoid convoluted notation, we will not indicate that the function F and its derivatives are to be calculated for the function $\overline{W}$. We will note, for example:

$$\frac{\partial F}{\partial W}(\overline{W}) \quad \text{by} \quad \frac{\partial F}{\partial W}.$$

Using the formulas of integration by parts over time and of Ostrogradski for the space variables, it follows:

$$\begin{aligned}
\delta_W \Lambda(\overline{W}, W^*) = & \int_{t_0}^{t_1}\int_S \left(\frac{\partial F}{\partial W} - \frac{\partial}{\partial t}\frac{\partial F}{\partial W_{,t}} - \frac{\partial}{\partial x_1}\frac{\partial F}{\partial W_{,1}} - \frac{\partial}{\partial x_2}\frac{\partial F}{\partial W_{,2}} \right. \\
& \left. + \frac{\partial^2}{\partial x_1^2}\frac{\partial F}{\partial W_{,11}} + \frac{\partial^2}{\partial x_1 \partial x_2}\frac{\partial F}{\partial W_{,12}} + \frac{\partial^2}{\partial x_2^2}\frac{\partial F}{\partial W_{,22}} \right) W^*\, dx_1\, dx_2\, dt \\
& + \int_{t_0}^{t_1}\int_{\overline{S}} W^* \left(n_1 \frac{\partial F}{\partial W_{,1}} + n_2 \frac{\partial F}{\partial W_{,2}} - n_1 \left(\frac{\partial}{\partial x_1}\frac{\partial F}{\partial W_{,11}} + \frac{\partial}{\partial x_2}\frac{\partial F}{\partial W_{,12}} \right) \right. \\
& \left. \qquad - n_2 \frac{\partial}{\partial x_2}\frac{\partial F}{\partial W_{,22}} \right) \\
& + \int_{t_0}^{t_1}\int_{\overline{S}} \left(W^*_{,1} \left(n_2 \frac{\partial F}{\partial W_{,12}} + n_1 \frac{\partial F}{\partial W_{,11}} \right) + W^*_{,2}\, n_2 \frac{\partial F}{\partial W_{,22}} \right) d\overline{s}dt.
\end{aligned} \quad [2.83]$$

The terms at the boundaries over time which appear during integration by parts are nil considering the hypothesis [2.80]. Quantities n_1 and n_2 are the direction cosines of the normal vector external to the contour $\overline{S}$ of the plate.

The normal and tangential derivative $W_{,n}$ and $W_{,s}$ are linked to the derivatives $W_{,x_1}$ and $W_{,x_2}$ by the following relations:

$$\begin{aligned}
& W_{,n} = W_{,1} n_1 + W_{,2} n_2 \quad \text{and} \quad W_{,s} = W_{,1} n_2 + W_{,2} n_1 , \\
& W_{,1} = n_1 W_{,n} - n_2 W_{,s} \quad \text{and} \quad W_{,2} = n_2 W_{,n} - n_1 W_{,s} .
\end{aligned} \quad [2.84]$$

By introducing $W_{,n}$ and $W_{,s}$ the third integral of the second member of [2.83] becomes:

$$\int_{t_0}^{t_1}\int_S \left(W^*_{,n}\left((n_1)^2 \frac{\partial F}{\partial W_{,11}} + n_1 n_2 \frac{\partial F}{\partial W_{,21}} + (n_2)^2 \frac{\partial F}{\partial W_{,22}} \right) + W^*_{,s}\left(n_1 n_2 \left(\frac{\partial F}{\partial W_{,22}} - \frac{\partial F}{\partial W_{,11}} \right) - n_2^2 \frac{\partial F}{\partial W_{,12}} \right)\right) d\bar{s}dt. \quad [2.85]$$

Observing that:

$$\int_{t_0}^{t_1}\int_S W^*_{,s}\left(n_1 n_2 \left(\frac{\partial F}{\partial W_{,22}} - \frac{\partial F}{\partial W_{,11}} \right) - n_2^2 \frac{\partial F}{\partial W_{,12}} \right) d\bar{s}dt = -\int_{t_0}^{t_1}\int_S W^* \left(n_1 n_2 \left(\frac{\partial F}{\partial W_{,22}} - \frac{\partial F}{\partial W_{,11}} \right) - n_2^2 \frac{\partial F}{\partial W_{,12}} \right)_{,s} d\bar{s}dt, \quad [2.86]$$

after calculation and suitable grouping of terms we obtain a new expression for [2.83]:

$$\begin{aligned}\delta_W \Lambda(\overline{W}, W^*) = & \int_{t_0}^{t_1}\int_S \left(\frac{\partial F}{\partial W} - \frac{\partial}{\partial t}\frac{\partial F}{\partial W_{,t}} - \frac{\partial}{\partial x_1}\frac{\partial F}{\partial W_{,1}} - \frac{\partial}{\partial x_2}\frac{\partial F}{\partial W_{,2}} \right. \\ & \left. + \frac{\partial^2}{\partial x_1^2}\frac{\partial F}{\partial W_{,11}} + \frac{\partial^2}{\partial x_1 \partial x_2}\frac{\partial F}{\partial W_{,12}} + \frac{\partial^2}{\partial x_2^2}\frac{\partial F}{\partial W_{,22}} \right) W^* \, dx_1 \, dx_2 \, dt \\ & + \int_{t_0}^{t_1}\int_S W^* \left(n_1 \frac{\partial F}{\partial W_{,1}} + n_2 \frac{\partial F}{\partial W_{,2}} - n_1\left(1 + n_2^2\right)\frac{\partial}{\partial x_1}\frac{\partial F}{\partial W_{,11}} \right. \\ & - n_2(1 + n_1^2)\frac{\partial}{\partial x_2}\frac{\partial F}{\partial W_{,22}} - n_1^3 \frac{\partial}{\partial x_2}\frac{\partial F}{\partial W_{,12}} - n_2^3 \frac{\partial}{\partial x_1}\frac{\partial F}{\partial W_{,12}} \\ & \left. + n_2 n_1^2 \frac{\partial}{\partial x_2}\frac{\partial F}{\partial W_{,11}} + n_1 n_2^2 \frac{\partial}{\partial x_1}\frac{\partial F}{\partial W_{,22}} \right) d\bar{s}dt \\ & + \int_{t_0}^{t_1}\int_S W^*_{,n} \left(n_1^2 \frac{\partial F}{\partial W_{,11}} + n_2^2 \frac{\partial F}{\partial W_{,22}} + n_1 n_2 \frac{\partial F}{\partial W_{,21}} \right) d\bar{s}dt. \end{aligned} \quad [2.87]$$

Note: to apply the variational method the replacement of the first member of [2.86] by the second member is obligatory since over $\overline{S}$ the tangential derivative $W^*_{,s}$ is completely determined by the given of the function W^*; these two quantities thus cannot vary separately and must be grouped.

The Euler equations associated with the extremalization of the functional [2.87] are obtained by writing:

$$\delta_W \Lambda(\overline{W}, W^*) = 0 \quad \forall\, W^*$$

After calculation follows the equation of motion [2.88] and the boundary conditions [2.89] – [2.92].

Equation of motion:

$$\frac{\partial F}{\partial W} - \frac{\partial}{\partial t}\frac{\partial F}{\partial W_{,t}} - \frac{\partial}{\partial x_1}\frac{\partial F}{\partial W_{,1}} - \frac{\partial}{\partial x_2}\frac{\partial F}{\partial W_{,2}} + \frac{\partial^2}{\partial x_1^2}\frac{\partial F}{\partial W_{,11}} + \frac{\partial^2}{\partial x_1 \partial x_2}\frac{\partial F}{\partial W_{,12}} + \frac{\partial^2}{\partial x_2^2}\frac{\partial F}{\partial W_{,22}} + \frac{\partial^2}{\partial x_2 \partial t}\frac{\partial F}{\partial W_{,2t}} + \frac{\partial^2}{\partial x_1 \partial t}\frac{\partial F}{\partial W_{,1t}} = 0 \qquad [2.88]$$

$$\forall\,(x_1, x_2) \in S \;,\quad \forall\, t \in \left]t_0, t_1\right[.$$

Boundary conditions to be verified $\forall\,(x_1, x_2) \in \overline{S}\;, \;\forall\, t \in \left]t_0, t_1\right[$:

either $\overline{W}(x_1, x_2, t) = \overline{d}(x_1, x_2, t)$, [2.89]

or

$$n_1 \frac{\partial F}{\partial W_{,1}} + n_2 \frac{\partial F}{\partial W_{,2}} - n_1\left(1 + n_2^2\right)\frac{\partial}{\partial x_1}\frac{\partial F}{\partial W_{,11}} - n_2(1 + n_1^2)\frac{\partial}{\partial x_2}\frac{\partial F}{\partial W_{,22}} - n_1^3 \frac{\partial}{\partial x_2}\frac{\partial F}{\partial W_{,12}} - n_2^3 \frac{\partial}{\partial x_1}\frac{\partial F}{\partial W_{,12}} + n_2\, n_1^2 \frac{\partial}{\partial x_2}\frac{\partial F}{\partial W_{,11}} + n_1\, n_2^2 \frac{\partial}{\partial x_1}\frac{\partial F}{\partial W_{,22}} = 0 \qquad [2.90]$$

and:

either $\overline{W}_{,n}(x_1, x_2, t) = \overline{c}(x_1, x_2, t)$, [2.91]

$$\text{or } n_1^2 \frac{\partial F}{\partial W_{,11}} + n_2^2 \frac{\partial F}{\partial W_{,22}} + n_1 n_2 \frac{\partial F}{\partial W_{,21}} = 0, \qquad [2.92]$$

where $\overline{d}(x_1, x_2, t)$ and $\overline{c}(x_1, x_2, t)$ are the displacements and the normal derivative of the displacements imposed on the boundary $\overline{S}$ of the plate. For a clamped boundary, these two functions are nil.

2.6.3. *Some results*

Providing the Euler equations in very general cases of functionals is difficult, taking into account the heaviness of the expressions that have to be handled, in particular, boundary conditions. It has, however, appeared necessary to us to gather the results which will be brought into use in the following chapters and which in fact cover nearly all the functionals interesting for our purposes.

We still consider that the functionals depend only on one function in order not to weigh down the writing since the case of functionals depending on several functions amounts applying the results, which we provide, to each function. We are still in the situation where the extremum is calculated over the set of functions verifying the initial and final conditions.

2.6.3.1. *Mechanical type functional of non-deformable solid*

The functions describing these systems depend only on time; they are most often generalized co-ordinates $q_i(t)$. The mechanical type functionals of not deformable solid are thus of the form:

$$L\left(q(t)\right) = \int_{t_0}^{t_1} F\left(q(t)\right) dt. \qquad [2.93]$$

The function F depends on the function q(t) and on its first derivative $q_{,t}(t)$.

The Euler equation associated with the extremalization of [2.93] is given by [2.94]; it is the simple form of the Lagrange equations:

$$\frac{\partial F}{\partial q} - \frac{d}{dt}\frac{\partial F}{\partial q_{,t}} = 0 \quad \forall t \in \left]t_0, t_1\right[. \qquad [2.94]$$

2.6.3.2. *Static beam type functional*

This case has been detailed in section 2.6.1, to which we refer the reader. The functional of the type [2.60] leads to the equation of motion [2.65] and to the boundary conditions [2.66] and [2.67].

2.6.3.3. *Beams vibration type functional*

The functions describing the vibratory behavior of beams depend on time and a variable of space. The corresponding functionals are of the type:

$$L\big(W(x,t)\big) = \int_{t_0}^{t_1}\int_0^L F\big(W(x,t)\big)\,dxdt\,. \qquad [2.95]$$

In the integrand of [2.95] we consider that the function F depends on W and on its partial derivative $\partial^{j+1}W/\partial t\partial x^j$ which we will note $W_{,tx^j}$, where the index j varies from 1 to n.

The calculation of extremum of the functional leads to verifying an equation in the domain and n boundary conditions:

Equation of motion:

$$\frac{\partial F}{\partial W} - \frac{\partial}{\partial t}\frac{\partial F}{\partial W_{,t}} - \frac{\partial}{\partial x}\frac{\partial F}{\partial W_{,x}} + \ldots + (-1)^n \frac{\partial^{n+1}}{\partial t\partial x^n}\frac{\partial F}{\partial W_{,tx^n}} = 0$$

$$\forall (x,t) \in \left]0, L\right[\times \left]t_0, t_1\right[, \qquad [2.96]$$

or in shortened form:

$$\sum_{i=0}^{1}\sum_{j=0}^{n}(-1)^{i+j}\frac{\partial^{i+j}}{\partial t^i\partial x^j}\frac{\partial F}{\partial W_{,t^i x^j}} = 0 \quad \forall (x,t) \in \left]0, L\right[\times \left]t_0, t_1\right[. \qquad [2.97]$$

Let us observe that a nil derivation index indicates that there is no derivation.

For example:

$$W_{,t^0x^3} = \frac{\partial^3 W}{\partial x^3} \quad \text{and} \quad W_{,t^0x^0} = W\,. \qquad [2.98]$$

The boundary conditions are given by the n alternatives to be verified in x = 0 and x = L at any moment $t \in \left]t_0, t_1\right[$

for $j = 0, \ldots, n-1$:

$$\text{either } \frac{\partial \overline{W}}{\partial x^j} = 0, \qquad [2.99]$$

$$\text{or } \sum (-1)^{i+1} \frac{\partial^{i-j-1}}{\partial x^{i-j-1}} \left(\frac{\partial F}{\partial W_{,x^i}} \right) = 0.$$

Let us take as an example the case n = 2. Equation [2.97] becomes:

$$\frac{\partial F}{\partial W} - \frac{\partial}{\partial t}\frac{\partial F}{\partial W_{,t}} - \frac{\partial}{\partial x}\frac{\partial F}{\partial W_{,x}} + \frac{\partial^2}{\partial x^2}\frac{\partial F}{\partial W_{,xx}} + \frac{\partial^2}{\partial x \partial t}\frac{\partial F}{\partial W_{,xt}} = 0$$

$$\forall (x, t) \in \left]0, L\right[\times \left]t_0, t_1\right[.$$

The two boundary conditions are obtained with [2.99]:

$$\text{either } \overline{W} = 0,$$

$$\text{or } \frac{\partial F}{\partial W_{,x}} - \frac{\partial}{\partial x}\left(\frac{\partial F}{\partial W_{,xx}} \right) = 0;$$

and:

$$\text{either } \frac{\partial \overline{W}}{\partial x} = 0,$$

$$\text{or } \frac{\partial F}{\partial W_{,xx}} = 0.$$

2.6.3.4. *Plates vibration type functional*

This case has been described in section 2.6.2.

2.6.3.5. *Three-dimensional medium vibration type functional*

The functions describing the vibratory behavior depend on time and three variables of space; the functionals are of the type:

$$L\big(W(x_1, x_2, x_3, t)\big) = \int_{t_0}^{t_1} \int_V F\big(W(x_1, x_2, x_3, t)\big) dx_1\, dx_2\, dx_3\, dt. \qquad [2.100]$$

We will suppose that the integrand depends on the function W and its first derivative with respect to time and the three variables of space.

The equation to verify $\forall (x_1, x_2, x_3) \in V$, $\forall t \in \left]t_0, t_1\right[$ is:

$$\frac{\partial F}{\partial W} - \frac{\partial}{\partial t}\frac{\partial F}{\partial W_{,t}} - \frac{\partial}{\partial x_1}\frac{\partial F}{\partial W_{,x_1}} - \frac{\partial}{\partial x_2}\frac{\partial F}{\partial W_{,x_2}} - \frac{\partial}{\partial x_3}\frac{\partial F}{\partial W_{,x_3}} = 0 . \qquad [2.101]$$

The boundary condition to verify $\forall (x_1, x_2, x_3) \in \overline{V}$, $\forall t \in \left]t_0, t_1\right[$ is given by the alternative:

either $\quad W = 0,$ [2.102]

or $\quad \frac{\partial F}{\partial W_{,1}} n_1 + \frac{\partial F}{\partial W_{,2}} n_2 + \frac{\partial F}{\partial W_{,3}} n_3 = 0,$ [2.103]

where n_1, n_2, n_3 are the direction cosines of the external normal vector.

2.7. Conclusion

The variational method that we have just presented transforms the local description in terms of equilibrium of force from Chapter 1 into a global description in terms of energy; they are two manifestations of the same phenomenon. Practically, the search for the solution of a problem is carried out by the calculation of stationary point values of a functional over the set of fields of displacements and/or admissible stresses. All the interest of the method consists in this particular way of obtaining the solutions. Indeed, the complexity of the phenomena is such that in general only approximations are possible. However, this search for approximate solutions is performed in a simple and natural manner using the variational method since it suffices to employ the same technique of calculation of extremum but on subsets of the fields of displacements and/or admissible stresses. This restriction of the fields is delicate because it is carried out *a priori*, taking into account the characteristics of the studied elastic solid medium (geometry, stress type, etc.); these are the assumptions of condensation, which owe their name to the fact that they often lead to mono or bi-dimensional continuous media.

Several other functionals built on the same basic idea could be proposed; we have limited ourselves here to the two principal ones, Reissner and Hamilton, which will be used alternatively in the following discussion.

From a mathematical point of view, the variational formulation is based on directional derivation; this operation generally takes rather a long time to carry out but can be curtailed considerably by the use of the Euler equations which stem from it. For a certain number of standard functionals we have provided the Euler equations. The rather heavy formalism requires a little training, for which the following chapters will provide plenty of opportunity.

Chapter 3

Equation of Motion for Beams

3.1. Objective of the chapter

The three-dimensional equations of a continuous solid elastic medium vibrations provided in Chapter 1 are of a great complexity and in general cannot be solved analytically. However, elastic solids used in the mechanical engineering present geometrical characteristics which simplify the mathematical analysis of their vibrations. These simplifications made *a priori* have led to the theories of beams, plates and shells. In the following chapters we will present the traditional simplifications; let us state here that with the use of the variational approach, this step will lead to "condensing" the three-dimensional continuous medium into a simpler, bi or mono-dimensional, continuous medium.

Theories of beams consist of constructing mono-dimensional models and in this sense represent the simplest continuous media. This simplicity is extremely useful since it leads to obtaining analytical solutions of the problem equations and, consequently, to studying the vibratory phenomena in a comprehensive fashion.

Research of the basic vibratory phenomena results in the identification of three elementary movements: longitudinal vibrations, vibrations of torsion and bending vibrations. Of course, such a decomposition of the beam movements is a simplification based on a decoupling linked to the excitation type and the frequency band. The study of coupled longitudinal movements, torsion and bending is possible, but with an increased difficulty of resolution.

The equations will be set with the use of Reissner's functional and, thus, of mixed variables: tensor of stress and of displacements. However, purely for purposes of comparison we will provide the results obtained with Hamilton's functional. The approach is based on the calculation of the extremum of functionals presented in Chapter 2 and more precisely on the use of the Euler equations. To benefit from this chapter the reader must have a good grasp of the variational techniques given in Chapter 2.

3.2. Hypotheses of condensation of straight beams

The defined mechanical medium is considered (Figure 3.1). Two of the dimensions (width b and thickness h) of this mechanical medium are small compared to the third (length L). Such a geometrical particularity leads to sides x_2 and x_3 of the points of the continuous medium that never move away considerably from the axis 1, which is the longitudinal axis of the beam passing through the center of the cross-section $x_2 = 0$ and $x_3 = 0$. To exploit the preceding observation mathematically we carry out a development of the components of displacement and components of the tensor of stress of the solid medium in a Taylor series:

$$W_i(x_1, x_2, x_3, t) = W_i(x_1, 0, 0, t) + x_2 \frac{\partial W_i}{\partial x_2}(x_1, 0, 0, t) + x_3 \frac{\partial W_i}{\partial x_3}(x_1, 0, 0, t) + \ldots \quad [3.1]$$

$$\sigma_{ij}(x_1, x_2, x_3, t) = \sigma_{ij}(x_1, 0, 0, t) + x_2 \frac{\partial \sigma_{ij}}{\partial x_2}(x_1, 0, 0, t) + x_3 \frac{\partial \sigma_{ij}}{\partial x_3}(x_1, 0, 0, t) + \ldots \quad [3.2]$$

Taking into account the field of variation of x_2 and x_3, it appears reasonable to truncate the developments [3.1] and [3.2] with linear terms; this is the beam hypothesis.

The problem, therefore, is not solved since the functions $W_i(x_1, 0, 0, t)$, $\frac{\partial W_i}{\partial x_2}(x_1, 0, 0, t)$, etc., to be calculated require knowledge of $W_i(x_1, x_2, x_3, t)$ and of $\sigma_{ij}(x_1, x_2, x_3, t)$ as well as their first derivatives in x_2 and x_3; however, these are precisely the unknowns of the problem. In fact, the developments [3.1] and [3.2] are interesting for the shape of displacements and the tensor of the stress which they suggest.

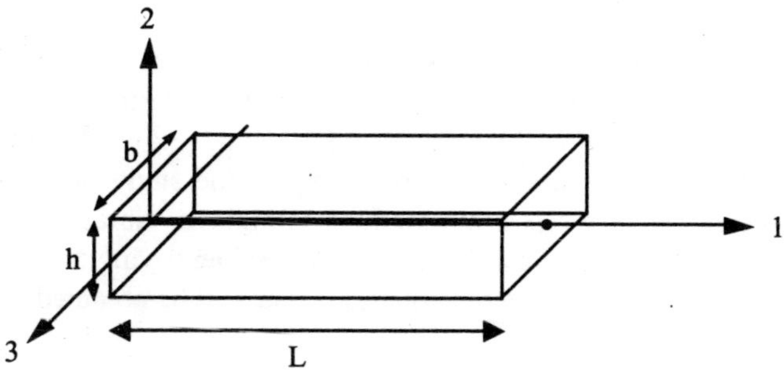

Figure 3.1. *Geometry of a beam*

It is this general form which we will retain while writing down the fields of displacements and of stress [3.3] and [3.4] as:

$$W_i(x_1, x_2, x_3, t) = W_i^0(x_1, t) + x_2 W_i^2(x_1, t) + x_3 W_i^3(x_1, t), \tag{3.3}$$

$$\sigma_{ij}(x_1, x_2, x_3, t) = \sigma_{ij}^0(x_1, t) + x_2\, \sigma_{ij}^2(x_1, t) + x_3\, \sigma_{ij}^3(x_1, t). \tag{3.4}$$

They are formally identical to [3.1] and [3.2], but the functions W_i^0, W_i^2, W_i^3, σ_{ij}^0, σ_{ij}^2 and σ_{ij}^3 are now independent and must be adjusted in order to verify as well as possible the equations of continuous medium vibrations.

Let us note that the displacement and stress fields [3.3] and [3.4] are too simplified to verify in all points the three-dimensional equations of the continuous solid elastic media vibrations provided equations [2.1] – [2.5] in Chapter 2 and are thus mere approximations of the 3D solutions.

For beams, we will verify the equations of elastic solid media vibrations only in the sense of an average over the cross-section. The equations which will result from it will depend only on the variable of space x_1 and of the time; they are thus characteristic of a mono-dimensional medium. This transformation of a three-dimensional medium into a mono-dimensional medium via developments [3.3] and [3.4] is sometimes called condensation and the hypotheses expressed by [3.3] and [3.4] are the hypotheses of condensation.

We thus define a beam as a continuous medium the displacement and tensor of stresses components of which can be tackled using the condensation hypotheses [3.3] and [3.4] with an acceptable precision.

Resolving the problem of vibrations of beams in general consists of determining the 27 unknown functions of the fields [3.3] and [3.4], that is, to solve 27 paired equations. This formidably complex task has not yet been performed. It is preferable to simplify the condensation hypotheses [3.3] and [3.4] based on particular excitations conditions. This amounts to breaking up the study of the vibrations of beams into three elementary cases: longitudinal vibrations, vibrations of torsion and bending vibrations. We will follow this procedure by identifying to the best possible extent the hypotheses that underlie the equations that will be obtained.

3.3. Equations of longitudinal vibrations of straight beams

3.3.1. *Basic equations with mixed variables*

Initially, we will define the hypotheses of condensation adapted to the study of the longitudinal vibrations of beams; in fact the issue is to preserve only the dominating terms in the fields [3.3] and [3.4]. This simplification of displacement and stress fields is carried out by a physical analysis of displacements and constraints associated with the type of vibration considered. The operation is not easy; in our opinion it constitutes the most delicate part of the modeling of dynamic behavior of elastic solids.

Longitudinal vibrations of beams bring about considerable displacements along axis 1 and weak displacements along axes 2 and 3; we accept, moreover, that the displacement along axis 1 is the same for all the points of the same cross-section, that is:

$$\begin{aligned} &W_1(x_1, x_2, x_3, t) = W_1^0(x_1, t), \\ &W_2(x_1, x_2, x_3, t) = 0, \\ &W_3(x_1, x_2, x_3, t) = 0. \end{aligned} \qquad [3.5]$$

The field of displacement [3.5] is of course a first approximation of the real movement; it must be noted that the Poisson effect, which describes the reduction of the cross-section when the beam lengthens, is not taken into account. We could consider a finer theory taking this effect into account by using the field of displacement [3.6]:

$$\begin{aligned} &W_1(x_1, x_2, x_3, t) = W_1^0(x_1, t), \\ &W_2(x_1, x_2, x_3, t) = x_2 W_2^2(x_1, t) \\ &W_3(x_1, x_2, x_3, t) = x_3 W_3^3(x_1, t) \end{aligned} \qquad [3.6]$$

To understand the physical significance of the hypotheses, the reader could proceed graphically in the following manner. Let us consider a rectangular cross-section and represent the displacements of its points induced by the term $W_1^0(x_1, t)$ (Figure 3.2). It is a displacement following axis 1 identical for all the points of a given cross-section; it is thus a translation of the cross-sections along axis 1.

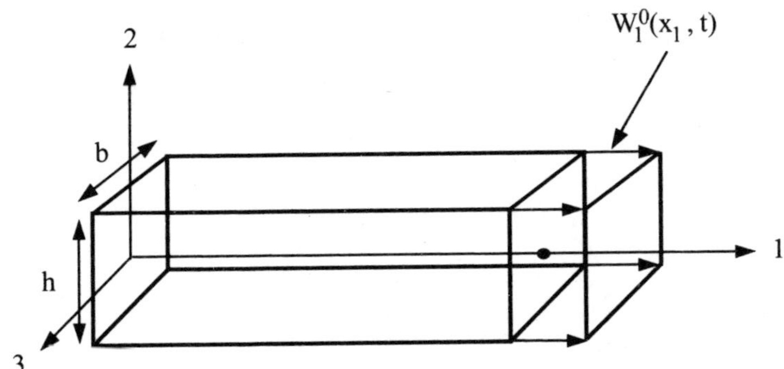

Figure 3.2. *Graphic representation of displacement* $W_1^0(x_1, t)$

Figure 3.3 depicts the displacement $x_2W_2^2(x_1, t)$ representing a movement of flattening and swelling of the cross-section along axis 2; the term $x_3W_3^3(x_1, t)$ represents the same type of movement following axis 3.

These two movements characterize the Poisson effect, that is, the reduction of the cross-section when the beam is extended by traction or the increase in the cross-section when the beam is subjected to a longitudinal compression.

We leave the task to interpret each term of the fields [3.3] and [3.4] to the reader by way of exercise.

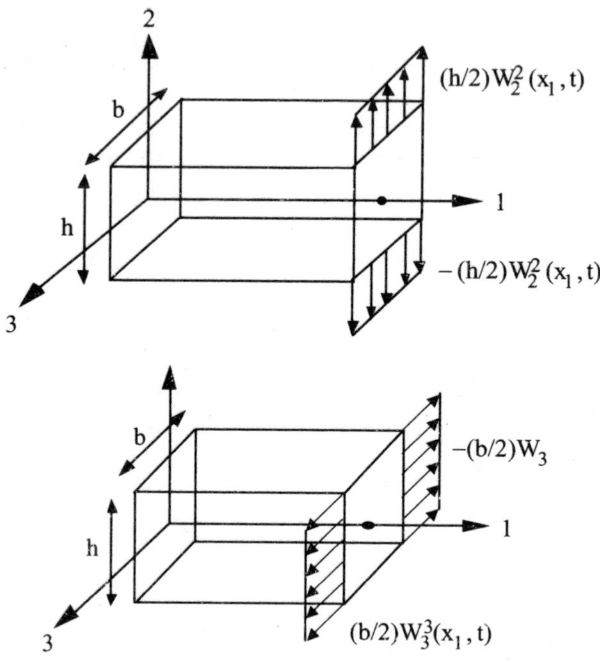

Figure 3.3. *Graphic representation of displacements* $x_2W_2^2(x_1, t)$ *and* $x_3W_3^3(x_1, t)$

In terms of stresses we suppose that the effect of traction-compression dominates, that is that longitudinal stress σ_{11} is considerably greater than other stresses. We pose:

$$\sigma_{11}(x_1, x_2, x_3, t) = \sigma_{11}^0(x_1, t),$$

$$\sigma_{ij}(x_1, x_2, x_3, t) = 0 \quad \text{if} \quad (i, j) \neq (1, 1). \qquad [3.7]$$

The fields [3.5] and [3.7] constitute the hypotheses of condensation which we consider for the study of longitudinal vibrations; we will thus neglect the Poisson effect in the rest of the discussion.

To obtain the equations of motion and the boundary conditions which must be verified by $W_1^0(x_1, t)$ and $\sigma_{11}^0(x_1, t)$ we use the Reissner's functional defined in Chapter 2 (equation [2.38]) supposing that the external surface forces are nil $F_i = 0$.

Generally:

$$R(W_i, \sigma_{ij}) = \int_{t_0}^{t_1}\int_V \left(\frac{1}{2}\rho\left(\frac{\partial W_i}{\partial t}\right)^2 - \sigma_{ij}\,\varepsilon_{ij} + \frac{1}{2}\sigma_{ij}\,S_{ijkl}\,\sigma_{kl} + f_i W_i \right) dvdt\,. \qquad [3.8]$$

When restricting W_i and σ_{ij} to [3.5] and [3.7] and replacing them, taking into account the fact that the force of volume f_i is null in free vibration, it follows:

$$R(W_1^0, \sigma_{11}^0) = \int_{t_0}^{t_1}\int_V \left(\frac{1}{2}\rho\left(\frac{\partial W_1^0}{\partial t}\right)^2 - \sigma_{11}^0\, W_{1,1}^0 + \frac{1}{2}\sigma_{11}^0\, S_{1111}\,\sigma_{11}^0 \right) dvdt\,. \qquad [3.9]$$

The dependence of the stress and displacements fields being fixed on x_2 and x_3 we can integrate over the cross-section of the beam; separating the integral of volume into an integral over the length and one over the cross-section, it follows:

$$R(W_1^0, \sigma_{11}^0) = \int_{t_0}^{t_1}\int_0^L \left(\frac{1}{2}\rho S\left(\frac{\partial W_1^0}{\partial t}\right)^2 - S\sigma_{11}^0 W_{1,1}^0 + \frac{1}{2}S(\sigma_{11}^0)^2\, S_{1111} \right) dx_1\, dt\,, \qquad [3.10]$$

where S is the cross-section of the beam, possibly a function of x_1 for a beam with a variable section.

Integration over the cross-section condenses the continuous medium since the unknown functions are dependent only on the variable of space x_1. From a physical point of view, this approach makes it possible to verify the equations of elastic solid continuous media globally over the cross-section and no longer in every point. The reader can realize this by introducing [3.5] and [3.7] into the three-dimensional equations given in Chapter 2 (equations [2.1] – [2.5]) and by noting that these cannot be verified.

To obtain the partial derivative equations characteristic of longitudinal vibrations of straight beams, it suffices to render the functional [3.10] stationary. Calculation is quite simple if the results of Chapter 2 for the Euler equations associated with a functional are used. For example, the application of equation [2.97] of Chapter 2, to our case, yields:

$$\frac{\partial F}{\partial W_1^0} - \frac{\partial}{\partial t}\frac{\partial F}{\partial W_{1,t}^0} - \frac{\partial}{\partial x_1}\frac{\partial F}{\partial W_{1,1}^0} = 0$$

and:

$$\frac{\partial F}{\partial \sigma_{11}^0} = 0 .$$

Upon calculation that gives the equation of motion [3.11] and the stress-strain relation [3.12]:

Stress-strain relation:

$$\rho S \frac{\partial^2 W_1^0}{\partial t^2} - \frac{\partial}{\partial x_1}(S\sigma_{11}^0) = 0 \quad \forall t , \forall x_1 \in \left]0, L\right[\qquad [3.11]$$

Relation of beam behavior:

$$- S \frac{\partial W_1^0}{\partial x_1} + S\sigma_{11}^0 S_{1111} = 0 \quad \forall t , \forall x_1 \in \left]0, L\right[. \qquad [3.12]$$

By application of equations [2.99] of Chapter 2, we obtain the boundary conditions:

$$\begin{aligned} &\text{either} \quad W_1^0(x_1, t) = 0 \quad \forall t , \; x_1 = 0 \;\text{ and }\; x_1 = L \\ &\text{or} \quad W_1^0(x_1, t) \neq 0 \Rightarrow S\sigma_{11}^0 = 0 \quad \forall t , \; x_1 = 0 \;\text{ and }\; x_1 = L . \end{aligned} \qquad [3.13]$$

The boundary conditions are given in the form of an alternative which is always interpreted as the nullity of a displacement or that of a constraint. In our case we will speak of a clamped end when displacement is imposed as nil and of a free end when the displacement is left free of all movement. Taking into account [3.13], these two boundary conditions will be translated mathematically by:

free end: $W_1^0 = 0$,

clamped end: $S\sigma_{11}^0 = 0$.

A beam in longitudinal vibration will thus take three types of basic boundary conditions:

$$\text{clamped-clamped:}\quad W_1^0\left(0,t\right)=0 \quad\text{and}\quad W_1^0\left(L,t\right)=0 \quad \forall\, t\,; \qquad [3.14]$$

$$\text{free-clamped:}\quad S\sigma_{11}^0\left(0,t\right)=0 \quad\text{and}\quad W_1^0\left(L,t\right)=0 \quad \forall\, t\,; \qquad [3.15]$$

$$\text{free-free:}\quad S\sigma_{11}^0\left(0,t\right)=0 \quad\text{and}\quad S\sigma_{11}^0\left(L,t\right)=0 \quad \forall\, t\,. \qquad [3.16]$$

3.3.2. *Equations with displacement variables*

In order to limit the number of unknown functions and equations, we often proceed by substitution in the equations in order to make the variables of stress disappear and thus to formulate the problem using only displacement variables.

Let us draw from [3.12] the value of $\sigma_{11}^0(x_1\,,t)$ according to $W_1^0(x_1\,,t)$:

$$\sigma_{11}^0(x_1\,,t) = \frac{1}{S_{1111}}\frac{\partial W_1^0}{\partial x_1}(x_1\,,t)\,. \qquad [3.17]$$

For an isotropic material with a Young modulus E, we have $S_{1111} = 1/E$ it follows:

$$\sigma_{11}^0(x_1\,,t) = E\frac{\partial W_1^0}{\partial x_1}(x_1\,,t)\,. \qquad [3.18]$$

Substituting [3.18] in [3.11], we obtain the equation of the free vibrations with displacement variables:

$$\rho S\frac{\partial^2 W_1^0}{\partial t^2} - \frac{\partial}{\partial x_1}\left(ES\frac{\partial W_1^0}{\partial x_1}\right) = 0 \quad \forall\, t\;,\; \forall\, x_1 \in \left]0\,,L\right[\,. \qquad [3.19]$$

Substituting [3.18] in [3.13] we obtain the boundary conditions in displacement variables:

$$\text{either } W_1^0(x_1, t) = 0 \quad \forall t, \; x_1 = 0 \text{ and } x_1 = L$$

[3.20]

$$\text{or } ES\frac{\partial W_1^0}{\partial x_1}(x_1, t) = 0 \quad \forall t, \; x_1 = 0 \text{ and } x_1 = L.$$

In the particular case of the homogenous beam (it is the simplest case, which is, in fact, one of the rare cases that can be solved without difficulty (see Chapter 5)) it is supposed that E, ρ and S are constant at every point of the beam which leads to the equations:

Equation of vibrations:

$$\rho S\frac{\partial^2 W_1^0}{\partial t^2} - ES\frac{\partial^2 W_1^0}{\partial x_1^2} = 0 \quad \forall t, \; \forall x_1 \in \left]0, L\right[. \qquad [3.21]$$

Boundary conditions:

$$\text{either } W_1^0(x_1, t) = 0 \quad \forall t, \; x_1 = 0 \text{ and } x_1 = L$$

[3.22]

$$\text{or } ES\frac{\partial W_1^0}{\partial x_1}(x_1, t) = 0 \quad \forall t, \; x_1 = 0 \text{ and } x_1 = L.$$

3.3.3. *Equations with displacement variables obtained by Hamilton's functional*

The formulations in displacements no longer consider stresses as variables independent of displacement, but as quantities related by the three-dimensional stress-strain relation:

$$\sigma_{ij} = C_{ijkl}\,\varepsilon_{kl}. \qquad [3.23]$$

The hypotheses of condensation should thus only relate to displacements, the values of the stresses being the direct consequence of respecting [3.23]. Taking again the field of displacements [3.5] we deduce with the help of [3.23] the associated field of stresses; for an isotropic material we have:

$$\sigma_{11} = C_{1111}\, W^0_{1,1}\ ,\ \sigma_{22} = C_{2211}\, W^0_{1,1}\ ,\ \sigma_{33} = C_{3311}\, W^0_{1,1}\ ,$$
$$\sigma_{12} = \sigma_{21} = 0\ ,\ \sigma_{13} = \sigma_{31} = 0\ ,\ \sigma_{32} = \sigma_{23} = 0\,. \qquad [3.24]$$

The field of stresses [3.24] is different from [3.7] since the components of the tensor of stresses σ_{22} and σ_{33} are non-nil since the coefficients C_{2211} and C_{3311} are non-nil. It is, in fact, less realistic since from a physical point of view σ_{22} and σ_{33} must be weak; indeed, the boundary conditions in any point of the external surface of the beam are that of a free surface:

$$\sigma_{ij}\, n_j = 0\,, \qquad [3.25]$$

where the quantities n_j are the direction cosines of the external normal vector $\vec{n}$ to the external surface of the beam. Their values are illustrated in Figure 3.4.

Let us place ourselves at the point $(x_1\,, h/2\,, x_3)$ on the external surface of the beam $(n_1 = 0\,, n_2 = 1\,, n_3 = 0)$. We deduce from [3.25] that:

$$\sigma_{22}(x_1\,, h/2\,, x_3) = 0\,.$$

Placing ourselves at other points of the surface, we would obtain in the same manner:

$$\sigma_{22}(x_1\,, -h/2\,, x_3) = 0\ ,\ \ \sigma_{33}(x_1\,, x_2\,, b/2) = 0\ ,\ \ \sigma_{33}(x_1\,, x_2\,, -b/2) = 0\,.$$

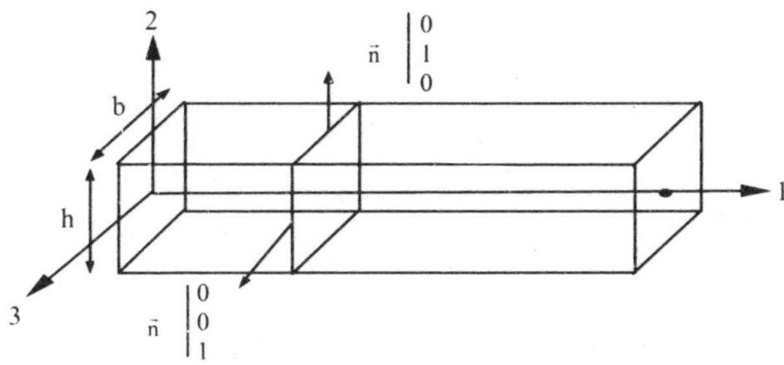

Figure 3.4. *Normal vector external to surface of the beam*

The thickness and the width of the beam are small by hypothesis and the stresses thus only vary a little throughout the cross-section, that is $\sigma_{22} \approx 0$ and $\sigma_{33} \approx 0$, which is contradictory with the hypotheses [3.24] but corresponds perfectly to [3.7].

We see here the great disadvantage of the formulation with displacements, which associates a much less realistic state of stresses to a realistic simplification of the field of displacements.

Using the field of displacements [3.5] in Hamilton's functional (equation [2.55], Chapter 2) it follows after integration over the cross-section:

$$H\left(W_1^0(x_1,t)\right) = \int_{t_0}^{t_1}\int_0^L \left[\frac{1}{2}\rho S\left(\frac{\partial W_1^0}{\partial t}\right)^2 - C_{1111}S\left(\frac{\partial W_1^0}{\partial x_1}\right)^2\right] dx_1\, dt . \qquad [3.26]$$

The calculation of the stationarity of the functional [3.26], taking into account the results of Chapter 2, leads to the equations:

Equation of vibrations:

$$\rho S\frac{\partial^2 W_1^0}{\partial t^2} - \frac{\partial}{\partial x_1}\left(C_{1111}\, S\frac{\partial W_1^0}{\partial x_1}\right) = 0 \quad \forall t\ ,\ \forall x_1 \in \left]0, L\right[. \qquad [3.27]$$

Boundary conditions:

$$\text{either}\quad W_1^0(x_1, t) = 0 \quad \forall t\ ,\ x_1 = 0 \text{ and } x_1 = L$$

$$[3.28]$$

$$\text{or}\quad C_{1111}\, S\frac{\partial W_1^0}{\partial x_1}(x_1, t) = 0 \quad \forall t\ ,\ x_1 = 0 \text{ and } x_1 = L .$$

Equations [3.27] and [3.28] correspond to [3.19] and [3.20]. There exists, however, a difference on the level of the equations coefficients since $C_{1111} \neq 1/S_{1111}$. For an isotropic material, for example, we have $S_{1111} = 1/E$ and $C_{1111} = E(1-\nu^2)/(1+\nu)(1-2\nu)$.

The comparison with experience shows that the results drawn from equations [3.19] and [3.20] are more satisfactory than those drawn from [3.27] and [3.28]. This established fact has led the users of formulations with displacements to amend the three-dimensional stress-strain relation in the case of beam or plate. We pose for the beams: $C_{1111} = E$. At this cost the formulations with displacements and mixed formulations lead to the same results.

3.4. Equations of vibrations of torsion of straight beams

3.4.1. *Basic equations with mixed variables*

Once again we adopt the methodology applied in the preceding section. It is thus necessary to define the hypotheses of condensation as the first step. The movement of torsion is characterized by a rotation of the cross-sections around the longitudinal axis of the beam; the stresses that result from it are of a shearing type. The fields of displacements and stresses of the beam are reduced under these conditions to:

$$
\begin{aligned}
&W_1(x_1, x_2, x_3, t) = 0, \\
&W_2(x_1, x_2, x_3, t) = -x_3\,\alpha(x_1, t), \\
&W_3(x_1, x_2, x_3, t) = +x_2\,\alpha(x_1, t)\ ;
\end{aligned}
\qquad [3.29]
$$

$$
\begin{aligned}
&\sigma_{11} = 0\ ,\ \sigma_{22} = 0\ ,\ \sigma_{33} = 0\ ,\ \sigma_{23} = \sigma_{32} = 0, \\
&\sigma_{12} = \sigma_{21} = -x_3\,\tau(x_1, t)\ ,\ \sigma_{13} = \sigma_{31} = +x_2\,\tau(x_1, t).
\end{aligned}
\qquad [3.30]
$$

To depict the field of displacements [3.29] we have traced the displacements in a cross-section in Figure 3.5. The quantity $\alpha(x_1, t)$ is the angle of torsion characteristic of the rotation of the cross-sections. The stresses associated with the movement of torsion are pure shear stresses.

Note: the hypotheses of condensation [3.29] and [3.30] are applicable to beams having a cross-section symmetrical with respect to axis 1 (in particular, circular). A non-symmetrical cross-section would introduce a coupling with the bending.

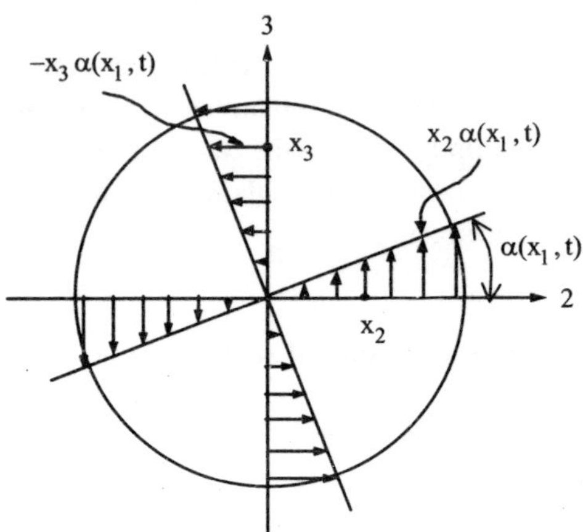

Figure 3.5. *Displacements in a cross-section corresponding to the field of displacement [3.29]*

Let us calculate the Reissner's functional of the problem by introducing [3.29] and [3.30] in equation [2.38] of Chapter 2. For a beam homogenous in the cross-section and consisting of an isotropic material, it follows (the calculation is left to the reader as an exercise):

$$R\big(\alpha(x_1,t),\tau(x_1,t)\big)=\int_{t_0}^{t_1}\int_0^L\left(\frac{1}{2}\rho\,I_0\left(\frac{\partial\alpha}{\partial t}\right)^2-I_0\,\tau\,\frac{\partial\alpha}{\partial x_1}+4\,S_{1212}\,I_0\,\tau^2\right)dx_1dt\,, \qquad [3.31]$$

with $I_0=\iint_S\left(x_2^2+x_3^2\right)\,dx_2\,dx_3$.

Let us now apply the results of Chapter 2 concerning the extremalization of the functional [3.31]. Noting $S_{1212}=1/4G$ where G is the Coulomb module of material, we obtain:

Equation of motion:

$$\rho I_0\frac{\partial^2\alpha}{\partial t^2}-\frac{\partial}{\partial x_1}(I_0\,\tau)=0\quad\forall t\;,\;\forall x_1\in\,]0,L[\,. \qquad [3.32]$$

Stress-strain relation:

$$I_0 \frac{\partial \alpha}{\partial x_1} = I_0 \frac{\tau}{G} \quad \forall t \ , \ \forall x_1 \in \left]0, L\right[. \qquad [3.33]$$

Boundary conditions:

$$\text{either} \quad \alpha\,(x_1\,,t) = 0 \quad \forall\, t \ , \ x_1 = 0 \ \text{ and } \ x_1 = L$$

$$[3.34]$$

$$\text{or} \quad \alpha\,(x_1\,,t) \neq 0 \Rightarrow I_0\ \tau\ (x_1\,,t) = 0 \quad \forall\, t \ , \ x_1 = 0 \ \text{ and } \ x_1 = L\ .$$

The term $I_0\,\tau\,(x_1\,,t)$ is homogenous with a torque, it is the torsional moment which is opposed to the rotation of torsion $\alpha\,(x_1\,,t)$.

Equations [3.32], [3.33] and [3.34] define the problem of pure torsion of isotropic beams; they result from a simplification of three-dimensional movement, realistic for beams with cross-sections symmetrical or nearly symmetrical with respect to axis 1.

3.4.2. *Equation with displacements*

It is enough to draw $\tau\,(x_1\,,t)$ on the basis of $\alpha\,(x_1\,,t)$ from equation [3.33]:

$$\tau = G \frac{\partial \alpha}{\partial x_1} \quad \forall t \ , \ \forall x_1 \in \left]0, L\right[, \qquad [3.35]$$

then to replace $\tau\,(x_1\,,t)$ with its expression resulting from [3.35] in the equation of motion [3.32], that is:

$$\rho I_0 \frac{\partial^2 \alpha}{\partial t^2} - \frac{\partial}{\partial x_1}\left(I_0 G \frac{\partial \alpha}{\partial x_1} \right) = 0 \quad \forall t \ , \ \forall x_1 \in \left]0, L\right[. \qquad [3.36]$$

Similarly, we obtain for the boundary conditions:

$$\begin{aligned}&\text{either}\quad \alpha\,(x_1\,,t)=0 \quad \forall\, t\ ,\ x_1=0\ ,\ x_1=L\\ &\text{or}\quad \alpha\,(x_1\,,t)\neq 0 \Rightarrow I_0\ G\frac{\partial\alpha}{\partial x_1}(x_1\,,t)=0 \quad \forall\, t\ ,\ x_1=0\ ,\ x_1=L\,.\end{aligned} \qquad [3.37]$$

The term $I_0\, G\dfrac{\partial\alpha}{\partial x_1}\left(x_1\,,t\right)$ is the torsional moment expressed on the basis of the angle of torsion.

Let us examine the particular case of the homogenous isotropic beam: it is the simplest case. It is supposed that G, ρ and I_0 are constants. The equations with displacements are simplified and become:

Equation of vibrations:

$$\rho I_0\frac{\partial^2\alpha}{\partial t^2}-I_0 G\frac{\partial^2\alpha}{\partial x_1^2}=0 \quad \forall\, t\ ,\ \forall\, x_1\in\left]0,L\right[. \qquad [3.38]$$

Boundary conditions:

$$\begin{aligned}&\text{either}\quad \alpha\,(x_1\,,t)=0 \quad \forall\, t\ ,\ x_1=0\ ,\ x_1=L\\ &\text{or}\quad \alpha\,(x_1\,,t)\neq 0 \Rightarrow I_0\ G\frac{\partial\alpha}{\partial x_1}(x_1\,,t)=0 \quad \forall\, t\ ,\ x_1=0\ ,\ x_1=L\,.\end{aligned} \qquad [3.39]$$

Note:

a) The boundary conditions of a beam in vibrations of torsion are of two types:

– clamped end: $\alpha\,(x_1\,,t)=0$ (zero rotation of the cross-section);

– free end: $I_0\, G\dfrac{\partial\alpha}{\partial x_1}(x_1\,,t)=0$ (zero torque).

A beam in vibration of torsion could thus be clamped-clamped, clamped-free or free-free.

b) The equations of vibrations of torsion of homogenous and isotropic beams [3.38] and [3.39] are formally identical to the equations of longitudinal vibrations of homogenous and isotropic beams [3.21] and [3.22]. This similarity is surprising enough taking into account the difference of the two movements considered; on the other hand, it is very practical since it makes it possible to study the two vibratory movements in the same way. In fact, equations [3.21], [3.22] and [3.38], [3.39] are also found in the two problems of sound pipes and of vibrating cords. Equation [3.21] or [3.38], representative of several vibratory phenomena, has received the name of the "waves equation".

c) The representations with displacements have the advantage of limiting the number of unknowns while making the components of the tensor of stresses disappear; however, many students use the equations with displacements without noting the relation of displacements with the components of the tensor of stresses. Such a presentation of the problem gives access to resonance frequencies, but can be misleading when we want to determine the most affected parts of the structure. It is indeed common to assimilate "strong stress" with "strong displacement", which in general is not true; according to the cases in question, a vibratory amplitude antinode corresponds to a stresses node or antinode.

d) The direct formulation with displacements using the hypotheses of condensation [3.29] in the Hamilton's functional again yields equations [3.38] and [3.39]. For the problem of torsion there is not only formal similarity between the direct formulation with displacements and that resulting from the formulation in mixed variables, as is the case for longitudinal vibrations, but there is a perfect correspondence of the two formulations.

3.5. Equations of bending vibrations of straight beams

3.5.1. *Basic equations with mixed variables: Timoshenko's beam*

The bending of straight beams represents a simultaneously transverse and longitudinal vibratory movement (rotation of the cross-sections), introducing longitudinal and shearing stresses. Setting out again the basic hypotheses of condensation of beams [3.3] and [3.4] and preserving only the terms prevalent in this vibratory state, we define the hypotheses of condensation of bending of beams: axis 1 coincides with the middle fiber of the beam, that is, the locations of the centers of gravity of the cross-sections.

Displacements:

$$\begin{aligned} &W_1(x_1, x_2, x_3, t) = x_2 W_1^2(x_1, t), \\ &W_2(x_1, x_2, x_3, t) = W_2^0(x_1, t), \\ &W_3(x_1, x_2, x_3, t) = 0. \end{aligned} \qquad [3.40]$$

Figure 3.6 shows the displacements of bending, which correspond to a dominant movement of translation of cross-sections along axis 2 and a rotation of the cross-sections with respect to axis 3.

Tensor of stresses:

$$\begin{aligned} &\sigma_{11}(x_1, x_2, x_3, t) = x_2\, \sigma_{11}^2(x_1, t) \quad , \quad \sigma_{22}(x_1, x_2, x_3, t) = 0, \\ &\sigma_{33}(x_1, x_2, x_3, t) = 0 \quad , \quad \sigma_{12}(x_1, x_2, x_3, t) = \sigma_{12}^0(x_1, t), \\ &\sigma_{13}(x_1, x_2, x_3, t) = 0 \quad , \quad \sigma_{23}(x_1, x_2, x_3, t) = 0. \end{aligned} \qquad [3.41]$$

Note:

a) Relations [3.40] and [3.41] introduce an effect of shearing via the term σ_{12}^0; taking it into account is characteristic of Timoshenko's hypothesis.

b) These hypotheses correspond to bending in the (1, 2) plane. We could also introduce bending into the (1, 3) plane by permutation of indices 2 and 3 in [3.40] and [3.41], and define the bending in space by summing up the fields.

c) These hypotheses are realistic if the cross-section presents symmetry with respect to axis 3.

Solving the problem of bending consists of determining the four unknown functions $W_1^2, W_2^0, \sigma_{11}^2, \sigma_{12}^0$. We will employ Reissner's functional to establish the equations which these unknown functions must verify.

We place ourselves within the framework of an orthotropic material whose orthotropism planes coincide with the planes defined by our system of axes (1, 2, 3). In this case the stress-strain relation is given by:

$$\begin{pmatrix} \varepsilon_{11} \\ \varepsilon_{22} \\ \varepsilon_{33} \\ \varepsilon_{12} \\ \varepsilon_{13} \\ \varepsilon_{23} \end{pmatrix} = \begin{pmatrix} S_{1111} & S_{1122} & S_{1133} & 0 & 0 & 0 \\ S_{2211} & S_{2222} & S_{2233} & 0 & 0 & 0 \\ S_{3311} & S_{3322} & S_{3333} & 0 & 0 & 0 \\ 0 & 0 & 0 & S_{1212} & 0 & 0 \\ 0 & 0 & 0 & 0 & S_{1313} & 0 \\ 0 & 0 & 0 & 0 & 0 & S_{2323} \end{pmatrix} \begin{pmatrix} \sigma_{11} \\ \sigma_{22} \\ \sigma_{33} \\ 2\,\sigma_{12} \\ 2\,\sigma_{13} \\ 2\,\sigma_{23} \end{pmatrix} \qquad [3.42]$$

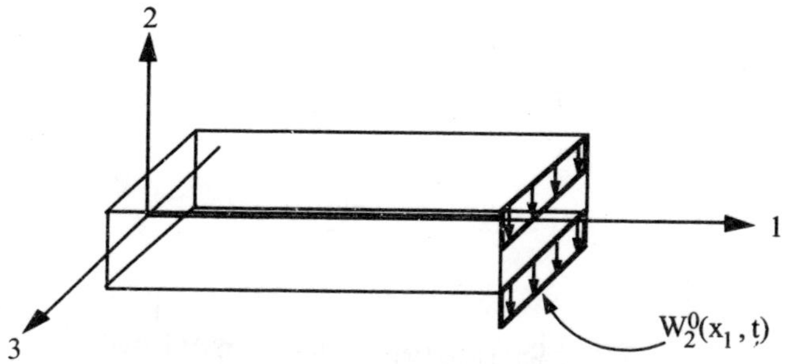

Figure 3.6a. *Transverse displacement* $W_2^0(x_1, t)$.
It is a translation of the cross-section along axis 2

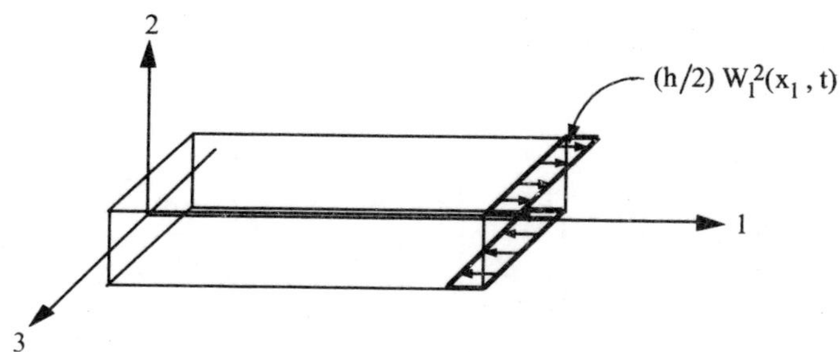

Figure 3.6b. *Longitudinal displacement* $x_2 W_1^2(x_1, t)$.
It is a rotation of the cross-section with respect to axis 3

The correspondence between the tensors S_{ijkl} and C_{ijkl} has been detailed in many works. However, to illustrate this correspondence, let us take the term ε_{12}. It comes with the 4th line from the preceding system: $\varepsilon_{12} = S_{1212}\, 2\sigma_{12}$. We also had: $\sigma_{12} = C_{1212}\, 2\varepsilon_{12}$. The identification leads to:

$$S_{1212} = \frac{1}{4C_{1212}}. \qquad [3.43]$$

Injecting the fields [3.42] and [3.43] into Reissner's functional [2.38], it follows:

$$R(W_1^2, W_2^0, \sigma_{11}^2, \sigma_{12}^0) = \int_{t_0}^{t_1}\int_V \left[\frac{1}{2}\rho\left(\left(x_2 \frac{\partial W_1^2}{\partial t}\right)^2 + \left(\frac{\partial W_2^0}{\partial t}\right)^2\right) - \sigma_{12}^0 \frac{\partial W_2^0}{\partial x_1} - \sigma_{12}^0 W_1^2 - x_2^2\,\sigma_{11}^2 \frac{\partial W_1^2}{\partial x_1} + \frac{1}{2}\left(S_{1111}(x_2\,\sigma_{11}^2)^2 + 4S_{1212}(\sigma_{12}^0)^2\right)\right] dvdt. \qquad [3.44]$$

For the type of material considered, $S_{1212} = 0$ (and all the symmetrical terms).

Noting $I_3 = \int_S x_2^2\, ds$ and $S = \int_S ds$ after integration on the cross-section, it follows:

$$R(W_1^2, W_2^0, \sigma_{11}^2, \sigma_{12}^0) = \int_{t_0}^{t_1}\int_0^L \left[\frac{1}{2}\rho\left(I_3\left(\frac{\partial W_1^2}{\partial t}\right)^2 + S\left(\frac{\partial W_2^0}{\partial t}\right)^2\right) - S\sigma_{12}^0 \frac{\partial W_2^0}{\partial x_1} - S\sigma_{12}^0 W_1^2 - I_3\sigma_{11}^2 \frac{\partial W_1^2}{\partial x_1} + \frac{1}{2}\left(S_{1111}\, I_3(\sigma_{11}^2)^2 + 4S_{1212}\, S(\sigma_{12}^0)^2\right)\right] dx_1 dt. \qquad [3.45]$$

The calculation of extremum of the functional [3.45] leads to the equations:

$$-S\sigma_{12}^0 - \rho I_3 \frac{\partial^2 W_1^2}{\partial t^2} + \frac{\partial}{\partial x_1}(I_3\,\sigma_{11}^2) = 0 \quad \forall t,\ \forall x_1 \in]0, L[, \qquad [3.46]$$

$$-\rho S \frac{\partial^2 W_2^0}{\partial t^2} + \frac{\partial}{\partial x_1}(S\sigma_{12}^0) = 0 \quad \forall t,\ \forall x_1 \in]0, L[, \qquad [3.47]$$

$$-S\left(W_1^2 + \frac{\partial W_2^0}{\partial x_1}\right) + S_{1212}\, S\sigma_{12}^0 = 0 \quad \forall t,\ \forall x_1 \in]0, L[, \qquad [3.48]$$

$$-I_3\left(\frac{\partial W_1^2}{\partial x_1} - S_{1111}\,\sigma_{11}^2\right) = 0 \quad \forall t,\ \forall x_1 \in]0, L[. \qquad [3.49]$$

Relations [3.46] and [3.47] are the equations of motion following axes 1 and 2; relations [3.48] and [3.49] are the stress-strain relation of the beam for σ_{12}^0 and σ_{11}^2.

The boundary conditions are given by the relations:

$$W_2^0 = 0 \text{ or } S\sigma_{12}^0 = 0 \text{ and } W_1^2 = 0 \text{ or } I_3\sigma_{11}^2 = 0$$
$$\forall\, t\ ,\ x_1 = 0 \text{ and } x_1 = L. \quad [3.50]$$

The term $S\sigma_{12}^0$ is homogenous with a force which is opposed to transverse displacement, which is called shearing force; it is introduced by the shearing σ_{12}^0. The term $I_3\sigma_{11}^2$ is homogenous to a torque opposed to the rotation of cross-sections and is called bending moment.

There will thus be a set of four boundary conditions possible for each end of the beam:

a) $W_2^0 = 0$ and $W_1^2 = 0$

The two movements along axes 1 and 2 are blocked. The end will be said to be clamped;

b) $W_2^0 = 0$ and $I_3\,\sigma_{11}^2 = 0$

Transverse movement is blocked, longitudinal movement is free. The end will be said to be supported;

c) $S\sigma_{12}^0 = 0$ and $I_3\,\sigma_{11}^2 = 0$

Both movements, longitudinal and transverse, are free, the constraints which correspond to them are nil, and the end is free;

d) $W_1^2 = 0$ and $S\sigma_{12}^0 = 0$

Longitudinal displacement is nil; transverse displacement is free. This boundary condition, difficult to realize in practice, is said to be guided.

A bending beam could thus be clamped-clamped, clamped-supported, etc.

3.5.2. *Equations with displacement variables: Timoshenko's beam*

It is enough to draw σ_{12}^0 and σ_{11}^2 from equations [3.48] and [3.49]:

$$\sigma_{12}^{0} = \frac{1}{4S_{1212}}\left(W_1^2 + \frac{\partial W_2^0}{\partial x_1}\right) \quad \forall t\ ,\ \forall x_1 \in \left]0, L\right[, \qquad [3.51]$$

$$\sigma_{11}^{2} = \frac{1}{S_{1111}}\frac{\partial W_1^2}{\partial x_1} \quad \forall t\ ,\ \forall x_1 \in \left]0, L\right[, \qquad [3.52]$$

then to introduce these expressions into equations [3.46] and [3.47] which become:

$$\rho I_3 \frac{\partial^2 W_1^2}{\partial t^2} - \frac{S}{4S_{1212}}\left(W_1^2 + \frac{\partial W_2^0}{\partial x_1}\right) + I_3 \frac{\partial}{\partial x_1}\left(\frac{1}{S_{1111}}\frac{\partial W_1^2}{\partial x_1}\right) = 0 \quad \forall t\ ,\ \forall x_1 \in \left]0, L\right[, \qquad [3.53]$$

$$-\rho S \frac{\partial^2 W_2^0}{\partial t^2} + \frac{\partial}{\partial x_1}\left(\frac{S}{4S_{1212}}\left(W_1^2 + \frac{\partial W_2^0}{\partial x_1}\right)\right) = 0 \quad \forall t\ ,\ \forall x_1 \in \left]0, L\right[. \qquad [3.54]$$

The boundary conditions [3.50] are given by:

$$\text{either } W_2^0 = 0, \quad \text{or } \frac{S}{4S_{1212}}\left(W_1^2 + \frac{\partial W_2^0}{\partial x_1}\right) = 0 \ \forall t,\ x_1 = 0,\ x_1 = L \qquad [3.55]$$

and

$$\text{either } W_1^2 = 0, \quad \text{or } \frac{I_3}{S_{1111}}\frac{\partial W_1^2}{\partial x_1} = 0 \ \forall t,\ x_1 = 0,\ x_1 = L. \qquad [3.56]$$

The writing with displacement variables reveals only two unknowns: the transverse displacement and the rotation of cross-sections. In practice, rotations of cross-sections are not accessible in experiments and the principal demonstration of the bending of beams is transverse displacement W_2^0. It is thus interesting to give only one equation function of this quantity. This is possible for homogenous beams by carrying out the following operations on equations [3.53] and [3.54]: we draw the value of $\partial W_1^2/\partial x_1$ from [3.54] according to W_2^0, and we then derive [3.53] with respect to x_1 and replace $\partial W_1^2 / \partial x_1$ by the expression obtained. This processing that we do not carry out in detail leads to the equation:

$$\frac{I_3}{S_{1111}}\frac{\partial^4 W_2^0}{\partial x_1^4}+\rho S\frac{\partial^2 W_2^0}{\partial t^2}+4\rho^2 I_3 S_{1212}\frac{\partial^4 W_2^0}{\partial t^4}-\rho I_3\left(1+4\frac{S_{1212}}{S_{1111}}\right)\frac{\partial^4 W_2^0}{\partial x_1^2\,\partial t^2}=0. \qquad [3.57a]$$

If we introduce the modules of Coulomb G and Young E with:

$E=1/S_{1111}$ and $G=1/4S_{1212}$,

while noting W_2^0 as W to reduce the writing, it follows:

$$EI_3\frac{\partial^4 W}{\partial x_1^4}+\rho S\frac{\partial^2 W}{\partial t^2}+\frac{\rho^2 I_3}{G}\frac{\partial^4 W}{\partial t^4}-\rho I_3\left(1+\frac{E}{G}\right)\frac{\partial^4 W}{\partial x_1^2\,\partial t^2}=0. \qquad [3.57b]$$

It is the most synthetic equation of beams with shearing and rotational inertia; let us recall that it is limited to homogenous beams.

A correction of the modulus of rigidity noted G' is often introduced with:

$G'=\alpha G$.

The multiplication coefficient α traditionally introduced to characterize the correction of shearing obviously does not have anything to do with the angle of torsion introduced in section 3.3.

This correction appears because the constant form of the shear stress throughout the thickness of the beam is very approximate. To realize the approximation, it is enough to note that to verify the boundary condition of the free surface, the shearing stress must be nil over these surfaces, which is not verified by the hypotheses.

We may clearly see that the form of the cross-section will have an influence on the correction that has to be applied. Various authors have been interested in this problem and have calculated the α corrections that have to be applied by comparing the solution with constant stress and a precise calculation of the shearing stress. We provide some corrections of the shearing modulus taken from the following references: Cowper [COW 66], Dharmarajan and McCutchen [DHA 73]:

a) beam with a circular cross-section made of an orthotropic material:

$$\alpha = \frac{6E_3}{7E_3 - 2\nu_{13}G_{13}} \; ;$$

E_3 is the Young modulus, ν_{13} is the Poisson's ratio, and G_{13} is the shearing modulus.

b) beam with a circular cross-section made of an isotropic material:

$$\alpha = \frac{6(1+\nu)}{7+6\nu} \; ;$$

c) beam with a rectangular cross-section made of an orthotropic material:

$$\alpha = \frac{5E_3}{6E_3 - \nu_{13}G_{13}} \; ;$$

d) beam with a rectangular cross-section made of an isotropic material:

$$\alpha = \frac{10(1+\nu)}{12+11\nu} \; ;$$

e) hollow tube with a circular cross-section (a and b are the interior and exterior diameters) made of an orthotropic material:

$$\alpha = \frac{6E_3(m^4-1)(1+m^2)}{E_3(7m^6+27m^4-27m^2-7)-\nu_{13}G_{13}(2m^6+18m^4-18m^2-2)}$$

with $m = b/a$.

For thin-walled tubes $m \approx 1$, the expression of α is simplified:

$$\alpha = \frac{E_3}{2E_3 - \nu_{13}G_{13}} \; ;$$

f) hollow tube with a circular cross-section made of an isotropic material (case of the thin wall $m \approx 1$):

$$\alpha = \frac{2(1+\nu)}{2+\nu} .$$

3.5.3. *Basic equations with mixed variables: Euler-Bernoulli beam*

The hypotheses that we will describe rest on the fact that shearing stress σ_{12}^0 is generally low and can thus be neglected in the first approximation. By using the relation [3.48] which in Timoshenko's model connects this shearing stress with the displacements W_1^2 and W_2^0, we note that if $\sigma_{12}^0 = 0$, we have $W_1^2 = -\partial W_2^0/\partial x_1$. These observations result in adopting the following hypotheses of condensation when transverse shearing is neglected:

Displacements:

$$\begin{aligned}
&W_1(x_1, x_2, x_3, t) = -x_2 \frac{\partial W_2^0}{\partial x_1}(x_1, t),\\
&W_2(x_1, x_2, x_3, t) = W_2^0(x_1, t),\\
&W_3(x_1, x_2, x_3, t) = 0.
\end{aligned} \qquad [3.58]$$

The displacements translate the equality of the rotation angle of the cross-sections with the slope of transverse displacement.

Tensors of stresses:

$$\begin{aligned}
&\sigma_{11}(x_1, x_2, x_3, t) = x_2\, \sigma_{11}^2(x_1, t)\ ,\ \sigma_{22}(x_1, x_2, x_3, t) = 0,\\
&\sigma_{33}(x_1, x_2, x_3, t) = 0\ ,\ \sigma_{12}(x_1, x_2, x_3, t) = \sigma_{12}^0(x_1, t),\\
&\sigma_{13}(x_1, x_2, x_3, t) = 0\ ,\ \sigma_{23}(x_1, x_2, x_3, t) = 0.
\end{aligned} \qquad [3.59]$$

Upon introducing these displacements and stress fields into the Reissner's functional (equation [2.38], Chapter 2), it follows after calculations:

$$\begin{aligned}
R\,(W_2^0, \sigma_{11}^2) = \frac{1}{2}\int_{t_0}^{t_1}\!\!\int_0^L \Bigg[& \rho I_3 \left(\frac{\partial^2 W_2^0}{\partial x_1 \partial t}\right)^2 + \rho S \left(\frac{\partial W_2^0}{\partial t}\right)^2 \\
& - I_3\, \sigma_{11}^2 \frac{\partial^2 W_2^0}{\partial x_1^2} + I_3\, S_{1111} (\sigma_{11}^2)^2 \Bigg] dx_1\, dt.
\end{aligned} \qquad [3.60]$$

The calculation of extremum of the functional [3.60] leads to the equations:

Equation of motion:

$$-\rho S\frac{\partial^2 W_2^0}{\partial t^2}+\frac{\partial}{\partial x_1}\left(\rho I_3\frac{\partial^3 W_2^0}{\partial x_1\partial t^2}\right)-\frac{\partial^2}{\partial x_1^2}(I_3\,\sigma_{11}^2)=0 \qquad [3.61]$$

$$\forall t\ ,\ \forall x_1\in\left]0,L\right[.$$

Stress-strain relation:

$$-I_3\frac{\partial^2 W_2^0}{\partial x_1^2}+I_3\,S_{1111}\,\sigma_{11}^2=0\quad \forall t\ ,\ \forall x_1\in\left]0,L\right[. \qquad [3.62]$$

Boundary conditions:

$$\text{let }\ W_2^0=0\ ,\ \text{let }\ \frac{\partial}{\partial x_1}(I_3\,\sigma_{11}^2)=0\quad \forall t\ ,\ x_1=0\ \text{ and }\ x_1=L \qquad [3.63]$$

and

$$\text{let }\ \frac{\partial W_2^0}{\partial x_1}=0\ ,\ \text{let }\ I_3\,\sigma_{11}^2=0\quad \forall t\ ,\ x_1=0\ \text{ and }\ x_1=L\,. \qquad [3.64]$$

3.5.4. *Equations of the Euler-Bernoulli beam with displacement variable*

To obtain the equations of Bernoulli's beam as a function of the transverse displacement of cross-sections variable $W_2^0(x_1, t)$, it suffices to draw $\sigma_{11}^2(x_1, t)$ from equation [3.62] and then to replace in [3.61], [3.63] and [3.64]:

$$\sigma_{11}^2=\frac{1}{S_{1111}}\frac{\partial^2 W_2^0}{\partial x_1^2} \qquad [3.65]$$

Equation of motion:

$$-\rho S\frac{\partial^2 W_2^0}{\partial t^2}+\frac{\partial}{\partial x_1}\left(\rho I_3\frac{\partial^3 W_2^0}{\partial x_1\,\partial t^2}\right)-\frac{\partial^2}{\partial x_1^2}\left(\frac{I_3}{S_{1111}}\frac{\partial^2 W_2^0}{\partial x_1^2}\right)=0 \qquad [3.66]$$

$$\forall t\ ,\forall x_1\in\left]0,L\right[.$$

Boundary conditions:

$$\text{either } W_2^0 = 0 \text{ , or } \frac{\partial}{\partial x_1}\left(\frac{I_3}{S_{1111}}\frac{\partial^2 W_2^0}{\partial x_1^2}\right) = 0 \qquad [3.67]$$

$$\forall t \text{ , } x_1 = 0 \text{ and } x_1 = L$$

and

$$\text{either } \frac{\partial W_2^0}{\partial x_1} = 0 \text{ , or } \frac{I_3}{S_{1111}}\frac{\partial^2 W_2^0}{\partial x_1^2} = 0 \quad \forall t \text{ , } x_1 = 0 \text{ and } x_1 = L \text{ .} \qquad [3.68]$$

To obtain the traditional equation of bending beams vibrations we must introduce an additional simplification neglecting the effects of rotational inertia of the beam.

We thus neglect $\frac{\partial}{\partial x_1}\left(\rho I_3 \frac{\partial^3 W_2^0}{\partial x_1 \partial t^2}\right)$ in [3.66], which while replacing $1/S_{1111}$ by E_1 (Young's modulus of material in the longitudinal direction) yields:

Equation of motion:

$$-\rho S \frac{\partial^2 W_2^0}{\partial t^2} - \frac{\partial^2}{\partial x_1^2}\left(I_3 E_1 \frac{\partial^2 W_2^0}{\partial x_1^2}\right) = 0 \quad \forall t \text{ , } \forall x_1 \in \left]0, L\right[\text{.} \qquad [3.69]$$

Boundary conditions:

$$\text{either } W_2^0 = 0 \text{ , or } \frac{\partial}{\partial x_1}\left(I_3 \, E_1 \frac{\partial^2 W_2^0}{\partial x_1^2}\right) = 0 \quad \forall t \text{ , } x_1 = 0 \text{ and } x_1 = L \qquad [3.70a]$$

and

$$\text{either } \frac{\partial W_2^0}{\partial x_1} = 0 \text{ , or } I_3 \, E_1 \frac{\partial^2 W_2^0}{\partial x_1^2} = 0 \quad \forall t \text{ , } x_1 = 0 \text{ and } x_1 = L \text{ .} \qquad [3.70b]$$

Note:

a) Vibratory movement is described by equations which no longer reveal the stresses, but the latter are of course always present and can be calculated with the expression [3.65] as long as $W_2^0(x_1, t)$ is known.

b) Equations with displacements [3.69] and [3.70] could be obtained directly with the Hamilton's functional, but would require a modification of the elastic constants of three-dimensional material, as for the longitudinal beams vibrations (see section 3.3.3). The functional to be used in this case is given by:

$$H\left(W_2^0(x_1,t)\right)=\int_{t_0}^{t_1}\int_0^L\left[\frac{\rho s}{2}\left(\frac{\partial W_2^0}{\partial t}\right)^2-\frac{EI}{2}\left(\frac{\partial^2 W_2^0}{\partial x_1^2}\right)^2\right]dx_1dt. \qquad [3.71]$$

c) In the case of isotropic material, equations [3.69], [3.70] and [3.71] of course remain valid; the longitudinal Young modulus is then simply the Young modulus of the isotropic material.

d) The quantity $\dfrac{\partial}{\partial x_1}\left(E_1 I_3 \dfrac{\partial^2 W_2^0}{\partial x_1^2}\right)$, appearing in [3.70] and placed in duality with the displacement W_2^0, is homogenous to a force: it is called the shearing force. The quantity $E_1 I_3 \dfrac{\partial^2 W_2^0}{\partial x_1^2}$, appearing in [3.71] and opposed to the rotation of the cross-sections $\partial W_2^0/\partial x_1$ is homogenous to a torque: it is called bending moment.

3.6. Complex vibratory movements: sandwich beam with a flexible inside

The hypotheses of condensation which we have used in the preceding sections describe the three elementary vibratory movements of beams homogenous in the cross-section. The non-homogenous beam has more complex states of stresses and displacements in its breadth, but they can be reconstituted on the basis of the elementary fields of bending, torsion, traction-compression. As an example we consider the case of a sandwich beam with a soft core.

The beam consists of three layers: the two on the outside are made of rigid materials (high elasticity modulus), while the core is made of a soft material (weak elasticity modulus).

Let us suppose that the beam is excited transversely; the rigid layers will have a movement of bending, and the soft internal layer will have a movement imposed by the displacement of the rigid layers, as shown in Figure 3.7. We will note the transverse displacement as $W(x, t)$ rather than $W_2^0(x_1, t)$ in order to avoid heaviness of writing and we use a local reference for each layer:

– layer 1: $x_2 \in \left[-\frac{e_1}{2}, \frac{e_1}{2}\right]$

$$\begin{aligned} W_1 &= -x_2 \frac{\partial W}{\partial x}(x, t), \\ W_2 &= W(x, t), \\ W_3 &= 0 ; \end{aligned} \qquad [3.72]$$

– layer 2: $x_2 \in \left[-\frac{e_2}{2}, \frac{e_2}{2}\right]$

$$\begin{aligned} W_1 &= g(x_2) \frac{\partial W}{\partial x}(x, t), \\ W_2 &= W(x, t), \\ W_3 &= 0, \end{aligned} \qquad [3.73]$$

with:

$$g(x_2) = x_2 \frac{e_1 + e_3}{e_2} + \frac{e_3 - e_1}{2} ; \qquad [3.74]$$

– layer 3: $x_2 \in \left[-\frac{e_3}{2}, \frac{e_3}{2}\right]$

$$\begin{aligned} W_1 &= -x_2 \frac{\partial W}{\partial x}(x, t), \\ W_2 &= W(x, t), \\ W_3 &= 0. \end{aligned} \qquad [3.75]$$

These fields of displacements make it possible to ensure the continuity of displacements at the interfaces; the reader can check it by calculating displacements in $x_2 = e_1/2$ for the first layer and $x_2 = -e_2/2$ for the layer 2, as well as in $x_2 = e_2/2$ and $x_2 = -e_3/2$ for layers 2 and 3.

Longitudinal displacements are represented in Figure 3.7; we can notice that layers 1 and 3 have neutral fibers (zero displacement) in the middle of the layers, whereas the flexible internal layer has a shifted neutral fiber.

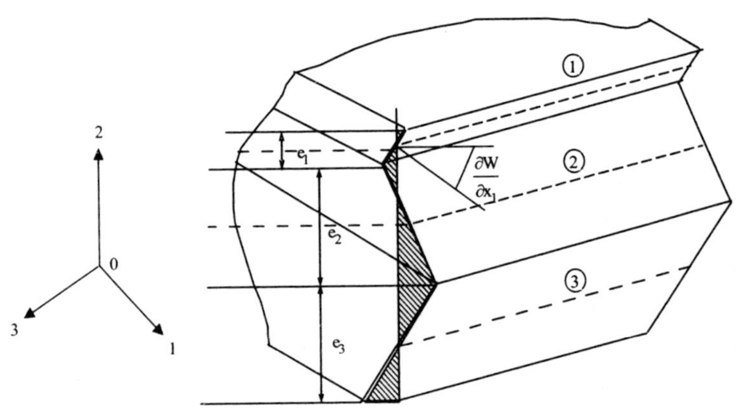

Figure 3.7. *Deformation of the cross-section of the sandwich beam*

To obtain the equation of vibrations of the sandwich beam, we will use Hamilton's functional rather than Reissner's functional, which will save us from introducing independent stress fields.

In expression [3.76], ρ_i is the density of material of layer i, S_i is the surface of section i, C^i_{1111} is the longitudinal module of layer i, and C^2_{1212} is the shearing modulus of layer 2.

After integrating over the cross-sections of the layers and regrouping terms, we obtain:

$$
\begin{aligned}
H\left(W(x,t)\right) = \int_{t_0}^{t_1}\int_0^L \Bigg[& (\rho_1 S_1 + \rho_2 S_2 + \rho_3 S_3)\left(\frac{\partial W}{\partial t}\right)^2 \\
& + (\rho_1 I_1 + \rho_2 I_2 + \rho_3 I_3)\left(\frac{\partial^2 W}{\partial x \partial t}\right)^2 \\
& - (C^1_{1111}\, I_1 + C^2_{1111}\, J_2 + C^3_{1111}\, I_3)\left(\frac{\partial^2 W}{\partial x^2}\right)^2 \\
& - C^2_{1212}\, K_2 \left(\frac{\partial W}{\partial x}\right)^2 \Bigg] \, dxdt
\end{aligned}
\qquad [3.76]
$$

where b is the width of the sandwich beam and e_i is the thickness of layer i:

$$
J_2 = \int_{S_2} g\,(x_2)^2\, dS_2 = \left(\frac{e_3 - e_1}{2}\right)^2 be_2 + \frac{e_1 + e_3}{12 e_2}\, be_2^3,
$$

$$
I_i = \int x_2^2\, dS_i = be_i^3/12,
$$

$$
K_2 = \int_{S_2} \left(\frac{e_1 + e_2 + e_3}{e_2}\right) dS_2 = \left(\frac{e_1 + e_2 + e_3}{e_2}\right) be_2 .
$$

The calculation of the extremum of the functional is performed thanks to the results of Chapter 2. The application to the functional [3.76] yields the equation of motion [3.77] and the boundary conditions [3.79] and [3.80]:

$$(\rho_1 S_1 + \rho_2 S_2 + \rho_3 S_3)\frac{\partial^2 W}{\partial t^2} - \frac{\partial}{\partial x}\left(C_{1212}^2 K_2 \frac{\partial W}{\partial x}\right) - \frac{\partial^2}{\partial x \partial t}\left((\rho_1 I_1 + \rho_2 J_2 + \rho_3 I_3)\frac{\partial^2 W}{\partial x \partial t}\right) + \frac{\partial^2}{\partial x^2}\left((C_{1111}^1 I_1 + C_{1111}^2 J_2 + C_{1111}^3 I_3)\frac{\partial^2 W}{\partial x^2}\right) = 0. \quad [3.77]$$

Using the note to section 3.3.3, we will replace C_{1111}^i by Young's modulus E_i; C_{1212}^2 is equal to the Coulomb modulus G_2. Moreover, if we suppose that the characteristics are constant with x and that rotational inertia is neglected, the equation becomes:

$$(\rho_1 S_1 + \rho_2 S_2 + \rho_3 S_3)\frac{\partial^2 W}{\partial t^2} - G_2 K_2 \frac{\partial^2 W}{\partial x^2} + (E_1 I_1 + E_2 J_2 + E_3 I_3)\frac{\partial^4 W}{\partial x^4} = 0. \quad [3.78]$$

The boundary conditions $\forall t$, $x_1 = 0$ and $x_1 = L$ to be verified are given by:

$$\text{either } W = 0 \text{ , or } G_2 K_2 \frac{\partial W}{\partial x} - (E_1 I_1 + E_2 J_2 + E_3 I_3)\frac{\partial^3 W}{\partial x^3} = 0 \quad [3.79]$$

and:

$$\text{either } \frac{\partial W}{\partial x} = 0 \text{ , or } -(E_1 I_1 + E_2 J_2 + E_3 I_3)\frac{\partial^2 W}{\partial x^2} = 0. \quad [3.80]$$

These equations describe the vibratory behavior of the sandwich beam; however, to be realistic, it is necessary that the hypotheses which have led to the fields of displacements [3.72], [3.73] and [3.74] are respected; that is, $E_2 << E_1$, $E_2 << E_3$, to have a soft internal layer with respect to the surface layers. From a practical point of view, this situation occurs when the internal layer consists of a viscoelastic material and the surface layers consist of rigid materials; equations [3.78], [3.79] and [3.80] are thus representations of the vibrations of beams with an internal damping layer. However, they are not usable for the sandwich beams with rigid cores.

The example that we have treated shows that using the fields of displacements associated with elementary vibratory states (bending, torsion, traction-compression), it is possible to reconstitute complex vibratory states, which appear in non-homogenous beams. The hypotheses of condensation adapted to the description of a complex movement are, however, not easy to determine without certain practice. The modeling of damping properties of multi-layer beams have brought about several models based on different hypotheses of condensation; some references are provided in the bibliography.

3.7. Conclusion

In this chapter we gave the equations governing longitudinal, transverse and torsion vibrations of straight beams. In the following chapters these equations will be used as a basis to describe the vibratory behavior of beams.

The method used for setting up the equations calls upon the functionals of Reissner and Hamilton presented in Chapter 2. We have also stuck to clearly specifying the hypotheses leading to the equations, which will make it possible for the reader to determine their limits of validity and thus to be critical of the results of an estimated calculation. From a practical point of view, the method breaks up into three parts:

a) establishing of condensation hypotheses which restrict the stresses and deformations field of the continuous medium, taking into account its form (truncated Taylor development) and external force applied (longitudinal, transverse, torsion);

b) calculating the functional taking into account the hypotheses of condensation;

c) calculating of the extremum of the functional.

The delicate part of work is undoubtedly establishing the hypotheses of condensation. An essential goal of this chapter is to show the reader how these hypotheses are established in the traditional cases and how calculations can be extended to more complex cases such as that of the sandwich beams with a soft core that we have addressed.

Chapter 4

Equation of Vibration for Plates

4.1. Objective of the chapter

Plates are continuous media with a more complicated mechanical behavior than that of beams. The greatest complexity comes from the fact that the description of plates' vibrations introduces functions with two variables of space. Thus, we have to deal with a two-dimensional (2D) medium.

The set up of equations is fundamentally identical to that of beams. We will use an energy formulation based on Reissner's functional with independent displacement and stress fields, and then kinematic hypotheses revealing two elementary movements: in plane vibration (membrane effect) and transverse vibrations. Various simplifying hypotheses will be presented leading to the model of Mindlin and then of Love-Kirchhoff. We will show, in particular, that the generally used equations are the result of very strong simplifying hypotheses and that these equations are often employed outside of their valid domain.

The plate being two-dimensional, we may sometimes find it beneficial to use polar rather than Cartesian co-ordinates. We will describe the passage between the two descriptions and will eventually arrive at the Love-Kirchhoff plate equations in polar co-ordinates.

4.2. Thin plate hypotheses

4.2.1. *General procedure*

We apply the same steps as for beams based on the fact that one of the dimensions of the structure (thickness h, direction x_3) is small compared to the width b and the length l. We can then develop the fields of displacements and stress in Taylor series and obtain a suitable approximation by truncating these fields to the first order:

$$W_i(x_1, x_2, x_3, t) \approx W_i(x_1, x_2, 0, t) + x_3 \frac{\partial W_i}{\partial x_3}(x_1, x_2, 0, t), \qquad [4.1]$$

$$\sigma_{ij}(x_1, x_2, x_3, t) \approx \sigma_{ij}(x_1, x_2, 0, t) + x_3 \frac{\partial \sigma_{ij}}{\partial x_3}(x_1, x_2, 0, t). \qquad [4.2]$$

The expressions [4.1] and [4.2] suggest seeking an approximation of the fields in the forms [4.3] and [4.4]:

$$W_i(x_1, x_2, x_3, t) = W_i^0(x_1, x_2, t) + x_3 W_i^3(x_1, x_2, t), \qquad [4.3]$$

$$\sigma_{ij}(x_1, x_2, x_3, t) = \sigma_{ij}^0(x_1, x_2, t) + x_3 \sigma_{ij}^3(x_1, x_2, t). \qquad [4.4]$$

As in theory of beams, the fields of displacements [4.3] and stresses [4.4] contain a set of vibratory states that are generally separated into independent movements so as to be able to study them easier. We separate the vibrations of plates into two elementary vibratory movements: vibrations in the plane of the plate and transverse vibrations, this second type of vibration being by far the most present in the problems encountered in practice.

4.2.2. *In plane vibrations*

This type of vibration corresponds to the longitudinal vibrations of beams. It is supposed that transverse displacement is nil:

$$W_3(x_1, x_2, x_3, t) = 0, \qquad [4.5a]$$

and those displacements in the directions 1 and 2 are constant throughout the thickness:

$$W_1(x_1, x_2, x_3, t) = W_1^0(x_1, x_2, t), \tag{4.5b}$$

$$W_2(x_1, x_2, x_3, t) = W_2^0(x_1, x_2, t). \tag{4.5c}$$

The field of tensors of stress is copied from that of beams and leads to:

$$\begin{cases} \sigma_{11}(x_1, x_2, x_3, t) = \sigma_{11}^0(x_1, x_2, t) \\ \sigma_{22}(x_1, x_2, x_3, t) = \sigma_{22}^0(x_1, x_2, t) \\ \sigma_{12}(x_1, x_2, x_3, t) = \sigma_{12}^0(x_1, x_2, t), \end{cases} \tag{4.6}$$

other stresses being supposed to be nil.

In the plate hypothesis, longitudinal vibrations in the plane are accompanied by longitudinal stresses but also by shearing (σ_{12}).

4.2.3. *Transverse vibrations: Mindlin's hypotheses*

This vibratory movement is counterpart beams bending movement. We will employ the hypotheses extrapolated from those described for beams in Chapter 3. The field of displacements is thus provided by:

$$\begin{cases} W_1(x_1, x_2, x_3, t) = x_3 W_1^3(x_1, x_2, t) \\ W_2(x_1, x_2, x_3, t) = x_3 W_2^3(x_1, x_2, t) \\ W_3(x_1, x_2, x_3, t) = W_3^0(x_1, x_2, t). \end{cases} \tag{4.7}$$

Transverse displacement is supposed to be constant throughout the thickness and is accompanied by longitudinal movements produced by rotations around axes 1 and 2.

Stresses associated to [4.7] are of the form [4.8]:

$$\begin{cases} \sigma_{11}(x_1, x_2, x_3, t) = x_3\sigma_{11}^3(x_1, x_2, t) \\ \sigma_{22}(x_1, x_2, x_3, t) = x_3\sigma_{22}^3(x_1, x_2, t) \\ \sigma_{12}(x_1, x_2, x_3, t) = x_3\sigma_{12}^3(x_1, x_2, t) \\ \sigma_{13}(x_1, x_2, x_3, t) = \sigma_{13}^0(x_1, x_2, t) \\ \sigma_{23}(x_1, x_2, x_3, t) = \sigma_{23}^0(x_1, x_2, t) \\ \sigma_{33}(x_1, x_2, x_3, t) = 0. \end{cases} \qquad [4.8]$$

The hypotheses [4.7] and [4.8] correspond to those of Mindlin; they are characterized by taking transverse shearing into account. This effect is negligible for isotropic materials at a low frequency. The Love-Kirchhoff theory, presented hereafter, is then sufficient. For anisotropic materials at a high frequency; taking into account transverse shearing is necessary for a good theory – experiment comparison. Let us note that the comment made in Chapter 3 on the incompatibility of shearing stresses σ_{13} and σ_{23} with the condition of free surface for $x_3 = \pm h/2$ is also present here. Indeed:

$$\sigma_{13}\left(x_1, x_2, \pm\frac{h}{2}, t\right) = \sigma_{13}^0(x_1, x_2, t)$$

$$\text{and } \sigma_{23}\left(x_1, x_2, \pm\frac{h}{2}, t\right) = \sigma_{23}^0(x_1, x_2, t),$$

whereas we should have zero constraints on the external surfaces of the plate. This comes from too strong a truncation of the developments [4.1] and [4.2]; the parabolic term would make it possible to get rid of the incompatibility but with very heavy calculations. The most frequently adopted procedure consists of keeping this simple model, the terms σ_{13}^0 and σ_{23}^0 appearing as the average (constant in the thickness) of variable stress in the thickness. This approach leads, as has been described for beams, to a correction of the shearing constants.

4.2.4. *Transverse vibrations: Love-Kirchhoff hypotheses*

These hypotheses are equivalent to the Euler-Bernoulli hypotheses for beams: they amount to supposing that transverse shearing is nil. For displacements it consists of equalizing rotations around axes 1 and 2 describing displacements in the thickness with the respective slopes of transverse displacement.

$$\begin{cases} W_1(x_1, x_2, x_3, t) = x_3 \dfrac{\partial W_3^0}{\partial x_1}(x_1, x_2, t) \\ W_2(x_1, x_2, x_3, t) = x_3 \dfrac{\partial W_3^0}{\partial x_2}(x_1, x_2, t) \\ W_3(x_1, x_2, x_3, t) = W_3^0(x_1, x_2, t). \end{cases} \quad [4.9]$$

The associated stress field is:

$$\begin{cases} \sigma_{11}(x_1, x_2, x_3, t) = x_3 \sigma_{11}^3(x_1, x_2, t) \\ \sigma_{22}(x_1, x_2, x_3, t) = x_3 \sigma_{22}^3(x_1, x_2, t) \\ \sigma_{12}(x_1, x_2, x_3, t) = x_3 \sigma_{12}^3(x_1, x_2, t) \\ \sigma_{13}(x_1, x_2, x_3, t) = 0 \\ \sigma_{23}(x_1, x_2, x_3, t) = 0 \\ \sigma_{33}(x_1, x_2, x_3, t) = 0. \end{cases} \quad [4.10]$$

4.2.5. *Plates which are non-homogenous in thickness*

Plates homogenous through thickness are the ones most generally found in practice; they present a remarkable property of decoupling vibrations in the plane and transverse vibrations, which can thus be studied separately. When mechanical characteristics vary through thickness of the plate, movements are coupled and must be studied together; we will clarify this point later on. The fields of displacements to be considered are provided by the superposition of the two fields describing movements in the plane and transverse movements. For example, for the Mindlin's plate, the field of displacements to be considered is the superposition of [4.7] and [4.5]:

$$\begin{cases} W_1(x_1, x_2, x_3, t) = W_1^0(x_1, x_2, t) + x_3 W_1^3(x_1, x_2, t) \\ W_2(x_1, x_2, x_3, t) = W_2^0(x_1, x_2, t) + x_3 W_2^3(x_1, x_2, t) \\ W_3(x_1, x_2, x_3, t) = W_3^0(x_1, x_2, t). \end{cases} \quad [4.11]$$

4.3. Equations of motion and boundary conditions of in plane vibrations

The equations are set up using the variational method based on Reissner's functional given in Chapter 2 (equation [2.38]) that we particularize here for the case of free vibrations, that is, without an external force of excitation:

$$R(W_i, \sigma_{ij}) = \int_{t_0}^{t_1}\int_V \left[\frac{1}{2}\rho\left(\frac{\partial W_i}{\partial t}\right)^2 - \sigma_{ij} W_{i,j} + \frac{1}{2}\sigma_{ij} S_{ijkl}\sigma_{kl}\right] dvdt \; . \qquad [4.12]$$

It suffices now to use the approximations [4.5] and [4.6] of the fields of displacements and stresses in the functional [4.12]:

$$R(W_i, \sigma_{ij}) = \int_{t_0}^{t_1} h \int_S \left[\frac{1}{2}\rho\left(\left(\frac{\partial W_1^0}{\partial t}\right)^2 + \left(\frac{\partial W_2^0}{\partial t}\right)^2\right) - \sigma_{11}^0 \frac{\partial W_1^0}{\partial x_1}\right.$$
$$+ \sigma_{12}^0\left(\frac{\partial W_1^0}{\partial x_2} + \frac{\partial W_2^0}{\partial x_1}\right) - \sigma_{22}^0 \frac{\partial W_2^0}{\partial x_2} + \frac{1}{2}(\sigma_{11}^0 S_{1111}\sigma_{11}^0 \qquad [4.13]$$
$$\left. + 2\sigma_{11}^0 S_{1122}\sigma_{22}^0 + \sigma_{22}^0 S_{2222}\sigma_{22}^0 + 4\sigma_{12}^0 S_{1212}\sigma_{12}^0)\right] dsdt \, .$$

In the preceding expression we have supposed that the material was homogenous, that the thickness of the plate h was constant and that the stress-strain relation of material had zero terms $(S_{1211} = S_{1222} = 0)$, which is verified for an isotropic material, or a material which is orthotropic, the orthotropy axes being 1 and 2. We have used S to denote the surface of the plate.

Taking into account integration over the thickness, the description of plate vibrations is performed by functions of a space with two dimensions defined over a surface S. For this reason the plate is compared to a 2D medium with surface S and contour $\overline{S}$.

The calculation of the extremum of the functional is carried out using the Euler equations which we have provided in Chapter 2 (equations [2.88] to [2.92]). The functional having only first derivatives, many of the terms of the Euler equations will be nil, in particular, only one group of boundary conditions will be considered ([2.89] and [2.90]).

There are five equations to verify on the surface of the plate S, since there are five unknown functions. They are obtained by the calculation of the Euler equation with respect to each unknown function W_1^0, W_2^0, σ_{11}^0, σ_{12}^0, σ_{22}^0 :

$$-\rho\frac{\partial^2 W_1^0}{\partial t^2}+\frac{\partial \sigma_{11}^0}{\partial x_1}+\frac{\partial \sigma_{12}^0}{\partial x_2}=0 \quad \forall\ (x_1,x_2)\in S\ ,\ \forall t, \qquad [4.14]$$

$$-\rho\frac{\partial^2 W_2^0}{\partial t^2}+\frac{\partial \sigma_{12}^0}{\partial x_1}+\frac{\partial \sigma_{22}^0}{\partial x_2}=0 \quad \forall\ (x_1,x_2)\in S\ ,\ \forall t, \qquad [4.15]$$

$$\frac{\partial W_1^0}{\partial x_1}=S_{1111}\,\sigma_{11}^0+S_{1221}\,\sigma_{22}^0 \quad \forall\ (x_1,x_2)\in S\ ,\ \forall t, \qquad [4.16]$$

$$\frac{\partial W_1^0}{\partial x_2}+\frac{\partial W_2^0}{\partial x_1}=4S_{1212}\,\sigma_{12}^0 \quad \forall\ (x_1,x_2)\in S\ ,\ \forall t, \qquad [4.17]$$

$$\frac{\partial W_2^0}{\partial x_2}=S_{1122}\,\sigma_{11}^0+S_{2222}\,\sigma_{22}^0 \quad \forall\ (x_1,x_2)\in S\ ,\ \forall t. \qquad [4.18]$$

Equations [4.14] and [4.15] are the equations of motion in directions 1 and 2, while equations [4.16], [4.17] and [4.18] represent the stress-strain relation associated with the movements in the plane of the plate.

The associated boundary conditions are given by [2.89] and [2.90] in any point of the boundary line $\overline{S}$ defining the plate where n_1 and n_2 are the direction cosines of the external normal vector (see Figure 4.1):

$\forall\ (x_1,x_2)\in \overline{S}$:

$$\begin{cases} \text{either} & : \quad W_1^0=0 \\ \text{or} & : \quad \sigma_{11}^0\ n_1+\sigma_{12}^0\ n_2=0 \end{cases} \qquad [4.19]$$

and:

$$\begin{cases} \text{either} & : \quad W_2^0 = 0 \\ \text{or} & : \quad \sigma_{12}^0\, n_1 + \sigma_{22}^0\, n_2 = 0\,. \end{cases} \qquad [4.20]$$

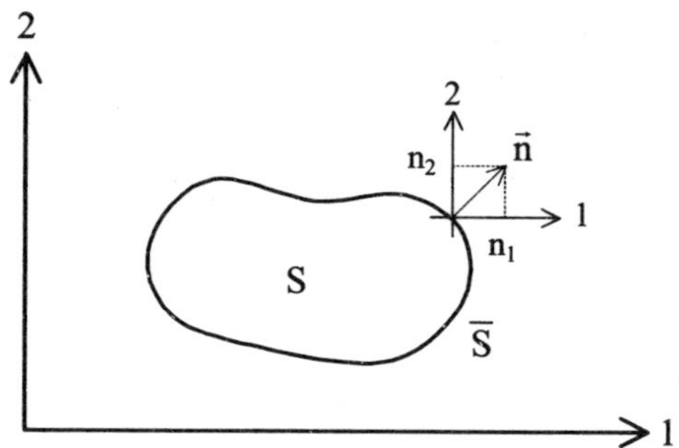

Figure 4.1. *Representation of a plate with a surface* S *and an external normal unit vector* $\vec{n}$ *in a point of its boundary line* $\overline{S}$

These boundary conditions are interpreted physically as the nullity of displacement (clamped) or of stresses placed in duality (free boundary).

From equations [4.14] – [4.20] a formulation with displacements can be drawn. It is enough to express the three components of the tensor of the stresses $\sigma_{11}^0, \sigma_{12}^0$ and σ_{22}^0 with respect to W_1^0 and W_2^0 thanks to the three relations [4.16], [4.17] and [4.18]:

$$\begin{pmatrix} \sigma_{11}^0 \\ \sigma_{22}^0 \\ \sigma_{12}^0 \end{pmatrix} = \begin{pmatrix} C_{1111}^0 & C_{1122}^0 & 0 \\ C_{2211}^0 & C_{2222}^0 & 0 \\ 0 & 0 & C_{1212}^0 \end{pmatrix} \begin{pmatrix} \partial W_1^0/\partial x_1 \\ \partial W_2^0/\partial x_2 \\ \partial W_1^0/\partial x_2 + \partial W_2^0/\partial x_1 \end{pmatrix}. \qquad [4.21]$$

The coefficients of the matrix appearing in [4.21] are easily identified in equations [4.16], [4.17] and [4.18]; we leave the calculation thereof for the general case to the reader.

For an isotropic material we have:

$$S_{1111} = S_{2222} = 1/E \ , \ S_{1122} = -\nu/E \ , \ S_{1212} = 1/(4G)$$

where E is the Young modulus, G is the shearing modulus and ν is the Poisson's material ratio.

The coefficients are drawn from this:

$$C^0_{1111} = C^0_{2222} = E/(1-\nu^2),$$
$$C^0_{1122} = \nu E/(1-\nu^2), \quad [4.22]$$
$$C^0_{1212} = G = E/(2(1+\nu)).$$

These elasticity coefficients do not correspond to those of the three-dimensional stress-strain relation; they constitute a law of two-dimensional behavior applicable to plates.

By changing [4.14] and [4.15] we obtain the two equations of motion to be verified in any point of S:

$$-\rho\frac{\partial^2 W^0_1}{\partial t^2} + \frac{\partial}{\partial x_1}\left(C^0_{1111}\frac{\partial W^0_1}{\partial x_1} + C^0_{1122}\frac{\partial W^0_2}{\partial x_2}\right) + \frac{\partial}{\partial x_2}\left(C^0_{1212}\left(\frac{\partial W^0_1}{\partial x_2} + \frac{\partial W^0_2}{\partial x_1}\right)\right) = 0, \quad [4.23]$$

$$-\rho\frac{\partial^2 W^0_2}{\partial t^2} + \frac{\partial}{\partial x_2}\left(C^0_{2222}\frac{\partial W^0_2}{\partial x_2} + C^0_{1122}\frac{\partial W^0_1}{\partial x_1}\right) + \frac{\partial}{\partial x_1}\left(C^0_{1212}\left(\frac{\partial W^0_1}{\partial x_2} + \frac{\partial W^0_2}{\partial x_1}\right)\right) = 0. \quad [4.24]$$

The boundary conditions are given by [4.25] and [4.26]; they must be verified in any point of the boundary $\bar{S}$:

either : $W^0_1 = 0$,

or : $\left(C^0_{1111}\frac{\partial W^0_1}{\partial x_1} + C^0_{1122}\frac{\partial W^0_2}{\partial x_2}\right) n_1 + C^0_{1212}\left(\frac{\partial W^0_1}{\partial x_2} + \frac{\partial W^0_2}{\partial x_1}\right) n_2 = 0$ [4.25]

and:

$$\text{either} \ : \ W_2^0 = 0,$$

$$\text{or} \ : \ \left(C_{2211}^0 \frac{\partial W_1^0}{\partial x_1} + C_{2222}^0 \frac{\partial W_2^0}{\partial x_2} \right) n_2 + C_{1212}^0 \left(\frac{\partial W_1^0}{\partial x_2} + \frac{\partial W_2^0}{\partial x_1} \right) n_1 = 0. \qquad [4.26]$$

The formulation with displacements reduces the number of unknowns, but increases the order of derivation of the equations; nonetheless, this form is the one generally used. Let us recall that setting up equations using Reissner's functional followed by substitution of stresses by their expressions according to the displacements, as we did while passing from equations [4.14] – [4.20] to [4.23] – [4.26], avoids the inconsistencies of the stress-strain relation which appear during the set up of equations with displacement variables using the Hamilton's functional (see Chapter 3). In the case of an isotropic material we can use the expressions [4.22] in [4.23] – [4.26] to arrive at the following equations, equations [4.23] and [4.24] being written in a more compact form:

$$-\rho \frac{\partial^2}{\partial t^2} \begin{pmatrix} W_1^0 \\ W_2^0 \end{pmatrix} + \frac{E}{1-\nu^2} \begin{pmatrix} \frac{\partial^2}{\partial x_1^2} + \frac{1-\nu}{2} \frac{\partial^2}{\partial x_2^2} & \frac{1+\nu}{2} \frac{\partial^2}{\partial x_1 \partial x_2} \\ \frac{1+\nu}{2} \frac{\partial^2}{\partial x_1 \partial x_2} & \frac{\partial^2}{\partial x_1^2} + \frac{1-\nu}{2} \frac{\partial^2}{\partial x_2^2} \end{pmatrix} \begin{pmatrix} W_1^0 \\ W_2^0 \end{pmatrix} = \begin{pmatrix} 0 \\ 0 \end{pmatrix}. \qquad [4.27]$$

The boundary conditions to verify $\forall (x_1, x_2) \in \overline{S}$ become:

$$\text{either} \ : \ W_1^0 = 0,$$

$$\text{or} \ : \ \frac{E}{1-\nu^2} \left(\left(\frac{\partial W_1^0}{\partial x_1} + \nu \frac{\partial W_2^0}{\partial x_2} \right) n_1 + \frac{1-\nu}{2} \left(\frac{\partial W_1^0}{\partial x_2} + \frac{\partial W_2^0}{\partial x_1} \right) n_2 \right) = 0 \qquad [4.28]$$

and:

$$\text{either} \ : \ W_2^0 = 0,$$

$$\text{or} \ : \ \frac{E}{1-\nu^2} \left(\left(\frac{\partial W_2^0}{\partial x_2} + \nu \frac{\partial W_1^0}{\partial x_1} \right) n_2 + \frac{1-\nu}{2} \left(\frac{\partial W_1^0}{\partial x_2} + \frac{\partial W_2^0}{\partial x_1} \right) n_1 \right) = 0. \qquad [4.29]$$

4.4. Equations of motion and boundary conditions of transverse vibrations

4.4.1. *Mindlin's hypotheses: equations with mixed variables*

The equations are set up by rendering the extremum of Reissner's functional [4.12] particularized for the fields of displacements and of stress [4.7] and [4.8], that is:

$$
\begin{aligned}
R(W_i, \sigma_{ij}) = \int_{t_0}^{t_1}\int_S \Bigg[& \frac{1}{2}\rho\left(I\left(\frac{\partial W_1^3}{\partial t}\right)^2 + I\left(\frac{\partial W_2^3}{\partial t}\right)^2 + h\left(\frac{\partial W_3^0}{\partial t}\right)^2 \right) \\
& - I\left(\sigma_{11}^3 \frac{\partial W_1^3}{\partial x_1} + \sigma_{22}^3 \frac{\partial W_2^3}{\partial x_2} + \sigma_{12}^3 \left(\frac{\partial W_1^3}{\partial x_2} + \frac{\partial W_2^3}{\partial x_1} \right) \right) \\
& - h\sigma_{13}^3 \left(W_1^3 + \frac{\partial W_3^0}{\partial x_1} \right) - h\sigma_{23}^3 \left(W_2^3 + \frac{\partial W_3^0}{\partial x_2} \right) \\
& + \frac{I}{2}\left(\frac{(\sigma_{11}^3)^2}{E} + \frac{(\sigma_{22}^3)^2}{E} + \frac{(\sigma_{12}^3)^2}{G} - 2\nu \frac{\sigma_{11}^3 \sigma_{22}^3}{E} \right) \\
& + \frac{h}{2}\left(\frac{(\sigma_{13}^3)^2}{G} + \frac{(\sigma_{23}^3)^2}{G} \right) \Bigg] dsdt.
\end{aligned}
\qquad [4.30]
$$

In this expression we considered that the thickness was constant and equal to h. Moreover, we posed $I = h^3/12$ and considered an isotropic material.

The calculation of extremum is carried out thanks to the Euler equations [2.88] – [2.92] of Chapter 2. The Euler equations stemming from displacement variables (W_1^3, W_2^3, W_3^0) yield the equations of motion:

$$
-\rho I \frac{\partial^2 W_1^3}{\partial t^2} + I \frac{\partial \sigma_{11}^3}{\partial x_1} + I \frac{\partial \sigma_{12}^3}{\partial x_2} - h\sigma_{13}^0 = 0 \quad \forall\ (x_1, x_2) \in S, \qquad [4.31]
$$

$$
-\rho I \frac{\partial^2 W_2^3}{\partial t^2} + I \frac{\partial \sigma_{22}^3}{\partial x_2} + I \frac{\partial \sigma_{12}^3}{\partial x_1} - h\sigma_{23}^0 = 0 \quad \forall\ (x_1, x_2) \in S, \qquad [4.32]
$$

$$- \rho h \frac{\partial^2 W_3^0}{\partial t^2} + h \frac{\partial \sigma_{23}^0}{\partial x_2} + I \frac{\partial \sigma_{13}^0}{\partial x_1} = 0 \quad \forall (x_1, x_2) \in S, \qquad [4.33]$$

those stemming from the stresses ($\sigma_{11}^3, \sigma_{12}^3, \sigma_{22}^3, \sigma_{13}^0, \sigma_{23}^0$) provide the stress-strain relation:

$$\begin{pmatrix} \sigma_{11}^3 \\ \sigma_{22}^3 \\ \sigma_{12}^3 \\ \sigma_{13}^0 \\ \sigma_{23}^0 \end{pmatrix} = \begin{pmatrix} \frac{E}{1-\nu^2} & \frac{\nu E}{1-\nu^2} & 0\,0\,0 \\ \frac{\nu E}{1-\nu^2} & \frac{E}{1-\nu^2} & 0\,0\,0 \\ 0 & 0 & G\,0\,0 \\ 0 & 0 & 0\,G\,0 \\ 0 & 0 & 0\,0\,G \end{pmatrix} \begin{pmatrix} \partial W_1^3/\partial x_1 \\ \partial W_2^3/\partial x_2 \\ \partial W_1^3/\partial x_2 + \partial W_2^3/\partial x_1 \\ W_1^3 + \partial W_3^0/\partial x_1 \\ W_2^3 + \partial W_3^0/\partial x_2 \end{pmatrix}. \qquad [4.34]$$

Equations [4.31] and [4.32] are representative of the rotational movement in the thickness; equation [4.33] governs the transverse movement. We use the set of equations stemming from the calculation of extremum with respect to the stress variables $\sigma_{11}^3, \sigma_{12}^3, \sigma_{22}^3, \sigma_{13}^0$ and σ_{23}^0 in order to write the stress-strain relation in matrix form [4.34].

The calculation of extremum also provides the boundary conditions with [2.89] and [2.90] from Chapter 2. The direction cosines of the external unit normal vector to the boundary line $\overline{S}$ of the plate are noted (n_1, n_2):

$$\begin{aligned} &\text{either} \quad : W_1^3 = 0 \quad \forall (x_1, x_2) \in \overline{S}, \\ &\text{or} \quad : I(\sigma_{11}^3 n_1 + \sigma_{12}^3 n_2) = 0 \quad \forall (x_1, x_2) \in \overline{S} \end{aligned} \qquad [4.35]$$

and:

$$\begin{aligned} &\text{either} \quad : W_2^3 = 0 \quad \forall (x_1, x_2) \in \overline{S}, \\ &\text{or} \quad : I(\sigma_{12}^3 n_1 + \sigma_{22}^3 n_2) = 0, \quad \forall (x_1, x_2) \in \overline{S} \end{aligned} \qquad [4.36]$$

and:

$$\begin{aligned} &\text{either} \quad : W_3^0 = 0 \quad \forall\,(x_1, x_2) \in \overline{S}, \\ &\text{or} \quad : h\,(\sigma_{13}^0\, n_1 + \sigma_{23}^0\, n_2) = 0 \quad \forall\,(x_1, x_2) \in \overline{S}. \end{aligned} \qquad [4.37]$$

4.4.2. *Mindlin's hypotheses: equations with displacement variables*

It is enough to draw the expression of stresses from the stress-strain relation [4.34] and to report them in equations [4.31] to [4.33] and in the boundary conditions [4.35] – [4.37].

Equations of the motion to verify $\forall\,(x_1, x_2) \in S$:

$$\rho I \frac{\partial^2 W_1^3}{\partial t^2} - \frac{EI}{1-\nu^2}\frac{\partial^2 W_1^3}{\partial x_1^2} - IG\frac{\partial^2 W_1^3}{\partial x_2^2} + hG\left(W_1^3 + \frac{\partial W_3^0}{\partial x_1}\right) - I\left(G + \frac{\nu E}{1-\nu^2}\right)\frac{\partial^2 W_2^3}{\partial x_1 \partial x_2} = 0, \qquad [4.38]$$

$$\rho I \frac{\partial^2 W_2^3}{\partial t^2} - \frac{EI}{1-\nu^2}\frac{\partial^2 W_2^3}{\partial x_2} - IG\frac{\partial^2 W_2^3}{\partial x_1^2} + hG\left(W_2^3 + \frac{\partial W_3^0}{\partial x_2}\right) - I\left(G + \frac{\nu E}{1-\nu^2}\right)\frac{\partial^2 W_1^3}{\partial x_1 \partial x_2} = 0, \qquad [4.39]$$

$$-\rho h \frac{\partial^2 W_3^0}{\partial t^2} + hG\left(\frac{\partial^2 W_3^0}{\partial x_1^2} + \frac{\partial^2 W_3^0}{\partial x_2^2}\right) + hG\left(\frac{\partial W_1^3}{\partial x_1} + \frac{\partial W_2^3}{\partial x_2}\right) = 0. \qquad [4.40]$$

Boundary conditions to verify $\forall\,(x_1, x_2) \in \overline{S}$:

$$\begin{cases} \text{either} \quad : W_1^3 = 0 \\ \text{or} \quad : \dfrac{EI}{1-\nu^2}\left(\dfrac{\partial W_1^3}{\partial x_1} + \nu\dfrac{\partial W_2^3}{\partial x_2}\right) n_1 + GI\left(\dfrac{\partial W_1^3}{\partial x_2} + \dfrac{\partial W_2^3}{\partial x_1}\right) n_2 = 0 \end{cases} \qquad [4.41]$$

and

$$\begin{cases} \text{either } : W_2^3 = 0 \\ \text{or} \quad : \dfrac{EI}{1-\nu^2}\left(\dfrac{\partial W_2^3}{\partial x_2} + \nu \dfrac{\partial W_1^3}{\partial x_1}\right) n_2 + GI\left(\dfrac{\partial W_1^3}{\partial x_2} + \dfrac{\partial W_2^3}{\partial x_1}\right) n_1 = 0 \end{cases} \quad [4.42]$$

and

$$\begin{cases} \text{either } : W_3^0 = 0 \\ \text{or} \quad : Gh\left(W_1^3 + \dfrac{\partial W_3^0}{\partial x_1}\right) n_1 + Gh\left(W_2^3 + \dfrac{\partial W_3^0}{\partial x_2}\right) n_2 = 0 . \end{cases} \quad [4.43]$$

4.4.3. *Love-Kirchhoff hypotheses: equations with mixed variables*

The calculation of Reissner's functional is carried out with the field of displacement [4.9] and the stress field [4.10]. To simplify calculations we also consider here a homogenous and isotropic material as well as a plate with constant thickness:

$$\begin{aligned} R(W_i, \sigma_{ij}) = \int_{t_0}^{t_1}\int_S \Bigg[& \frac{1}{2}\rho\left(I\left(\frac{\partial^2 W_3^0}{\partial x_1 \partial t}\right)^2 + I\left(\frac{\partial^2 W_3^0}{\partial x_2 \partial t}\right)^2 + h\left(\frac{\partial W_3^0}{\partial t}\right)^2 \right) \\ & + I\left(\sigma_{11}^3 \frac{\partial^2 W_3^0}{\partial x_1^2} + \sigma_{22}^3 \frac{\partial^2 W_3^0}{\partial x_2^2} + 2\sigma_{12}^3 \frac{\partial^2 W_3^0}{\partial x_2 \partial x_1}\right) \\ & + \frac{1}{2}\left(\frac{I}{E}\left(\left(\sigma_{11}^3\right)^2 + \left(\sigma_{22}^3\right)^2 - 2\nu\sigma_{11}^3\sigma_{22}^3\right) + \frac{I}{G}\left(\sigma_{12}^3\right)^2\right)\Bigg] dsdt . \end{aligned} \quad [4.44]$$

The equation of motion and the stress-strain relation are calculated here also thanks to Euler's equations [2.88] – [2.92] from Chapter 2. They are more complicated to apply than in the preceding cases since the functional [4.44] reveals a second derivative. Upon calculation it follows:

Equation of the motion to verify $\forall (x_1, x_2) \in S$:

$$\rho h \frac{\partial^2 W_3^0}{\partial t^2} - \rho I \left(\frac{\partial^4 W_3^0}{\partial x_1^2 \partial t^2} + \frac{\partial^4 W_3^0}{\partial x_2^2 \partial t^2} \right) - I \left(\frac{\partial^2 \sigma_{11}^3}{\partial x_1^2} + \frac{\partial^2 \sigma_{22}^3}{\partial x_2^2} + 2 \frac{\partial^2 \sigma_{12}^3}{\partial x_1 \partial x_2} \right) = 0. \qquad [4.45]$$

Stress-strain relation to verify $\forall (x_1, x_2) \in S$:

$$\begin{pmatrix} \sigma_{11}^3 \\ \sigma_{22}^3 \\ \sigma_{12}^3 \end{pmatrix} = \begin{pmatrix} \frac{E}{1-\nu^2} & \frac{\nu E}{1-\nu^2} & 0 \\ \frac{\nu E}{1-\nu^2} & \frac{E}{1-\nu^2} & 0 \\ 0 & 0 & \frac{E}{1+\nu} \end{pmatrix} \begin{pmatrix} \frac{\partial^2 W_3^0}{\partial x_1^2} \\ \frac{\partial^2 W_3^0}{\partial x_2^2} \\ \frac{\partial^2 W_3^0}{\partial x_1 \partial x_2} \end{pmatrix} \qquad [4.46]$$

Boundary conditions to verify $\forall (x_1, x_2) \in \overline{S}$:

either : $W_3^0 = 0$, [4.47]

$$\text{or} \quad : I \left(\frac{\partial \sigma_{11}^3}{\partial x_1}(n_1 + n_1\, n_2^2) + \frac{\partial \sigma_{22}^3}{\partial x_2}(n_2 + n_2\, n_1^2) + 2\frac{\partial \sigma_{12}^3}{\partial x_2} n_1^3 + \frac{\partial \sigma_{11}^3}{\partial x_2} n_2\, n_1^2 + \frac{\partial \sigma_{22}^3}{\partial x_1} n_1\, n_2^2 - 2\frac{\partial \sigma_{12}^3}{\partial x_1} n_2^3 - \rho \frac{\partial^3 W_3^0}{\partial t^2 \partial x_1} n_1 - \rho \frac{\partial^3 W_3^0}{\partial t^2 \partial x_2} n_2 \right) = 0 \qquad [4.48]$$

and

either : $\frac{\partial W_3^0}{\partial n} = 0$, [4.49]

or : $I \left(\sigma_{11}^3\, n_1^2 + 2\sigma_{12}^3\, n_1\, n_2 + \sigma_{22}^3\, n_2^2 \right) = 0$. [4.50]

The number of unknowns is smaller than in the case of the Mindlin's hypotheses, but the equations have a higher order of derivation and are, therefore, rather complicated.

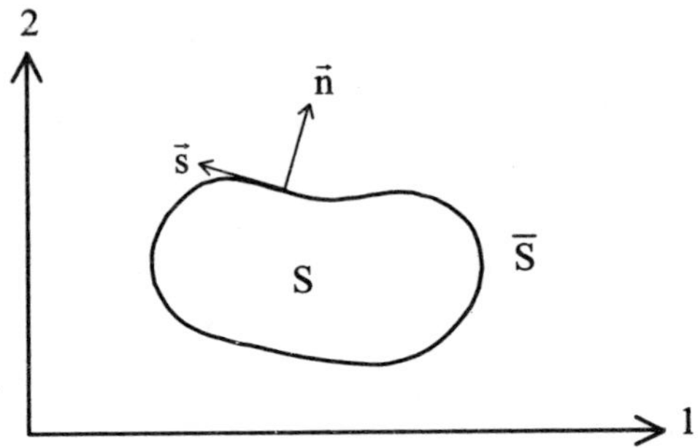

Figure 4.2. *Representation of the unit vectors in a point of the boundary line* $\overline{S}$ *of the plate.* $\vec{n}$ *normal* $\vec{s}$ *tangent*

We can give another form of more compact boundary conditions by introducing the normal $\partial/\partial n$ and tangential $\partial/\partial s$ derivatives connected to the derivatives following x_1 and x_2 by the relations:

$$\frac{\partial}{\partial x_1} = n_1 \frac{\partial}{\partial n} - n_2 \frac{\partial}{\partial s} \qquad [4.51]$$

and $$\frac{\partial}{\partial x_2} = n_2 \frac{\partial}{\partial n} + n_1 \frac{\partial}{\partial s} ;$$

that is also:

$$\frac{\partial}{\partial n} = n_1 \frac{\partial}{\partial x_1} + n_2 \frac{\partial}{\partial x_2} \qquad [4.52]$$

and $$\frac{\partial}{\partial s} = -n_2 \frac{\partial}{\partial x_1} + n_1 \frac{\partial}{\partial x_2} .$$

Let us denote by bending moment M the quantity placed in duality with the normal derivative of transverse displacement $\left(\frac{\partial W_3^0}{\partial n}\right)$ by the boundary condition [4.50]:

$$M = I\,(\sigma_{11}^3\, n_1^2 + 2\sigma_{12}^3\, n_1\, n_2 + \sigma_{22}^3\, n_2^2) .$$

Let us denote by shearing force T the quantity placed in duality with transverse displacement W_3^0 by the boundary condition [4.48]. After a long but not difficult calculation, it is shown that this quantity is also written in the form [4.53]:

$$T = -\frac{\partial}{\partial n}\left(M + \rho I \frac{\partial^2 W_3^0}{\partial t^2} \right) + I \frac{\partial}{\partial s}\left((\sigma_{11}^3 - \sigma_{22}^3)\, n_1\, n_2 + 2\sigma_{12}^3 (n_2^2 - n_1^2) \right). \quad [4.53]$$

The boundary conditions [4.49] and [4.50] are thus written $\forall\, (x_1, x_2) \in \overline{S}$:

$$\begin{cases} \text{either} & : W_3^0 = 0 \\ \text{or} & : T = 0\,. \end{cases} \quad [4.54]$$

and:

$$\begin{cases} \text{either} & : \dfrac{\partial W_3^0}{\partial n} = 0 \\ \text{or} & : M = 0\,. \end{cases} \quad [4.55]$$

4.4.4. *Love-Kirchhoff hypotheses: equations with displacement variables*

In this case it is also possible to substitute the tensors of stresses with their expressions according to the transverse displacement W_3^0 provided by the stress-strain relation [4.46]. Substituting in [4.45] we obtain the equation of motion:

$$-\rho h \frac{\partial^2 W}{\partial t^2} + \rho I \left(\frac{\partial^4 W}{\partial x_1^2\, \partial t^2} + \frac{\partial^4 W}{\partial x_2^2\, \partial t^2} \right) - D\left(\frac{\partial^4 W}{\partial x_1^4} + 2\frac{\partial^4 W}{\partial x_1^2\, \partial x_2^2} + \frac{\partial^4 W}{\partial x_2^4} \right) = 0 \quad [4.56]$$

with $D = EI/(1 - \nu^2)$ bending stiffness. [4.56']

We also obtain by substitution the boundary conditions [4.54] and [4.55] where the bending moment and the shearing force are given by:

$$M = D\left[\left(\frac{\partial^2 W}{\partial x_1^2} + \nu \frac{\partial^2 W}{\partial x_2^2}\right) n_1^2 + 2(1-\nu)\frac{\partial^2 W}{\partial x_1 x_2} n_1 n_2 + \left(\frac{\partial^2 W}{\partial x_2^2} + \nu \frac{\partial^2 W}{\partial x_1^2}\right) n_2^2\right] \quad [4.57]$$

and:

$$T = \frac{\partial}{\partial n}\left(M + \rho I \frac{\partial^2 W}{\partial t^2}\right) + \frac{\partial}{\partial s}\left(D\left(\left(\frac{\partial^2 W}{\partial x_1^2} - \frac{\partial^2 W}{\partial x_2^2}\right)(1+\nu)\, n_1 n_2 + 4(1-\nu)\frac{\partial^2 W}{\partial x_1 \partial x_2}(n_2^2 - n_1^2)\right)\right) \quad [4.58]$$

In order to be concise, in these expressions we have replaced W_3^0 with W.

These are the most frequently employed equations, but an additional simplification is introduced by neglecting the effect of rotational inertia (we will see, on the basis of the case of beams in Chapter 6, that this simplification is acceptable for a low frequency). We then have the standard equations:

$$-\rho h \frac{\partial^2 W}{\partial t^2} - D\left(\frac{\partial^4 W}{\partial x_1^4} + 2\frac{\partial^4 W}{\partial x_1^2 \partial x_2^2} + \frac{\partial^4 W}{\partial x_2^4}\right) = 0 \quad \forall (x_1, x_2) \in S \quad [4.59]$$

and:

$$\begin{cases} \text{either} & : W = 0 \\ & \qquad \forall (x_1, x_2) \in \overline{S} \\ \text{or} & : T = 0 \end{cases} \quad [4.60]$$

and:

$$\begin{cases} \text{either} & : \dfrac{\partial W}{\partial n} = 0 \\ \text{or} & : M = 0 \end{cases} \quad \forall\ (x_1, x_2) \in \overline{S}\ . \qquad [4.61]$$

M is given by [4.57] and T by [4.58]. In the expression of the shearing force, the term $\rho I \partial^2 W / \partial t^2$ associated with rotational inertia is neglected.

4.4.5. *Love-Kirchhoff hypotheses: equations with displacement variables obtained using Hamilton's functional*

In this chapter we have until now made exclusive use of Reissner's functional, the equations with displacement variables being obtained by substitution of stresses by displacements in the equations with mixed variables.

It is also possible to directly obtain the equations with displacements by using Hamilton's functional. The two approaches were detailed in Chapter 3 in the case of beams, thus, we will present this approach for plates only briefly by limiting ourselves to the most often encountered case of the Love-Kirchhoff hypotheses.

The field of displacements considered is the one described by equation [4.9]. It suffices then to replace general displacements by those from expression [4.9] in the general expression of Hamilton's functional given in equation [2.55] in Chapter 2. Upon calculation we obtain after integration over thickness for a plate with constant thickness h made of an isotropic material homogenous in the thickness:

$$\begin{aligned} H\big(W(x_1, x_2, t)\big) = \int_{t_0}^{t_1}\!\!\int_S \Bigg(& \rho h/2(\partial W/\partial t)^2 - (D/2)\bigg((\partial^2 W/\partial x_1^2)^2 \\ & + (\partial^2 W/\partial x_2^2)^2 \bigg) + 2\nu(\partial^2 W/\partial x_1^2)(\partial^2 W/\partial x_2^2) \qquad [4.62] \\ & + 2(1-\nu)(\partial^2 W/\partial x_1 \partial x_2) \Bigg) dx_1\, dx_2\, dt\,. \end{aligned}$$

In this expression, D is the bending stiffness of the plate defined in [4.56'] and rotational inertia has been neglected in the expression of kinetic energy. It is the classical functional used in the problems of plates, in particular, within the framework of the Rayleigh-Ritz method.

We calculate Euler's equations [2.88] with respect to the variable W to obtain the equation of motion and equations [2.89], [2.90], [2.91] and [2.92] to determine the boundary conditions. After all the calculations we receive equations identical to those obtained in section 4.4.4. There is, thus, an equivalence between the set-up of equations using Reissner's functional with two independent fields and Hamilton's functional with only one field of displacements. However, this equivalence is obtained by using the moduli of elasticity C^0_{ijkl} given by [4.21] and not the moduli C_{ijkl} of the three-dimensional stress-strain relation in Hamilton's functional. Here also, as for beams in Chapter 3, the two formulations are equivalent only if the elastic moduli are modified and adapted to the case under consideration.

4.4.6. *Some comments on the formulations of transverse vibrations*

There are several possible formulations of transverse plates vibrations. The Mindlin's hypotheses are adapted to the description of anisotropic plates and high frequencies. The Love-Kirchhoff hypotheses result from the statics of isotropic plates and constitute the most severe approximation; they are realistic only for low frequency forecasting of isotropic plates vibrations. However, the simplicity of equations resulting from the Love-Kirchhoff hypotheses often leads to their use outside of their actual field of validity. A first approximation of the result is then obtained.

In general, formulations with displacements are preferred to mixed formulations because they decrease the number of equations to be processed. We can, however, wonder whether the concentration of derivations over a restricted number of equations is not damaging in the end, especially for physical interpretation. The expression of the boundary shearing force [4.58] is completely explicit on the subject of the difficulty of interpretation.

4.5. Coupled movements

We have defined two elementary movements to describe the vibrations of plates: vibrations in the plane of the plate and transverse vibrations. These vibratory states are uncoupled for homogenous plates but are coupled as soon as the plate has variable characteristics in the thickness.

Let us consider a vibratory movement of the plate resulting from a combination of transverse vibrations and vibrations in the plane. The fields of displacements and stresses describing the movement are defined by the superposition of [4.5] and [4.7] and of [4.6] and [4.8], that is:

$$\begin{cases} W_1(x_1, x_2, x_3, t) = W_1^0(x_1, x_2, t) + x_3\, W_1^3(x_1, x_2, t) \\ W_2(x_1, x_2, x_3, t) = W_2^0(x_1, x_2, t) + x_3\, W_2^3(x_1, x_2, t) \\ W_3(x_1, x_2, x_3, t) = W_3^0(x_1, x_2, t), \end{cases}$$

$$\sigma_{11}(x_1, x_2, x_3, t) = \sigma_{11}^0(x_1, x_2, t) + x_3\, \sigma_{11}^3(x_1, x_2, t),$$

$$\sigma_{22}(x_1, x_2, x_3, t) = \sigma_{22}^0(x_1, x_2, t) + x_3\, \sigma_{22}^3(x_1, x_2, t),$$

$$\sigma_{12}(x_1, x_2, x_3, t) = \sigma_{12}^0(x_1, x_2, t) + x_3\, \sigma_{12}^3(x_1, x_2, t),$$

$$\sigma_{13}(x_1, x_2, x_3, t) = \sigma_{13}^0(x_1, x_2, t),$$

$$\sigma_{23}(x_1, x_2, x_3, t) = \sigma_{23}^0(x_1, x_2, t),$$

$$\sigma_{33}(x_1, x_2, x_3, t) = 0.$$

The Reissner's functional of the problem is rather bulky and we do not write it down as a whole but limit ourselves only to the terms revealing the variables W_1^3 and W_1^0 in order to write the two equations of motion which result from it:

$$\begin{aligned} R(W_i, \sigma_{ij}) = \int_{t_0}^{t_1}\int_S \Bigg(& \frac{1}{2}\int_{-h/2}^{+h/2} \rho \left(x_3^2 \left(\frac{\partial W_1^3}{\partial t}\right)^2 + 2x_3 \frac{\partial W_1^3}{\partial t}\frac{\partial W_1^0}{\partial t} \right. \\ & \left. + \left(\frac{\partial W_1^0}{\partial t}\right)^2 \right) dx_3 - \int_{-h/2}^{+h/2} \left(\sigma_{11}^0 \frac{\partial W_1^0}{\partial x_1} + x_3 \left(\sigma_{11}^0 \frac{\partial W_1^3}{\partial x_1} + \sigma_{11}^3 \frac{\partial W_1^0}{\partial x_1} \right) \right. \\ & \left. + x_3^2\, \sigma_{11}^3 \frac{\partial W_1^3}{\partial x_1} \right) dx_3 - \int_{-h/2}^{+h/2} \left(\sigma_{12}^0 \frac{\partial W_1^0}{\partial x_2} + x_3 \left(\sigma_{12}^0 \frac{\partial W_1^3}{\partial x_2} + \sigma_{12}^3 \frac{\partial W_1^0}{\partial x_2} \right) \right. \\ & \left. + x_3^2\, \sigma_{12}^3 \frac{\partial W_1^3}{\partial x_2} \right) dx_3 - \int_{-h/2}^{+h/2} \sigma_{13}^0\, W_1^3\, dx_3 + \ldots \Bigg) dsdt. \end{aligned} \qquad [4.63]$$

The coupling of plane and transverse movements comes from the terms in x_3 of [4.63]; after integration over the thickness these terms are eliminated in the 2nd and 3rd lines because:

$$\int_{-h/2}^{+h/2} x_3\, dx_3 = 0 \quad [4.64]$$

but they generally remain in the first line if the density varies in the thickness:

$$\int_{-h/2}^{+h/2} \rho x_3\, dx_3 \neq 0. \quad [4.65]$$

If the density is constant in thickness, it can stem form the integral [4.65], which is then nil.

The integration of [4.63] over the thickness yields:

$$\begin{aligned} R(W_i, \sigma_{ij}) = \int_{t_0}^{t_1}\int_S \Bigg(& \frac{1}{2}\int_{-h/2}^{+h/2} \rho x_3^2\, dx_3 \left(\frac{\partial W_1^3}{\partial t}\right)^2 + \int_{-h/2}^{+h/2} \rho x_3\, dx_3 \frac{\partial W_1^3}{\partial t}\frac{\partial W_1^0}{\partial t} \\ & + \frac{1}{2}\int_{-h/2}^{+h/2} \rho\, dx_3 \left(\frac{\partial W_1^0}{\partial t}\right)^2 - h\,\sigma_{11}^0 \frac{\partial W_1^0}{\partial x_1} - I\,\sigma_{11}^3 \frac{\partial W_1^3}{\partial x_1} \\ & - h\,\sigma_{12}^0 \frac{\partial W_1^0}{\partial x_2} - I\,\sigma_{12}^3 \frac{\partial W_1^3}{\partial x_2} - h\,\sigma_{13}^0\, W_1^3 + \dots \Bigg) ds dt. \end{aligned} \quad [4.66]$$

We are able to calculate the equations of motion coming from the calculation of extremum with respect to the two functions W_1^3 and W_1^0:

$$\int_{-h/2}^{+h/2} \rho x_3^2\, dx_3 \frac{\partial^2 W_1^3}{\partial t^2} + \int_{-h/2}^{+h/2} \rho x_3\, dx_3 \frac{\partial^2 W_1^0}{\partial t^2} - I\frac{\partial \sigma_{11}^3}{\partial x_1} - I\frac{\partial \sigma_{12}^3}{\partial x_2} + h\sigma_{13}^0 = 0, \quad [4.67]$$

$$\int_{-h/2}^{+h/2} \rho\, dx_3 \frac{\partial^2 W_1^0}{\partial t^2} + \int_{-h/2}^{+h/2} \rho\, x_3\, dx_3 \frac{\partial^2 W_1^3}{\partial t^2} - h\frac{\partial \sigma_{11}^0}{\partial x_1} - h\frac{\partial \sigma_{12}^0}{\partial x_2} = 0. \quad [4.68]$$

Equation [4.67] can be compared to [4.31]; equation [4.68] can be compared to [4.14]. The coupling of movement in the plane W_1^0 and transverse movement W_1^3

is related to the value of the integral [4.65], which is nil if the density is constant through the thickness; we then find the equations uncoupled since [4.67] coincides exactly with [4.31] and [4.68] coincides exactly with [4.14]. The examination of the whole of Reissner's functional and the equations which result from it would show that the property highlighted for the two equations [4.67] and [4.68] can be generalized and that for a material with constant characteristics in the thickness, transverse vibrations and vibrations in the plane are uncoupled.

4.6. Equations with polar co-ordinates

4.6.1. *Basic relations*

In certain problems we may find it beneficial to use polar co-ordinates rather than Cartesian co-ordinates which we have employed up until now. The transformation of equations written in Cartesian co-ordinates into equations written in polar co-ordinates is rather simple. We will exploit it here.

Let us introduce r and θ, the polar co-ordinates of a point of the plate. They are connected to the Cartesian co-ordinates x_1 and x_2 by the relations:

$$\begin{cases} x_1 = r\sin\theta \\ x_2 = r\cos\theta, \end{cases}$$

$$\begin{cases} r = \sqrt{x_1^2 + x_2^2} \\ \theta = \text{Arctg}\left(\dfrac{x_1}{x_2}\right). \end{cases} \quad [4.69]$$

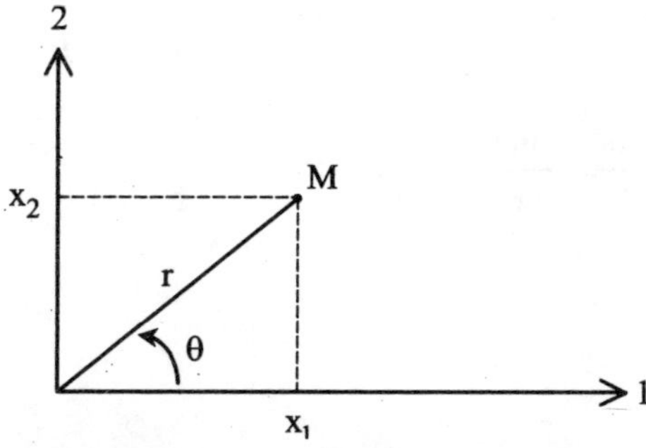

Figure 4.3. *Cartesian and polar co-ordinates of the point M*

By observing the rule of chain derivation, we deduce:

$$\begin{cases} \dfrac{\partial}{\partial x_1} = \dfrac{\partial}{\partial r}\dfrac{\partial r}{\partial x_1} + \dfrac{\partial}{\partial \theta}\dfrac{\partial \theta}{\partial x_1} \\ \dfrac{\partial}{\partial x_2} = \dfrac{\partial}{\partial r}\dfrac{\partial r}{\partial x_2} + \dfrac{\partial}{\partial \theta}\dfrac{\partial \theta}{\partial x_2}. \end{cases} \quad [4.70]$$

A rather simple calculation with equations [4.69] yields the following results:

$$\frac{\partial r}{\partial x_1} = \sin\theta \ ; \ \frac{\partial \theta}{\partial x_1} = \frac{\cos\theta}{r} \ ; \ \frac{\partial r}{\partial x_2} = \cos\theta \ ; \ \frac{\partial \theta}{\partial x_2} = -\frac{\sin\theta}{r}.$$

With [4.70] we deduce from it:

$$\begin{cases} \dfrac{\partial}{\partial x_1} = \sin\theta\dfrac{\partial}{\partial r} + \dfrac{\cos\theta}{r}\dfrac{\partial}{\partial \theta} \\ \dfrac{\partial}{\partial x_2} = \cos\theta\dfrac{\partial}{\partial r} - \dfrac{\sin\theta}{r}\dfrac{\partial}{\partial \theta}. \end{cases} \quad [4.71]$$

Equations [4.71] are at the basis of the transformation of equations with Cartesian co-ordinates into polar co-ordinates. Thanks to these expressions, derivatives of higher orders can be calculated. For example, let us calculate the second derivative:

$$\begin{cases} \dfrac{\partial^2}{\partial x_1^2} = \left(\sin\theta\dfrac{\partial}{\partial r} + \dfrac{\cos\theta}{r}\dfrac{\partial}{\partial \theta}\right)\left(\sin\theta\dfrac{\partial}{\partial r} + \dfrac{\cos\theta}{r}\dfrac{\partial}{\partial \theta}\right) \\ \dfrac{\partial^2}{\partial x_2^2} = \left(\cos\theta\dfrac{\partial}{\partial r} - \dfrac{\sin\theta}{r}\dfrac{\partial}{\partial \theta}\right)\left(\cos\theta\dfrac{\partial}{\partial r} - \dfrac{\sin\theta}{r}\dfrac{\partial}{\partial \theta}\right). \end{cases} \quad [4.72]$$

Calculation provides the following expressions:

$$\frac{\partial^2}{\partial x_1^2} = \sin^2\theta \frac{\partial^2}{\partial r^2} - \frac{2\sin\theta\cos\theta}{r^2}\frac{\partial}{\partial\theta} + \frac{2\sin\theta\cos\theta}{r}\frac{\partial^2}{\partial r\partial\theta} + \frac{\cos^2\theta}{r}\frac{\partial}{\partial r} + \frac{\cos^2\theta}{r^2}\frac{\partial^2}{\partial\theta^2},$$

$$\frac{\partial^2}{\partial x_1^2} = \cos^2\theta \frac{\partial^2}{\partial r^2} + \frac{2\sin\theta\cos\theta}{r^2}\frac{\partial}{\partial\theta} - \frac{2\sin\theta\cos\theta}{r}\frac{\partial^2}{\partial r\partial\theta} + \frac{\sin^2\theta}{r}\frac{\partial}{\partial r} + \frac{\sin^2\theta}{r^2}\frac{\partial^2}{\partial\theta^2}.$$ [4.73]

It is then remarkable to note that the Laplace operator has the simple form:

$$\Delta = \frac{\partial^2}{\partial x_1^2} + \frac{\partial^2}{\partial x_2^2} = \frac{\partial^2}{\partial r^2} + \frac{1}{r}\frac{\partial}{\partial r} + \frac{1}{r^2}\frac{\partial^2}{\partial\theta^2}.$$ [4.74]

4.6.2. *Love-Kirchhoff equations of the transverse vibrations of plates*

We are interested in the standard equation [4.59], which neglects rotational inertia. The operator appearing in plate bending equation of motion is also called bilaplacian:

$$\Delta^2 = \frac{\partial^4}{\partial x_1^4} + 2\frac{\partial^4}{\partial x_1^2\partial x_2^2} + \frac{\partial^4}{\partial x_2^4}.$$ [4.75]

By using the result [4.74] we note that:

$$\Delta^2 = \left(\frac{\partial^2}{\partial r^2} + \frac{1}{r}\frac{\partial}{\partial r} + \frac{1}{r^2}\frac{\partial^2}{\partial\theta^2}\right)\left(\frac{\partial^2}{\partial r^2} + \frac{1}{r}\frac{\partial}{\partial r} + \frac{1}{r^2}\frac{\partial^2}{\partial\theta^2}\right)$$ [4.76]

After calculation we obtain:

$$\Delta^2 = \frac{\partial^4}{\partial r^4} + \frac{2}{r^2}\frac{\partial^4}{\partial r^2\,\partial\theta^2} + \frac{1}{r^4}\frac{\partial^4}{\partial\theta^4} + \frac{2}{r}\frac{\partial^3}{\partial r^3} - \frac{2}{r^3}\frac{\partial^3}{\partial\theta^2\,\partial r} - \frac{1}{r^2}\frac{\partial^2}{\partial r^2} + \frac{4}{r^4}\frac{\partial^2}{\partial\theta^2} + \frac{1}{r^3}\frac{\partial}{\partial r}. \tag{4.77}$$

In the case of axisymmetric movement $W(r,\theta)$ is independent of θ and derivations with respect to this variable are nil; the operator is greatly simplified:

$$\Delta^2 = \frac{\partial^4}{\partial r^4} + \frac{2}{r}\frac{\partial^3}{\partial r^3} - \frac{1}{r^2}\frac{\partial^2}{\partial r^2} + \frac{1}{r^3}\frac{\partial}{\partial r}. \tag{4.78}$$

The equation of the transverse vibrations of plates retains the form:

$$-\rho h\frac{\partial^2 W}{\partial t^2} - D\Delta^2(W) = 0 \tag{4.79}$$

where the bilaplacian Δ^2 is given by [4.77] in general and by [4.78] for axisymmetric movements. In Cartesian co-ordinates, the bilaplacian is equal to [4.75].

The boundary conditions associated with equation [4.79] were provided in [4.60] and [4.61] in Cartesian co-ordinates; we may of course provide an expression thereof in polar co-ordinates. For that it is necessary to express the normal derivative, the bending moment and the shearing force in these co-ordinates.

Let us examine the normal derivative:

$$\frac{\partial}{\partial n} = n_1\frac{\partial}{\partial x_1} + n_2\frac{\partial}{\partial x_2}.$$

With [4.71] it follows:

$$\frac{\partial}{\partial n} = (n_1\sin\theta + n_2\cos\theta)\frac{\partial}{\partial r} + \frac{1}{r}(n_1\cos\theta - n_2\sin\theta)\frac{\partial}{\partial\theta}. \tag{4.80}$$

The expression of bending moment has been given in [4.57], while the expression of shearing force is in [4.58]. To obtain the expressions in polar co-ordinates, the procedure is commonplace since it is a question of replacing the derivative with respect to the Cartesian variables by the expressions [4.71]; however, calculation is extremely long and the expressions obtained are very heavy. In addition we would find ourselves in the case of an axisymmetric problem, which implies the circular shape of the plate. Under these conditions the direction cosines of the external unit normal vector take the form:

$$n_2 = \cos\theta, \; n_1 = \sin\theta. \qquad [4.81]$$

We note with [4.80] and [4.81] that in this case:

$$\frac{\partial}{\partial n} = \frac{\partial}{\partial r}.$$

To calculate the bending moment and the shearing force, let us take the particular expressions resulting from [4.73] in the axisymmetric case where the displacement of the plate is independent of θ. It follows:

$$\frac{\partial^2}{\partial x_1^2} = \sin^2\theta \frac{\partial^2}{\partial r^2} + \frac{\cos^2\theta}{r}\frac{\partial}{\partial r}, \; \frac{\partial^2}{\partial x_2^2} = \cos^2\theta \frac{\partial^2}{\partial r^2} + \frac{\sin^2\theta}{r}\frac{\partial}{\partial r}.$$

A similar calculation also leads to:

$$\frac{\partial^2}{\partial x_1 \partial x_2} = \sin\theta \; \cos\theta \frac{\partial^2}{\partial r^2} - \frac{\cos\theta \sin\theta}{r}\frac{\partial}{\partial r}.$$

Once all the calculations have been done, the two expressions are obtained:

$$M = D\left(\frac{\partial^2 W}{\partial r^2} + \nu \frac{\partial W}{\partial r}\right), \qquad [4.82]$$

$$T = D\left(\frac{\partial^3 W}{\partial r^3} + \nu \frac{\partial^2 W}{\partial r^2}\right). \qquad [4.83]$$

We have neglected the effect of rotational inertia in the shearing force.

In the case of the circular plate with axisymmetric vibratory movement, the equations of the problem are greatly simplified, in particular, the expressions of bending moment and of the shearing force. These expressions are valid only in one very particular case of circular plate vibrations since the vector is independent of θ, which occurs for an excitation that is also axisymmetric as a point transverse force applied to the center of the plate. For an offset transverse force, the simplified equations are no longer usable since transverse displacement will depend on θ.

4.7. Conclusion

In this chapter we have established the equations of the vibrations of thin plates. The variational set-up of equations is based on Reissner's functional and we have systematically obtained equations with mixed variables, then by substitution the equations with displacements. The case of movements in the plane and then of transverse movements have been tackled. For transverse vibrations of plates, the two traditional hypotheses were exposed (Mindlin and Love-Kirchhoff). Finally, the equations in polar co-ordinates were provided in the simple case of the Love-Kirchhoff operator.

We have attempted to show the methods and thus provide the reader with a general procedure to establish equations of plates motion. As for beams, which have been covered in Chapter 3, all the information is contained in the hypotheses of condensation adopted for displacements and stresses. For plates, these hypotheses result from a development of the various functions describing the vibrations of the plate in Taylor series over the thickness. For thin plates that we consider, these developments are truncated of the first order taking into account the low thickness.

In addition to the technique of setting up equations, the suggested procedure makes it possible to determine the domain of applicability of the established theory thanks to the physical interpretation which arises from the hypotheses of condensation employed.

Chapter 5

Vibratory Phenomena Described by the Wave Equation

5.1. Introduction

The wave equation is a partial derivative equation which we highlighted in Chapter 3 during the study of longitudinal and torsion vibrations of beams. This equation is also representative of two other vibratory phenomena which are often encountered: vibrations of cords and fluctuations of acoustic pressure in pipes.

The study of the wave equation is particularly interesting because its relative simplicity makes it possible to easily find a solution and describe many basic concepts.

In the first section we present the problem, and more precisely we recapitulate the set of applications of the wave equation; then we demonstrate the uniqueness of the solution. In the following section, we provide a solution by the method of propagation, which will lead us to notion of the image source to take into account the boundary conditions. Resolution by separation of variables will then be carried out, which will lead to the key concept of the natural mode of vibration, from which will result the general form of the response by modal decomposition. Finally a summary table of the modal system for the case of standard boundary conditions is drawn up.

The last section will give a detailed presentation of two applications with some of the most remarkable physical tendencies of vibratory behavior. Moreover, they will provide practical examples of calculations, modal system and vibratory response, with displacements as well as with stresses.

5.2. Wave equation: presentation of the problem and uniqueness of the solution

5.2.1. *The wave equation*

The wave equation is the following partial derivative equation:

$$\frac{\partial^2 y}{\partial t^2}(x, t) - c^2 \frac{\partial^2 y}{\partial x^2} = 0 . \qquad [5.1]$$

The function y(x, t) represents vibratory movement, the constant c is characteristic of the studied medium; it is called celerity or waves propagation velocity.

This equation is representative of longitudinal vibratory and torsion movements of homogenous beams, as we have shown in Chapter 3. In fact, vibrations of cords and fluctuations in pressure in pipes are also governed by this equation. The correspondence between the general equation [5.1] and the four physical situations that it describes is presented in Table 5.1.

To properly present the problem, that is, in fact, to ensure the uniqueness of the solution, it is necessary to provide equation [5.1] with boundary conditions and initial conditions.

Boundary conditions:

$$y(0, t) = -\frac{1}{\alpha_0}\frac{\partial y}{\partial x}(0, t) , \qquad [5.2]$$

$$y(L, t) = \frac{1}{\alpha_L}\frac{\partial y}{\partial x}(L, t) . \qquad [5.3]$$

Initial conditions:

$$y(x,0) = d_0(x) , \qquad [5.4]$$

$$\frac{\partial y}{\partial t}(x,0) = v_0(x) . \qquad [5.5]$$

with d_0 initial displacement and v_0 initial speed.

Physical situation	y(x , t)	c
Longitudinal vibrations	$U_1^0(x, t)$: longitudinal displacement of cross-sections	$c_L = \sqrt{\frac{E}{\rho}}$, celerity of longitudinal waves. E : Young modulus ρ : density
Vibrations of torsion of beams	$\alpha(x, t)$: rotation of cross-sections	$c_T = \sqrt{\frac{G}{\rho}}$, celerity of the waves of torsion. G : Coulomb modulus ρ : density
Vibrating cords	y(x, t) : transverse displacement of the cord	$c_T = \sqrt{\frac{T}{\rho S}}$ T : tension of the cord ρ : density S : cross-section of the cord
Sound pipes	p(x, t) : fluctuation in pressure	c : speed of sound in fluid (air 340 m/s).

Table 5.1. *The applicability of the wave equation*

Initial conditions are taken at the moment $t = 0$ to simplify the writing. Boundary conditions are particular: they correspond to resilient mounting characterized by α_0 and α_L for each end. From a physical point of view, our definition shows that these quantities are homogenous to lengths and representative of the limits impedance. Indeed, we find the classical boundary conditions of clamped end for $\alpha_0 = 0$ (or $\alpha_L = 0$), and of a free end making α_0 (or α_L) tend towards infinity:

clamped end: $y(0,t) = 0$.

free end: $\frac{\partial y}{\partial x}(0,t) = 0$.

These two types of boundary conditions thus appear as a borderline case of [5.2] and [5.3].

5.2.2. *Equation of energy and uniqueness of the solution*

5.2.2.1. *Equation of energy*

Let us multiply the two members of equation [5.1] by $\frac{\partial y}{\partial t}$; it follows:

$$\frac{\partial y}{\partial t}\frac{\partial^2 y}{\partial t^2}(x,t) - \frac{\partial y}{\partial t}c^2\frac{\partial^2 y}{\partial x^2} = 0. \qquad [5.6]$$

Let us take the integral of equation [5.6] between two fixed positions a and b:

$$\int_a^b \frac{\partial y}{\partial t}\frac{\partial^2 y}{\partial t^2}(x,t)\,dx - \int_a^b \frac{\partial y}{\partial t}c^2\frac{\partial^2 y}{\partial x^2}dx = 0. \qquad [5.7]$$

Let us integrate by parts the second term of the first member of [5.7]; we obtain:

$$\int_a^b \frac{\partial y}{\partial t}\frac{\partial^2 y}{\partial t^2}(x,t)\,dx + \int_a^b \frac{\partial y}{\partial x}c^2\frac{\partial^2 y}{\partial x\partial t}dx$$
$$= c^2\left(\frac{\partial y}{\partial x}(b,t)\frac{\partial y}{\partial t}(b,t) - \frac{\partial y}{\partial x}(a,t)\frac{\partial y}{\partial t}(a,t)\right). \qquad [5.8]$$

Observing that:

$$\frac{d}{dt}\int_a^b \frac{1}{2}\left(\frac{\partial y}{\partial t}\right)^2 dx = \int_a^b \frac{1}{2}\frac{\partial}{\partial t}\left(\frac{\partial y}{\partial t}\right)^2 dx = \int_a^b \frac{\partial y}{\partial t}\frac{\partial^2 y}{\partial t^2}dx, \qquad [5.9]$$

equation [5.8] is written:

$$\frac{1}{2}\frac{d}{dt}\int_a^b \left(\left(\frac{\partial y}{\partial t}\right)^2 + c^2\left(\frac{\partial y}{\partial x}\right)^2\right)dx$$
$$= c^2\left(\frac{\partial y}{\partial x}(b,t)\frac{\partial y}{\partial t}(b,t) - \frac{\partial y}{\partial x}(a,t)\frac{\partial y}{\partial t}(a,t)\right). \qquad [5.10]$$

Equation [5.10] is in fact nothing but the expression of conservation of energy, except for a multiplicative constant $(\mu = \text{constant})$. Indeed, let us multiply the two members by the linear density μ of the medium considered; it follows:

$$\frac{1}{2}\frac{d}{dt}\int_a^b\left(\mu\left(\frac{\partial y}{\partial t}\right)^2+c^2\mu\left(\frac{\partial y}{\partial x}\right)^2\right)dx$$
$$=\mu c^2\left(\frac{\partial y}{\partial x}(b,t)\frac{\partial y}{\partial t}(b,t)-\frac{\partial y}{\partial x}(a,t)\frac{\partial y}{\partial t}(a,t)\right). \qquad [5.11]$$

The first member represents the energy variation of the section [a, b] over time. Indeed, we recognize the sum of the densities of kinetic and deformation energies under the integral sign of the first member.

The second member is the difference of the powers introduced at the two ends a and b of the section.

The principle of conservation of energy is thus entirely contained in the wave equation.

If we now apply equation [5.11] to points 0 and L, ends of the beam, taking into account the boundary conditions [5.2] and [5.3] and after integration over time, it follows:

$$\frac{1}{2}\int_0^L\left(\mu\left(\frac{\partial y}{\partial t}\right)^2+c^2\,\mu\left(\frac{\partial y}{\partial x}\right)^2\right)dx+\frac{1}{2}\mu\,c^2\,\alpha_0\,y(0,t)^2$$
$$+\frac{1}{2}\mu\,c^2\,\alpha_L\,y(L,t)^2=E \qquad [5.12]$$

where E is a constant.

Equation [5.12] means that the energy of the total system, that is taking account the energy of the boundaries, is constant over time. In the 3rd and 4th term of the first member of [5.12] we recognize the energies of the boundaries. Calculating equation [5.12] at the initial moment we obtain the constant value of energy over time which is equal to the value taken at the initial moment, that is to say, taking into account [5.4] and [5.5]:

$$\frac{1}{2}\int_0^L\left(\mu\,(v_0)^2+c^2\,\mu\left(\frac{\partial d_0}{\partial x}\right)^2\right)dx+\frac{1}{2}\mu\,c^2\,\alpha_0\,d_0^2(0)$$
$$+\frac{1}{2}\mu\,c^2\,\alpha_L\,d_0^2(L)=E. \qquad [5.13]$$

The equation of energy [5.13] is the foundation of the demonstration of the solution uniqueness which is proposed in the following section.

5.2.2.2. *Uniqueness of the solution*

Let us consider two solutions $y_1(x, t)$ and $y_2(x, t)$ which verify equations [5.1] to [5.5] and their difference $Y(x, t)$.

$$Y(x, t) = y_1(x, t) - y_2(x, t) . \tag{5.14}$$

The linearity of these equations implies that the difference of the two solutions $Y(x, t)$ verifies the wave equation [5.1] as well as the boundary and initial conditions:

$$Y(0, t) = -\alpha_0 \frac{\partial Y}{\partial x}(0, t) , \tag{5.15}$$

$$Y(L, t) = \alpha_L \frac{\partial Y}{\partial x}(L, t) , \tag{5.16}$$

$$Y(x,0) = 0 , \tag{5.17}$$

$$\frac{\partial Y}{\partial t}(x,0) = 0 . \tag{5.18}$$

The function $Y(x, t)$ verifying the wave equation and the boundary conditions [5.15] and [5.16] also satisfies the integral form [5.12], that is:

$$\frac{1}{2}\int_0^L \left(\mu \left(\frac{\partial Y}{\partial t} \right)^2 + c^2 \mu \left(\frac{\partial Y}{\partial x} \right)^2 \right) dx + \frac{1}{2} \mu\, c^2\, \alpha_0 Y(0, t)^2 + \frac{1}{2} \mu\, c^2\, \alpha_L Y(L, t)^2 = E . \tag{5.19}$$

It is now enough to take equation [5.19] at $t = 0$ to deduce from it that the constant is nil, taking into account [5.17] and [5.18]. Thus, at any moment:

$$\frac{1}{2}\int_0^L \left(\mu \left(\frac{\partial Y}{\partial t} \right)^2 + c^2 \mu \left(\frac{\partial Y}{\partial x} \right)^2 \right) dx + \frac{1}{2} \mu\, c^2\, \alpha_0 Y(0, t)^2 + \frac{1}{2} \mu\, c^2\, \alpha_L Y(L, t)^2 = 0 . \tag{5.20}$$

Relation [5.20] expresses the nullity of the sum of positive quantities, which must thus also be nil on their own. From that we deduce, on the one hand, that

$Y(0,t) = 0$ and $Y(L,t) = 0$ and, on the other hand, that $\frac{\partial Y}{\partial x}$ and $\frac{\partial Y}{\partial t}$ must be nil almost everywhere in the open interval $]0, L[$. Since the function $Y(x,t)$ also verifies the wave equation, it must be continuously derivable twice and consequently nil. The two solutions $y_1(x,t)$ and $y_2(x,t)$ are combined. Thus, the uniqueness of the solution of the problem defined by equations [5.1] – [5.5] is proven.

5.3. Resolution of the wave equation by the method of propagation (d'Alembert's methodology)

5.3.1. *General solution of the wave equation*

Let us consider the wave equation [5.1] and seek a general solution by carrying out the change of variables:

$$u = x + ct, \tag{5.21}$$

$$v = x - ct. \tag{5.22}$$

By using the chain rule of derivation, we can show that:

$$\frac{\partial^2}{\partial x^2} = \frac{\partial^2}{\partial u^2} + 2\frac{\partial^2}{\partial u \partial v} + \frac{\partial^2}{\partial v^2} \tag{5.23}$$

and:

$$\frac{\partial^2}{\partial t^2} = c^2\left(\frac{\partial^2}{\partial u^2} - 2\frac{\partial^2}{\partial u \partial v} + \frac{\partial^2}{\partial v^2}\right); \tag{5.24}$$

then by replacing in the wave equation we obtain in the system of variables (u,v):

$$\frac{\partial^2 y}{\partial u \partial v} = 0. \tag{5.25}$$

This equation is solved in two stages:

$$\text{a) } \frac{\partial y}{\partial u} = F(u) \tag{5.26}$$

where $F(u)$ is an arbitrary function of u.

b) $y = f(u) + g(v)$ [5.27]

where f is the primitive of F and g is an arbitrary function.

Returning to the variables x and t via the definitions [5.21] and [5.22], the general solution of the wave equation is:

$$y(x,t) = f(x + ct) + g(x - ct). \quad [5.28]$$

The functions f and g are arbitrary provided that they are continuously derivable twice. This property of the solutions is absolutely remarkable. To understand its physical meaning, we will examine the first part of the solution $f(x + ct)$, at various moments. Let us propose an arbitrary movement at the moment t and observe the evolution over time (Figure 5.1). Displacement associated with the point x at the moment t is associated the point x' the moment t' if:

$$x + ct = x' + ct'. \quad [5.29]$$

From that we deduce:

$$x' = x - c(t - t'). \quad [5.30]$$

When time increases, displacement is relocated towards negative x without deformation; it is said that it is propagated.

The remarkable aspect of the solutions thus results in the fact that any initially imposed displacement (which is arbitrary) completely describes the displacement at any later moment just by translation of the initial shape.

By considering infinitesimal increases $x' = x + dx$ and $t' = t + dt$, we demonstrate with [5.30] that:

$$\frac{dx}{dt} = -c. \quad [5.31]$$

This quantity c is homogenous to speed: it characterizes the propagation. It is called celerity of the waves or propagation velocity. For the part of that solution that we have just studied, speed is negative and characterizes a propagation towards decreasing x.

A similar reasoning for the second part of the solution $g(x - ct)$ would show the same phenomenon but with an opposite propagation velocity $+c$ and thus a propagation towards growing x. The general solution is thus the superposition of two displacements propagating at the same speed but in opposite directions.

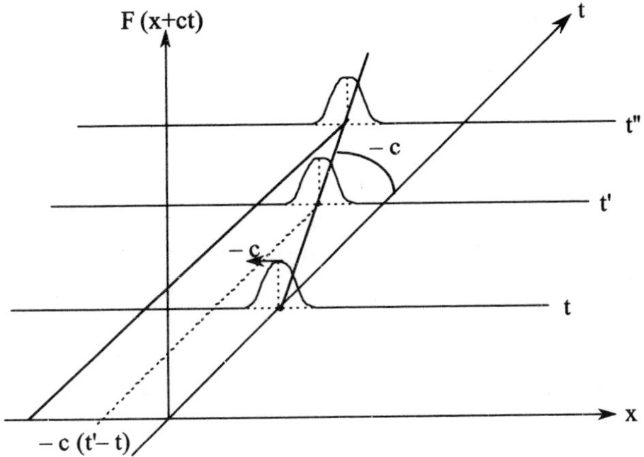

Figure 5.1. *Propagation of displacement without deformation. Displacement at the moment t is translated towards negative x at the moment t' with a propagation velocity equal to –c*

5.3.2. *Taking initial conditions into account*

Let us consider the initial conditions [5.4] and [5.5] and seek the general solution verifying them; it follows:

$$f(x) + g(x) = d_0(x), \tag{5.32}$$

$$c\,f'(x) - cg'(x) = v_0(x). \tag{5.33}$$

By ' in [5.33] we have denoted derivation. By transforming the two preceding equations we obtain:

$$f(x) = d_0(x) - g(x) \tag{5.34}$$

and:

$$c\left(d_0'(x) - 2g'(x)\right) = v_0(x). \tag{5.35}$$

By integrating the second equation and then using the result in the first we obtain:

$$g(\xi) = \frac{d_0(\xi)}{2} - \frac{d_0(0)}{2} - \frac{1}{2c}\int_0^{\xi} v_0(x)\,dx + g(0) \tag{5.36}$$

and:

$$f(\xi) = \frac{d_0(\xi)}{2} + \frac{d_0(0)}{2} + \frac{1}{2c}\int_0^{\xi} v_0(x)\,dx - g(0)\,. \tag{5.37}$$

These two expressions, taken respectively for $\xi = x - ct$ and $\xi = x + ct$, make it possible to find the general form of the solution verifying the initial conditions [5.4] and [5.5]:

$$y(x,t) = \frac{d_0(x + ct)}{2} + \frac{d_0(x - ct)}{2} + \frac{1}{2c}\int_{x-ct}^{x+ct} v_0(\xi)\,d\xi\,. \tag{5.38}$$

To illustrate this result let us consider the following example:

$$d_0(x) = \sin^2\left(\frac{2\pi}{L}\left(x - \frac{L}{2}\right)\right)\left\{H\left(x + \frac{L}{2}\right) - H\left(x - \frac{L}{2}\right)\right\} \tag{5.39}$$

with:

$$H(u) = 0 \ \text{ if } \ u < 0\,, \qquad H(u) = 1 \ \text{ if } \ u > 0 \tag{5.40}$$

and:

$$v_0(x) = 0\,. \tag{5.41}$$

We immediately deduce from it that for every $t \geq 0$:

$$\begin{aligned} y(x,t) = {} & \frac{1}{2}\sin^2\left(\frac{2\pi}{L}\left(x + ct - \frac{L}{2}\right)\right)\left(H\left(x + ct + \frac{L}{2}\right) - H\left(x + ct - \frac{L}{2}\right)\right) \\ & + \frac{1}{2}\sin^2\left(\frac{2\pi}{L}\left(x - ct - \frac{L}{2}\right)\right)\left(H\left(x - ct + \frac{L}{2}\right) - H\left(x - ct - \frac{L}{2}\right)\right). \end{aligned} \tag{5.42}$$

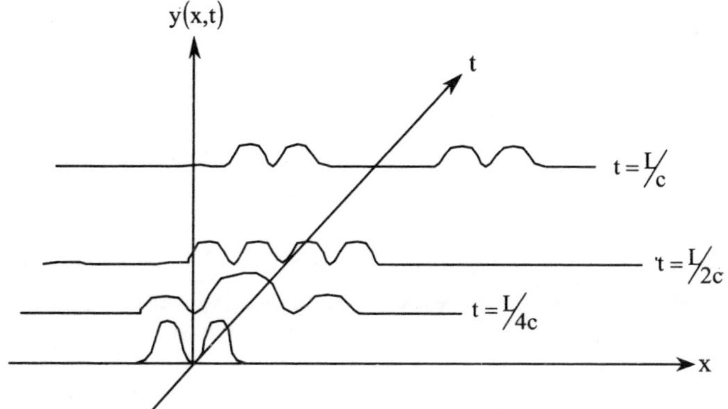

Figure 5.2. *Propagation of initial displacement*

Figure 5.2 illustrates the vibratory state at various moments after the initial moment. A certain number of characteristic phenomena can be observed in this figure:

a) the initial condition generates two identical displacement shapes that propagate in opposite directions;

b) each displacement has the same shape as the one imposed initially but with half the amplitude;

c) the two displacements do not become deformed during their propagation. This property is characteristic of a non-dispersive medium;

d) the two displacements are propagated with respective velocities $-c$ and $+\mathrm{c}$ for the two terms of [5.41].

Let us take as a second example the initial conditions of imposed velocity:

$$\begin{aligned} & \mathrm{d}_0(x) = 0 \\ & \mathrm{v}_0(x) = 0 \quad \text{if} \quad \mathrm{x} < 0\ ,\ \mathrm{x} > \mathrm{L} \\ \text{and} \quad & \mathrm{v}_0(x) = \sin\left(\frac{\pi}{\mathrm{L}}\mathrm{x}\right) \quad \text{if} \quad 0 < \mathrm{x} < \mathrm{L}. \end{aligned} \tag{5.43}$$

From that we deduce:

$$\mathrm{y(x,t)} = \frac{1}{2\mathrm{c}} \int_{\mathrm{x-ct}}^{\mathrm{x+ct}} \mathrm{v}_0(\xi)\,\mathrm{d}\xi\,, \tag{5.44}$$

that is:

$$\mathrm{y(x,t)} = \frac{1}{2\mathrm{c}} \left[\mathrm{D}_0(\xi)\right]_{\mathrm{x-ct}}^{\mathrm{x+ct}} \tag{5.45}$$

$$D_0(\xi) = -\frac{1}{\pi} \text{ if } \xi < 0,$$

$$D_0(\xi) = -\frac{1}{\pi}\cos\left(\frac{\pi\xi}{L}\right) \text{ if } 0 < x < L, \qquad [5.46]$$

$$D_0(\xi) = \frac{1}{\pi} \text{ if } \xi > L.$$

After calculations we can draw up Table 5.2, which gives the expression of vibratory displacement y(x, t) according to the respective values of x + ct and x – ct.

	$-\infty < x + ct < 0$	$0 < x + ct < L$	$L < x + ct < \infty$
$0 > X - ct > -\infty$	0	$-\frac{1}{2\pi c}\left[\cos\left(\frac{\pi(x+ct)}{L}\right) - 1\right]$	$\frac{L}{\pi c}$
$L > X - ct > 0$	$-\frac{1}{2\pi c}\left[1 - \cos\frac{(x-ct)}{L}\right]$	$-\frac{1}{2\pi c}\left[\cos\left(\frac{\pi(x+ct)}{L}\right) - \cos\left(\frac{\pi(x-ct)}{L}\right)\right]$	$\frac{1}{2\pi c}\left[1 + \cos\left(\frac{\pi(x-ct)}{L}\right)\right]$
$\infty > X - ct > L$	$-\frac{L}{\pi c}$	$-\frac{1}{2\pi c}\left[\cos\left(\frac{\pi(x+ct)}{L}\right) + 1\right]$	0

Table 5.2. *Values of y (x, t) according to x + ct and x – ct*

Figure 5.3 shows vibratory displacement at several characteristic moments.

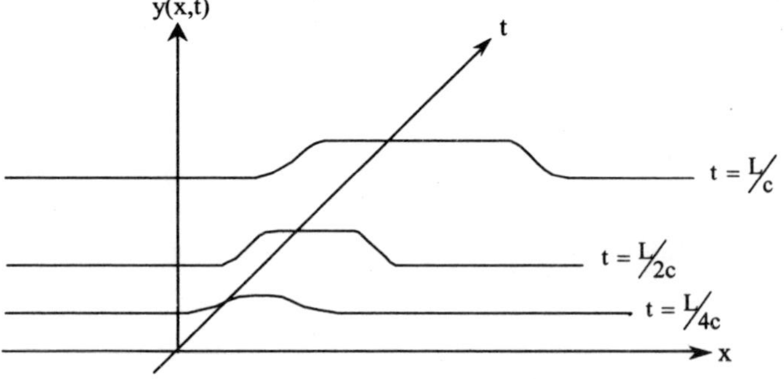

Figure 5.3. *Consecutive displacements under an initial condition of imposed velocity*

Vibratory behavior following an initial imposed velocity condition is different from behavior following an initial imposed displacement condition, which we have described previously.

Indeed, the initial velocity generated between 0 and L extends in the course of time to the whole of the beam with an evolution of form over time. Propagation velocities ± c remain present since they characterize the displacement of the displacement front (see Figure 5.4).

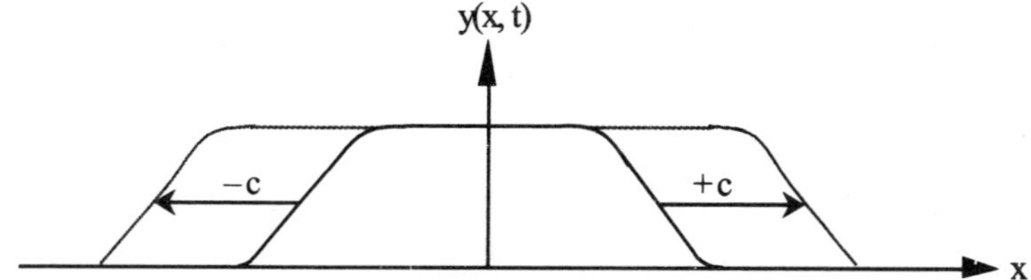

Figure 5.4. *Propagation velocity of displacements following an initial condition of imposed velocity*

5.3.3. *Taking into account boundary conditions: image source*

Taking into account boundary conditions leads to the concept of image source. To begin with, let us consider a semi-infinite medium $]-\infty, L]$, clamped at point $x = L$, that is, verifying:

$$y(L,t) = 0 \quad \forall t. \tag{5.47}$$

Let us further consider that the beam is subjected to the initial conditions [5.39] and [5.40]. The solution is then given by:

$$y(x,t) = \left(\frac{d_0(x+ct)}{2} + \frac{d_0(x-ct)}{2}\right) - \left(\frac{d_0(2L-x+ct)}{2} + \frac{d_0(2L-x-ct)}{2}\right). \tag{5.48}$$

The first term of the right-hand member is the direct wave, while the second is the wave reflected by the clamped end.

It is clearly a solution of the wave equation since it is a function of the two variables $x + ct$ and $x - ct$. We may also note that when $x = L$:

$$y(L,t) = 0. \tag{5.49}$$

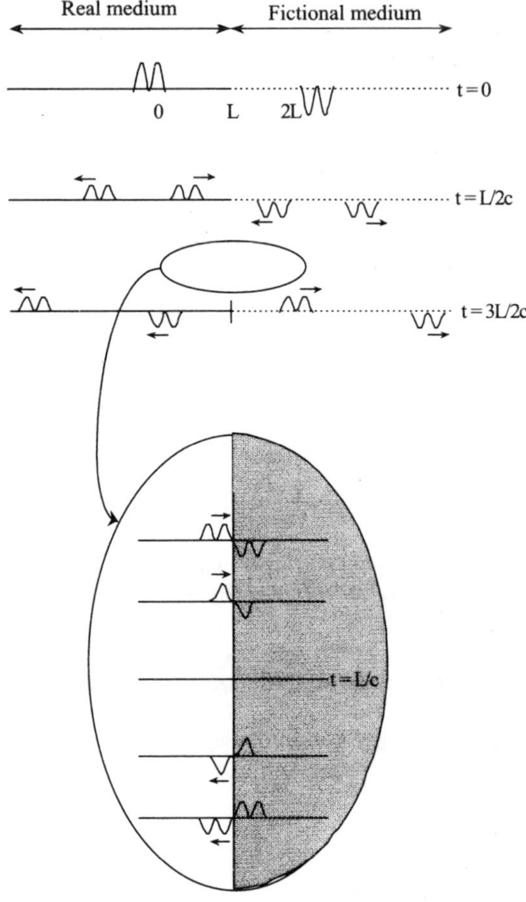

Figure 5.5. *Displacement of the real and fictional media at different point in time*

It is interesting to examine movement over time while introducing a fictional continuous medium in the continuity of the real medium (Figure 5.5). At the initial moment, in addition to real displacement, the solution reveals a mirror image displacement in the fictitious medium laid out antisymmetrically with respect to the clamped end. At the later moments we observe four disturbances occurring in the real and fictional media: two of them propagate with velocity c in the direction of increasing x and the two others towards decreasing x. In the real medium the clamped end reveals a reflection of the disturbance with an inversion of the sign of the displacements. The real source creates the direct field; the image source creates the field reflected by the boundary.

A boundary condition can thus be taken into account by introducing an image of the initial displacement into a fictional medium extending from the real medium.

This concept can extend to more complex configurations, in particular, with a finite beam clamped at the ends:

$$y(L,t)=0\,, \tag{5.50}$$

$$y(-L,t)=0\,. \tag{5.51}$$

Let us consider the initial conditions [5.39] and [5.40], as previously, and position in Figure the 5.6 the different images, which are now unlimited in number because it is necessary to introduce the image of an image:

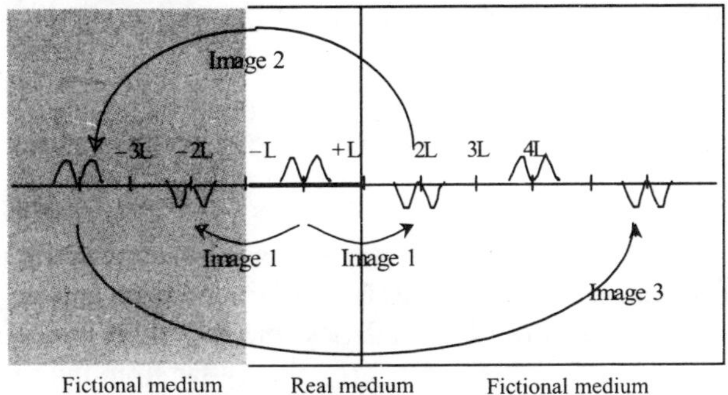

Figure 5.6. *Propagation in a finite medium*
Initial displacements in the real and fictional media
Image 1: images of the real source with respect to the two boundaries
Image 2: images of the first images with respect to the two boundaries
Image 3: images of the second images with respect to the two boundaries

The initial displacement $d_0(x)$ being defined as between $-L$ and $+L$, we associate an infinite number of images to it over the entire x axis. The initial displacement defined in the real and fictional media is written:

$$\begin{aligned} g(x,0) = \quad & d_0(x) \\ & - d_0(2L-x) - d_0(-2L-x) \qquad \text{(first images } I_1\text{)}\,, \\ & + d_0(4L-x) + d_0(-4L-x) \qquad \text{(second images } I_2\text{)}\,, \\ & - d_0(6L-x) - d_0(-6L-x) \qquad \text{(third images } I_3\text{)}\,. \end{aligned} \tag{5.52}$$

Displacement [5.52] is written out only until the third image but, naturally, it comprises an infinity thereof.

The vibratory response of the beam is obtained by applying the result [5.38] when initial speed is nil and initial displacement is given by [5.52]:

$$\begin{aligned} y(x,t) &= \frac{d_0(x+ct)}{2} + \frac{d_0(x-ct)}{2} \\ &\quad + \sum_{i=1}^{\infty} \big(d_0(2iL-(x+ct)) + d_0(2iL-(x-ct))\big)\left(\frac{(-1)^i}{2}\right) \\ &\quad + \sum_{i=1}^{\infty} \big(d_0(-2iL-(x+ct)) + d_0(-2iL-(x-ct))\big)\left(\frac{(-1)^i}{2}\right). \end{aligned} \qquad [5.53]$$

We can easily notice that the solution [5.53] really verifies the boundary conditions [5.50] and [5.51].

This technique of resolution by image source is applicable in many other problems. In the physical sense, it makes boundary conditions appear as more or less deforming mirrors. For absorbing boundaries, the reflection is accompanied by a weakening, and images of higher orders are initially far from the real medium and, when reaching it, have very low amplitude. The solution $y(x,t)$ then tends towards 0 when t tends towards infinity.

5.4. Resolution of the wave equation by separation of variables

5.4.1. *General solution of the wave equation in the form of separate variables*

A second method of resolving the equation is possible: it is based on the separation of variables. In general, the method that we are going to expose is the one preferred over the previous method because it highlights the concept of natural vibration modes.

Let us consider the solutions of the wave equation separated into the product of two functions f(x) and g(t):

$$y(x,t) = f(x)\ g(t). \qquad [5.54]$$

Introducing the form [5.54] into the wave equation [5.1], it follows:

$$\frac{d^2g}{dt^2}(t)\ f(x) = c^2\, g(t)\frac{d^2f}{dx^2}(x). \qquad [5.55]$$

Let us separate the variables in [5.55]:

$$\frac{\dfrac{d^2g}{dt^2}}{g(t)} = c^2 \frac{\dfrac{d^2f}{dx^2}}{f(x)} = \lambda(x,t)\,. \qquad [5.56]$$

The first member of [5.56] is independent of x, while the second member is independent of t; their equality implies that the function $\lambda(x,t)$ that we have introduced in the third member is simultaneously independent of x and t, i.e. equal to a constant a. Equation [5.56] is thus separated into two equations:

$$\frac{d^2g}{dt^2}(t) - ag(t) = 0\,, \qquad [5.57]$$

$$\frac{d^2f}{dx^2}(t) - \frac{a}{c^2} f(x) = 0\,. \qquad [5.58]$$

The constant a can be negative, positive or zero. Let us consider these three possibilities.

1) if $a < 0$, we will pose $a = -\omega^2$, $\omega \neq 0$.

Equations [5.57] and [5.58] become:

$$\frac{d^2g}{dt^2}(t) + \omega^2 g(t) = 0 \qquad [5.59]$$

and:

$$\frac{d^2f}{dx^2}(t) + k^2 f(x) = 0 \qquad [5.60]$$

with: $k = \omega/c$. [5.61]

The solutions of [5.59] and [5.60] are given by:

$$g(t) = A\cos(\omega t) + B\sin(\omega t)\,, \qquad [5.62]$$

$$f(x) = C\cos(kx) + D\sin(kx)\,. \qquad [5.63]$$

The time behavior is described by [5.62], the movement occurs with the angular frequency ω. The space aspect is described by [5.63], it is characterized by a wave number k. Angular frequency and wave number are linked by the relation [5.61] called the dispersion relation.

2) $a = 0 \quad (\omega = 0)$

Equations [5.59] and [5.60] become:

$$\frac{d^2 g}{dt^2}(t) = 0 \qquad [5.64]$$

and:

$$\frac{d^2 f}{dx^2}(t) = 0 . \qquad [5.65]$$

The solutions to [5.64] and [5.65] are:

$$g(t) = At + B \qquad [5.66]$$

and:

$$f(x) = Cx + D . \qquad [5.67]$$

These uniform movements are of a different nature to those described by [5.62] and [5.63]. This particular behavior would be representative of a zero angular frequency $(\omega = 0)$.

3) $a > 0$, we will pose $a = \delta^2$, $\delta > 0$.

Equations [5.57] and [5.58] become:

$$\frac{d^2 g}{dt^2}(t) - \delta^2 g(t) = 0 \qquad [5.68]$$

and:

$$\frac{d^2 f}{dx^2}(t) - \frac{\delta^2}{c^2} f(x) = 0 . \qquad [5.69]$$

The solutions to [5.68] and [5.69] are given by:

$$g(t) = Ae^{\delta t} + Be^{-\delta t} \qquad [5.70]$$

$$f(x) = Ce^{(\delta/c)x} + De^{-(\delta/c)x} \qquad [5.71]$$

In short, there are three different types of solution to the wave equation obtained by separation of variables:

$$a < 0 : y(x,t) = \left(A\cos(\omega t) + B\sin(\omega t)\right)\left(C\cos(kx) + D\sin(kx)\right), \qquad [5.72]$$

$$a = 0 : y(x,t) = \left(At + B\right)\left(Cx + D\right), \qquad [5.73]$$

$$a > 0 : y(x,t) = \left(Ae^{\delta t} + Be^{-\delta t}\right)\left(Ce^{(\delta/c)x} + De^{-(\delta/c)x}\right). \qquad [5.74]$$

5.4.2. *Taking boundary conditions into account*

Let us take the case of boundary conditions of the free type, that is, let us impose:

$$\frac{\partial y}{\partial x}(0,t) = 0 \text{ and } \frac{\partial y}{\partial x}(L,t) = 0 \cdot \qquad [5.75]$$

Taking into account the separation of variables, equations [5.75] become:

$$\frac{df}{dx}(0) = 0 \text{ and } \frac{df}{dx}(L) = 0. \qquad [5.76]$$

Solutions of the wave equation [5.72] – [5.74] must also verify [5.76] in order to authorize a physical movement. Let us examine whether this is possible.

Taking into account [5.76], the form [5.72] leads to equations:

$$D = 0 \text{ and } C\,k\sin(kL) = 0. \qquad [5.77]$$

That is to say, either to the trivial solution $C = D = 0$, or to the solution $D = 0$, $C \neq 0$ and:

$$\sin(kL) = 0. \qquad [5.78]$$

Equation [5.78] is called an equation with normal (or eigen-) frequencies; indeed, only certain values of k can satisfy it. These are normal wave numbers noted k_n:

$$k_n = \frac{n\pi}{L} \text{ for } n = 1, \ldots, \infty \cdot \quad [5.79]$$

Taking into account the relation of dispersion [5.61], an angular frequency ω_n corresponds to each normal wave number:

$$\omega_n = c\frac{n\pi}{L} \text{ for } n = 1, \ldots, \infty \cdot \quad [5.80]$$

Consequently, there is an infinite number of possible solutions resulting from the form [5.72]:

$$y_n(x, t) = \left(A_n \cos(\omega_n t) + B_n \sin(\omega_n t)\right) f_n(x) \quad [5.81]$$

$$\text{with: } f_n(x) = C_n \cos\left(\frac{n\pi}{L} x\right). \quad [5.82]$$

The function $f_n(x)$ is the mode shape. It is defined with an arbitrary multiplicative constant C_n, which we can normalize to one without losing the generality.

Now let us consider the solutions resulting from [5.73]. Taking into account [5.76] after calculation we have: C = 0 and unspecified D. Consequently, there is only one possible solution resulting from [5.73]:

$$y(x, t) = A_0 t + B_0 .$$

In the preceding expression, we normalized the constant D to a singular unit. This movement characterizes a displacement of a rigid body, also called a movement of a rigid solid.

To finish, let us take the third form of the solution given by [5.74]; the introduction into [5.76] gives the following relations:

$$C + D = 0 \text{ and } Ce^{L\delta/c} + De^{-L\delta/c} = 0,$$

that is:

$$C = -D \text{ and } C \sinh\left(\delta \frac{L}{c}\right) = 0.$$

Respecting these equations is only possible with the trivial solution $C = D = 0$. There are thus no solutions resulting from [5.74].

The most general movement is that resulting from the superposition of the solutions which we have highlighted:

$$y(x,t) = A_0 t + B_0 + \sum_{n=1}^{\infty} \left(A_n \cos(\omega_n t) + B_n \sin(\omega_n t)\right) \cos\left(\frac{n\pi}{L} x\right) \quad [5.83]$$

with: $\omega_n = c\frac{n\pi}{L}$.

Vibratory movement is the combination of a uniform movement of a rigid solid and of an infinity of vibratory movements with normal pulsations ω_n, and mode shapes characterized by the wave numbers of k_n.

The presence of movement of rigid solid is linked to the free-free boundary conditions; all other boundary conditions would eliminate this type of movement with only the solutions resulting from the form [5.72] remaining. In the clamped-free case, for example, we would obtain:

$$y(x,t) = \sum_{n=1}^{\infty} \left(A_n \cos(\omega_n t) + B_n \sin(\omega_n t)\right) \sin(k_n x) \quad [5.84]$$

$$\text{with: } \omega_n = ck_n \text{ and } k_n = \frac{2n-1}{2} \frac{\pi}{L}. \quad [5.85]$$

The analysis of [5.84] or [5.83] shows that movement with deformation is the sum of independent movements characterized by a normal angular frequency ω_n and a mode shape $f_n(x)$. Each pair of angular frequency ω_n and mode shape $f_n(x)$ constitutes a mode. Certain authors apply the name of normal mode to the mode shape $f_n(x)$ alone. The set of normal angular frequencies and mode shapes pairs constitutes the modal system of the beam. We can provide this modal system for the cases that we have treated:

$$\text{free-free: } \left(c\frac{n\pi}{L}, \cos\left(\frac{n\pi}{L} x\right)\right), \quad [5.86]$$

$$\text{clamped-free: } \left(c\frac{2n-1}{2}\frac{\pi}{L}, \sin\left(\frac{2n-1}{2}\frac{\pi}{L} x\right)\right). \quad [5.87]$$

Free vibratory movement thus occurs with normal angular frequency for the system considered.

Mode shapes of modes 1, 2 and 3, for the embedded-free case are examined in Figures 5.7, 5.8 and 5.9.

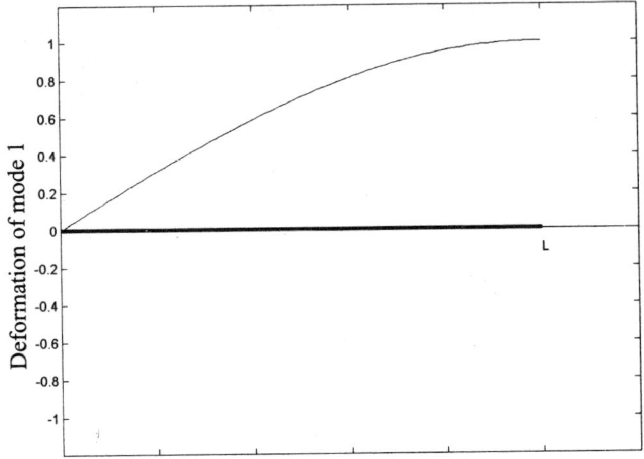

Figure 5.7. *Mode shape of mode 1 of the clamped-free medium*

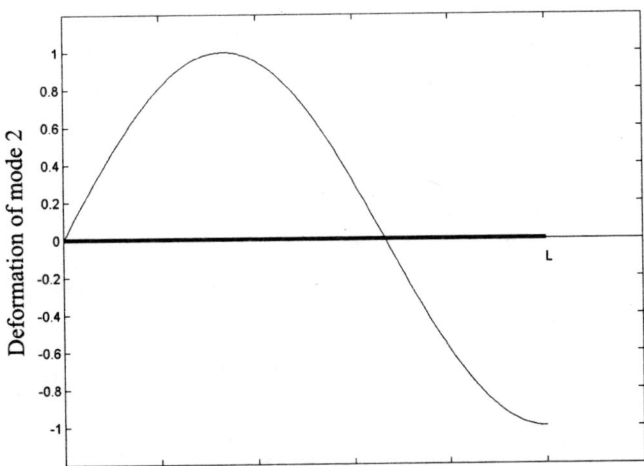

Figure 5.8. *Mode shape of mode 2 of the clamped-free medium*

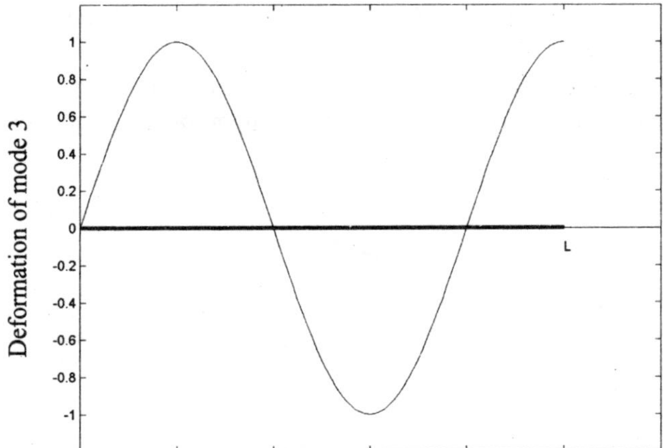

Figure 5.9. *Mode shape of mode 3 of the clamped-free medium*

The point characteristic of modal movement, that is, of the part of movement associated with an index n, is the appearance of nodes and antinodes of vibration.

Let us take the modal movement of the 1st order of the clamped-free beam $y_1(x, t)$:

$$y_1(x, t) = \left(A_1 \cos(\omega_1 t) + B_1 \sin(\omega_1 t)\right) \sin\left(\frac{\pi}{2L} x\right).$$

It is the product of the mode shape $f_1(x)$ by a sinusoidal function of time; at several consecutive moments, the medium traces the spindle defined in Figure 5.10.

The point of zero amplitude $(x = 0)$ is a node of vibration, while the point $x = L$ is an antinode of vibration; it corresponds to the maximum amplitude of the vibratory displacement of mode 1.

Let us examine the modal movement of the 2nd order:

$$y_2(x, t) = \left(A_2 \cos(\omega_2 t) + B_2 \sin(\omega_2 t)\right) \sin\left(\frac{3\pi}{2L} x\right).$$

Displacements over time are given in Figure 5.11. This movement presents two nodes in $x = 0$ and $x = 2L/3$ and two antinodes in $x = L/3$ and $x = L$.

The number of nodes (or antinodes) is characteristic of a mode; an additional node appears when we pass from mode n to mode n + 1. The total movement which is the superposition of these modal movements does not present a node in the strict sense of the term since the points of zero displacement are different for each mode. However, for conditions of excitation favoring a mode, that is those with amplitude much larger than of the other modes, we will find a low vibratory amplitude in the vicinity of the nodes of the favored mode often comparable to a node by extension.

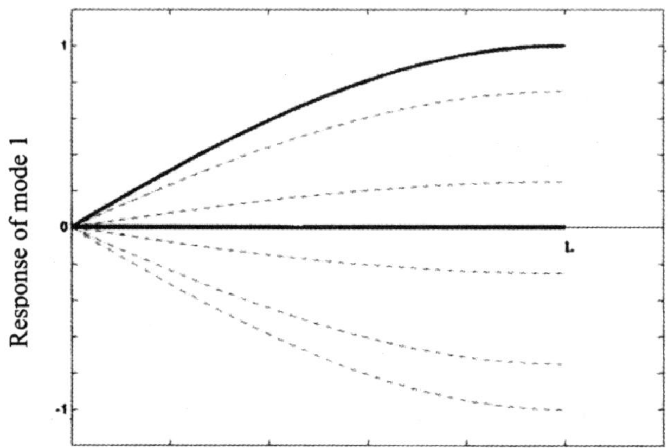

Figure 5.10. *Vibratory response of mode 1 of the clamped-free beam, at different moments*

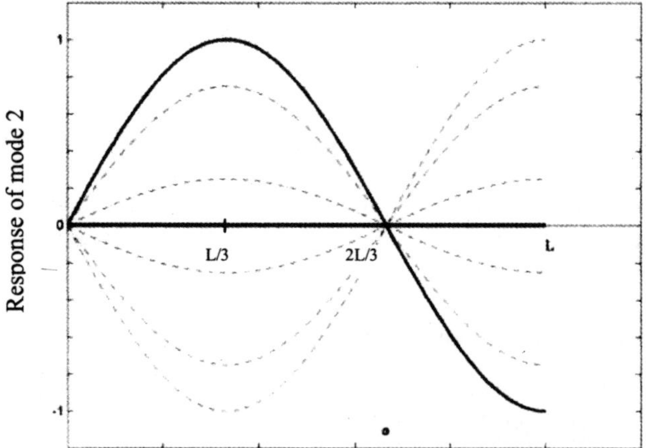

Figure 5.11. *Vibratory response of mode 2 of the clamped-free beam*

Expressions [5.81] and [5.84] give the general forms of the solutions verifying the wave equation and the free-free and clamped-free boundary conditions. There remain some unknowns since the terms A_i and B_i are arbitrary; in fact, the initial conditions will fix these values and thus ensure the uniqueness of the solution.

Finally, let us note that the modal system results from the verifying of the wave equation and the boundary conditions. It is the essential characteristic of vibrating systems, since this modal system defines all of the movements that the medium is likely to have. Table 5.3 summarizes the modal systems associated with the wave equation for different boundary conditions.

Notes:

a) The free-free boundary condition allows a movement without deformation which results in a zero normal angular frequency, that is, in a uniform movement. The mode shape equals 1, that is, the displacement is the same in any point of the structure.

b) The vibration normal angular frequencies in the clamped-clamped and free-free cases are identical. This situation is surprising since the systems are different, but does not have to lead us to thinking that the vibrations are the same ones because the mode shapes are completely different.

5.4.3. *Taking initial conditions into account*

The finite medium is now subject to the initial conditions [5.4] and [5.5] that we recollect:

$$y(x,0) = d_0(x),$$

$$\frac{\partial y}{\partial t}(x,0) = v_0(x).$$

Boundary conditions	**Characteristic equation**	**Normal angular frequency**	**Mode shapes**
Clamped-clamped $y(0,t) = 0$ $y(L,t) = 0$	$\sin\left(\frac{\omega L}{c}\right) = 0$	$\omega_n = \frac{n\pi}{L}c$ n=1,..., ∞	$\sin\left(\frac{n\pi}{L}x\right)$ n=1,..., ∞
Clamped-free $y(0,t) = 0$ $\frac{\partial y}{\partial x}(L,t) = 0$	$\cos\left(\frac{\omega L}{c}\right) = 0$	$\omega_n = \frac{2n-1}{2}\frac{\pi}{L}c$ n=1,..., ∞	$\sin\left(\frac{2n-1}{2}\frac{\pi}{L}x\right)$ n=1,..., ∞
Free-free $\frac{\partial y}{\partial x}(0,t) = 0$ $\frac{\partial y}{\partial x}(L,t) = 0$	Elastic modes $\sin\left(\frac{\omega L}{c}\right) = 0$ .... Solid mode	$\omega_n = \frac{n\pi}{L}c$ n=1,..., ∞ .... $\omega_0 = 0$	$\cos = \left(\frac{n\pi}{L}x\right)$ n=1,..., ∞ 1

Table 5.3. *Summary of the vibration modes, for the free or clamped boundary conditions*

Let us take the example of the free-free beam whose solution is given by [5.81] and [5.82]. Imposing the respect of the initial conditions, it follows:

$$y(x,0) = B_0 + \sum_{n=1}^{\infty} A_n \cos\left(\frac{n\pi}{L}\right) = d_0(x), \tag{5.88}$$

$$\frac{\partial y}{\partial t}(x,0) = A_0 + \sum_{n=1}^{\infty} B_n\, \omega_n \cos\left(\frac{n\pi}{L} x\right) = v_0(x). \tag{5.89}$$

Equation [5.88] shows that the terms B_0 and A_n correspond to the coefficients of development of the function $d_0(x)$ in Fourier series; similarly A_0 and B_n are the coefficients of development of $v_0(x)$. After all the calculations, it follows:

$$B_0 = \frac{1}{L}\int_0^L d_0(x)\, dx, \tag{5.90}$$

$$A_0 = \frac{1}{L}\int_0^L v_0(x)\, dx, \tag{5.91}$$

$$A_n = \frac{2}{L}\int_0^L \cos\left(\frac{n\pi}{L} x\right) d_0(x)\, dx, \tag{5.92}$$

$$B_n = \frac{2}{\omega_n L}\int_0^L \cos\left(\frac{n\pi}{L} x\right) v_0(x)\, dx. \tag{5.93}$$

The use of [5.90] – [5.93] in the general form of vibratory displacement [5.83] provides the solution to the problem. This solution is expressed by a series just as for the technique of preceding resolution using source-image (expression [5.52]). Each method has its advantages and its disadvantages; however, the separation of variables is generally preferred because it reveals the key concept of normal angular frequency and normal displacement.

The introduction of initial conditions consisted of developing functions $d_0(x)$ and $v_0(x)$ in Fourier series. This procedure is general but Fourier series will not be forcing the traditional developments into sine and cosine. Let us consider another case of boundary conditions. The general form of the solution is given by:

$$y(x,t) = \sum_{n=1}^{\infty} \left(A_n \cos(\omega_n t) + B_n \sin(\omega_n t)\right) f_n(x). \tag{5.94}$$

Respecting the initial conditions imposes:

$$\sum_{n=1}^{\infty} A_n f_n(x) = d_0(x), \qquad [5.95]$$

$$\sum_{n=1}^{\infty} B_n \omega_n f_n(x) = v_0(x). \qquad [5.96]$$

It is thus in general a decomposition into a series of normal functions which should be carried out. These functions have the property of orthogonality demonstrated in section 5.4.4.

$$\int_0^L f_n(x) f_p(x)\, dx = N_n \delta_{np} \qquad [5.97]$$

where N_n is the norm of mode n, δ_{np} is the Kroneker symbol.

Let us take the equality [5.95], multiply the two members by $f_p(x)$ and integrate over the length of the beam. Taking into account the property of orthogonality [5.97], it follows:

$$A_p = \frac{1}{N_p} \int_0^L d_0(x) f_p(x)\, dx. \qquad [5.98]$$

This expression relates to [5.92]. Similarly with [5.96] we obtain the expression B_p relating to [5.93]:

$$B_p = \frac{1}{\omega_p N_p} \int_0^L v_0(x) f_p(x)\, dx. \qquad [5.99]$$

5.4.4. *Orthogonality of mode shapes*

The essential property for the calculation of vibratory response following initial conditions is the orthogonality of mode shapes. To demonstrate it, let us, again base ourselves on equation [5.60] which must be verified for each mode shape.

For the oscillatory modes $\omega_n \neq 0$, mode shapes verify [5.100]:

$$\frac{d^2 f_n}{dx^2} + \frac{\omega_n^2}{c^2} f_n = 0. \qquad [5.100]$$

If rigid movement is possible, the associated mode shapes $f_0(x)$ must verify [5.65], that is:

$$\frac{d^2 f_0}{dx^2} = 0. \qquad [5.101]$$

We can assemble [5.100] and [5.101] into the single form [5.100] by introducing a zero normal angular frequency for the rigid movement.

Let us multiply [5.100] by the mode shape of the mode p, integrate between 0 and L and cleverly group:

$$-\int_0^L f_p(x) \frac{d^2 f_n}{dx^2}(x)\, dx = \frac{\omega_n^2}{c^2} \int_0^L f_p(x)\ f_n(x)\, dx. \qquad [5.102]$$

We also have a symmetrical equation by permuting the indices:

$$-\int_0^L f_n(x) \frac{d^2 f_p}{dx^2}(x)\, dx = \frac{\omega_p^2}{c^2} \int_0^L f_p(x)\ f_n(x)\, dx. \qquad [5.103]$$

Let us consider the first member of [5.103] and carry out integration by parts; after regrouping we obtain:

$$-\int_0^L f_p(x) \frac{d^2 f_n}{dx^2}(x)\, dx + \int_0^L f_n(x) \frac{d^2 f_p}{dx^2}(x)\, dx = \left[f_n(x) \frac{df_p}{dx}(x) \right]_0^L - \left[f_p(x) \frac{df_n}{dx}(x) \right]_0^L. \qquad [5.104]$$

Since the mode shapes verify the boundary conditions, it follows that the second member of [5.104] is nil. Indeed, for the traditional conditions, clamped or free, we have either $f_n(x)$, or $\frac{df_n}{dx}(x)$ (either $f_p(x)$ or $\frac{df_p}{dx}(x)$), which are nil at each end. From this it follows that the product $f_n(x)\frac{df_p}{dx}(x)$ is nil when $x = 0$ and $x = L$.

For the boundary conditions which we have considered at the beginning of the chapter (equations [5.2] and [5.3]) we have:

$$f_n(0) = \alpha_0 \frac{df_n}{dx}(0) \qquad [5.105]$$

and:

$$f_n(L) = \alpha_0 \frac{df_n}{dx}(L)\,. \qquad [5.106]$$

Let us replace $f_p(0)$, $f_p(L)$, $f_n(0)$, $f_n(L)$ by their expressions drawn from [5.105] and [5.106]; the second member of [5.104] becomes:

$$\alpha_L \frac{df_n}{dx}(L)\frac{df_p}{dx}(L) - \alpha_0 \frac{df_n}{dx}(0)\frac{df_p}{dx}(0) - \alpha_L \frac{df_n}{dx}(L)\frac{df_p}{dx}(L) + \alpha_0 \frac{df_n}{dx}(0)\frac{df_p}{dx}(0) = 0\,.$$

The second member of [5.104] is thus also nil for yield strengths. The relation [5.104] is consequently reduced to:

$$-\int_0^L f_p(x)\frac{d^2 f_n}{dx^2}(x)\,dx + \int_0^L f_n(x)\frac{d^2 f_p}{dx^2}(x)\,dx = 0\,. \qquad [5.107]$$

This equation shows the symmetry of the second derivative operator, which is at the base of the orthogonality of the mode shapes. Indeed, let us introduce this relation into [5.102] and [5.103]; it follows:

$$(\omega_n^2 - \omega_p^2)\int_0^L f_p(x)\ f_n(x)\,dx = 0\,. \qquad [5.108]$$

There is thus the alternative:

$$\omega_n = \omega_p \text{ and } \int_0^L f_n(x)\ f_n(x)\ dx = N_n \qquad [5.109]$$

or:

$$\omega_n \neq \omega_p \text{ and } \int_0^L f_p(x)\, f_n(x)\, dx = 0. \qquad [5.110]$$

This shows the orthogonality of the mode shapes. Let us note that [5.110] and [5.104] show that there is also a second property of orthogonality:

$$\int_0^L f_n(x) \frac{d^2 f_p}{dx^2}(x)\, dx = 0 \text{ if } \omega_n \neq \omega_p. \qquad [5.111]$$

In problems of vibration, there is always a double orthogonality of the mode shapes: orthogonality with respect to the operator of mass [5.110] and orthogonality with respect to the operator of stiffness [5.111] of the problem considered.

5.5. Applications

5.5.1. *Longitudinal vibrations of a clamped-free beam*

We consider a clamped-free beam subjected to a static force at its free end until the moment $t = 0$ when the force is suddenly canceled.

Figure 5.12. *Initial conditions of the beam*

To calculate the response of the beam at the moment $T > 0$ it is necessary, first of all, to calculate the deformation of the beam at the moment $T = 0$, then to introduce it as the initial condition of free vibratory displacement following the sudden cancellation of force.

Static deformation U(x) corresponding to the compression of the beam is obtained by resolving the problem of static equilibrium of the beam. It is thus a question of solving equation [5.112], which results from the equation of longitudinal vibrations of beams when displacement does not depend on time. We then force the solution to respect the boundary conditions [5.113].

$$ES\frac{d^2U}{dx^2} = 0 \quad \forall x \in]0, L[, \tag{5.112}$$

$$U(0) = 0 \text{ and } ES\frac{dU}{dx}(L) = -F. \tag{5.113}$$

The general solution of equation [5.112] is:

$$U(x) = C_0 + D_0 x. \tag{5.114}$$

The two boundary conditions [5.113] lead to the solution:

$$U(x) = -\frac{F}{ES}x. \tag{5.115}$$

The initial conditions to apply to the vibratory problem are deduced from this:

$$U(x,0) = -\frac{F}{ES}x \quad \text{and} \quad \frac{\partial U}{\partial t}(x,0) = 0. \tag{5.116}$$

Let us consider the solution of the problem obtained by modal decomposition that we have outlined in section 5.4.2, equations [5.83] and [5.87]:

$$U(x,t) = \sum_{n=1}^{\infty}\left(A_n \cos(\omega_n t) + B_n \sin(\omega_n t)\right)\sin\left(\frac{2n-1}{2}\frac{\pi}{L}x\right). \tag{5.117}$$

Taking into account of the initial conditions yields:

$$B_n = 0$$

$$\text{and} \quad A_n = -\frac{2}{L}\int\frac{F}{ES}x\sin\left(\frac{2n-1}{2}\frac{\pi}{L}x\right)dx = \frac{(-1)^n}{\left(\frac{2n-1}{2}\frac{\pi}{L}\right)^2}\frac{2}{L}\frac{F}{ES} \tag{5.118}$$

From it we deduce:

$$U(x,t) = \frac{2F}{ESL}\sum_{n=1}^{\infty}\frac{(-1)^n}{(2n-1)^2(\pi/8)^2}\cos(\omega_n t)\sin\left(\frac{2n-1}{2}\frac{\pi}{L}x\right). \tag{5.119}$$

Equation [5.119] gives vibratory displacement in any point x and at any moment t following the cancellation of the static force at the moment t = 0.

A second characteristic is key in the study of longitudinal vibrations of beams, that is the stress of traction-compression, which is connected to the displacement U(x, t) by the relation [3.18] provided in Chapter 3.

$$\sigma_{11}(x, t) = E \frac{\partial U}{\partial x}(x, t), \tag{5.120}$$

that is:

$$\sigma_{11}(x, t) = \frac{4F}{S\pi} \sum_{n=1}^{\infty} \frac{(-1)^n}{(2n-1)} \cos(\omega_n t) \cos\left(\frac{2n-1}{2} \frac{\pi}{L} x\right). \tag{5.121}$$

The analysis of the longitudinal displacement and the stress of traction-compression can be carried out mode by mode in order to break up the vibratory state into simple elements. First of all, let us consider the modal amplitudes of displacement and stress and carry them over to Figure 5.13:

Amplitude of displacement of the mode n: $\dfrac{8F}{ESL} \dfrac{(-1)^n}{(2n-1)^2 (\pi/L)^2}$,

Amplitude of stress of the mode n: $\dfrac{4F}{S\pi} \dfrac{(-1)^n}{(2n-1)}$.

We observe that the amplitudes of modal displacements decrease much more quickly than those of modal stresses; it will thus be easier to converge in displacement than in stress. Another consequence of this distortion of amplitude is the difficulty in extrapolating a visual feeling resulting from displacements from the state of stress of the beam. This is all the more true since the space mode shapes for displacements and stresses are radically different. Let us trace the mode shapes and stresses of the first three modes (Figure 5.14): the correspondence between a node of displacement and an antinode of stress and vice versa is remarkable. It is thus necessary to expect to record strong stresses at the places of low amplitude for beams in longitudinal vibrations as well as in torsion.

The “engineering rules” stipulate that mode 1 is dominating in the problems free of vibrations. We may clearly observe in our example the truthfulness of this assertion but it should not, however, be forgotten that, although widespread, this property is not true in general. In fact, the amplitude of various modes is related to the space form of the initial conditions, as shown by the expressions [5.100] and [5.101]. We can say that the modes whose mode shapes will be similar to the form of initial displacement and/or initial speed will respond strongly. In the case analyzed in this section, initial displacement is given in Figure 5.15; it is unarguably closer to the deformation of mode 1 than of those of the following modes which produce sign changes.

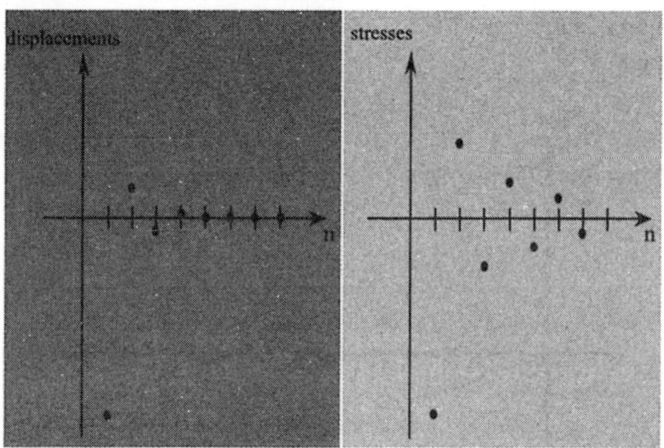

Figure 5.13. *Modal amplitudes of displacements and stresses*

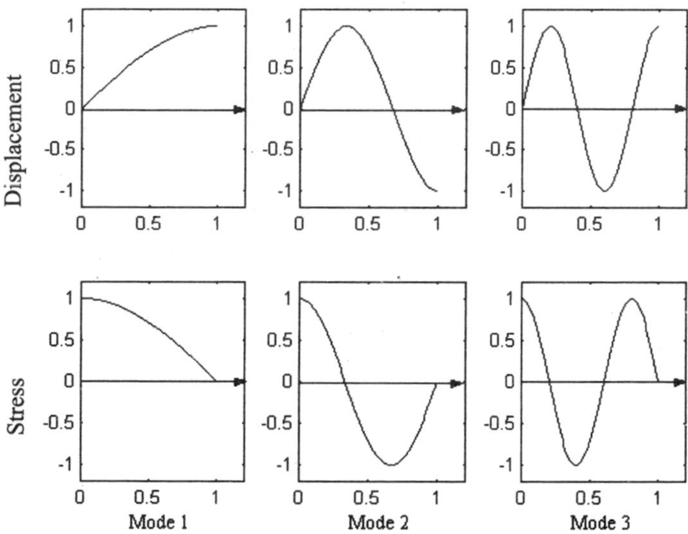

Figure 5.14. *Mode shapes of displacements and stresses for the first three modes*

This is the reason for the preponderance of mode 1 in vibratory displacement of beams that we have studied. It is also the reason for the great frequency of occurrence of this situation, because it is rare to create in practice the initial conditions close, for example, to mode 2. This would require the creation of an opposition to the initial phase, which is the application of two opposed forces in L and L/3. In general, favoring mode n requires the use of n static forces adjusted in sign and position.

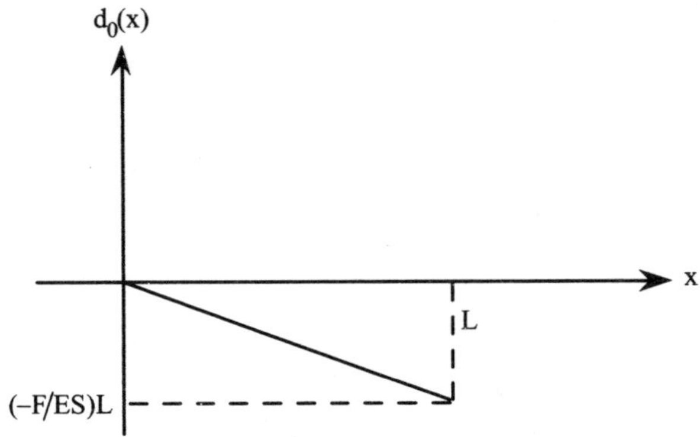

Figure 5.15. *Static deformation of the beam*

5.5.2. *Torsion vibrations of a line of shafts with a reducer*

We consider a reducer made up of two shafts coupled by a set of gears. To study the vibrations of torsion of this unit we model the system as defined in Figure 5.16.

The quantities R_1 and R_2 are respectively the radii of gears linked to shafts 1 and 2, I_1 and I_2 are polar inertias of the cross-sections of the two shafts, G_1 and G_2 are the moduli of materials rigidity, finally L_1 and L_2 the lengths.

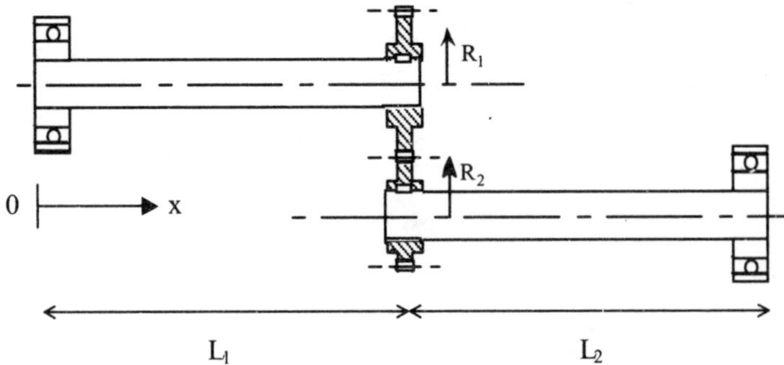

Figure 5.16. *Line of shafts with a reducer*

Noting as $\alpha_1(x, t)$ and $\alpha_2(x, t)$ the respective angles of torsion of shafts 1 and 2, and introducing local references into each beam, we can write the boundary conditions of the two shafts as follows:

– Torque of shaft 1 nil in 0 (free boundary):

$$G_1 I_1 \frac{\partial \alpha_1}{\partial x}(0, t) = 0 . \qquad [5.122]$$

– Equality of displacements at the point of contact of the teeth:

$$R_1 \alpha_1 (L_1, t) = -R_2 \alpha_2 (0, t) . \qquad [5.123]$$

Equilibrium of forces in contact with the teeth expressed according to the torques of the two shafts:

$$\frac{G_1 I_1}{R_1} \frac{\partial \alpha_1}{\partial x}(L_1, t) = -\frac{G_2 I_2}{R_2} \frac{\partial \alpha_2}{\partial x}(0, t) . \qquad [5.124]$$

Torque of shaft 2 nil in L_2 (free boundary):

$$G_2 I_2 \frac{\partial \alpha_2}{\partial x}(L_2, t) = 0 . \qquad [5.125]$$

Modeling employed to describe the behavior of the reducer neglects, on the one hand, the masses of the gears and, on the other hand, the elasticity of the teeth, all in all we are considering here a low frequency simplification.

We will calculate the total modal system of the set of two shafts, that is, we will find the solutions of the free vibratory problems.

The angles of torsion $\alpha_1(x, t)$ and $\alpha_2(x, t)$ must verify the equations of motion of shafts 1 and 2 and the boundary conditions [5.122] – [5.125] respectively.

Respecting the equations of motion involves:

$$\alpha_1(x, t) = \left(A_1 \cos(\omega_1 t) + B_1 \sin(\omega_1 t)\right)\left(C_1 \cos(k_1 x) + D_1 \sin(k_1 x)\right) \qquad [5.126]$$

with : $k_1 = \omega_1 \Big/ \sqrt{\dfrac{G_1}{\rho_1}}$,

$$\alpha_2(x,t) = \left(A_2 \cos(\omega_2 t) + B_2 \sin(\omega_2 t)\right) \left(C_2 \cos(k_2 x) + D_2 \sin(k_2 x)\right) \tag{5.127}$$

with: $k_2 = \omega_2 \Big/ \sqrt{\dfrac{G_2}{\rho_2}}$.

At this level of writing the solutions of the equations of motion the time functions are different. In order to satisfy the equations of connection [5.124] and [5.125] at any moment, it is, however, necessary to pose the equality of the temporal functions:

$$A_i \cos(\omega_i t) + B_i \sin(\omega_i t) = A \cos(\omega t) + B \sin(\omega t) \quad \text{for} \quad i = 1,2 .$$

The coupling between the two beams clearly leads to only one single system vibrating as a whole with the pulsation ω. The wave numbers and thus the wavelengths are, however, different for the two beams. We may observe that for a given angular frequency, the wavelength is shorter for greater velocities of waves of torsion.

$$k_1 = \omega \Big/ \sqrt{\frac{G_1}{\rho_1}} \ , \ \lambda_1 = 2\pi \sqrt{\frac{G_1}{\rho_1}} \Big/ \omega , \tag{5.128}$$

$$k_2 = \omega \Big/ \sqrt{\frac{G_2}{\rho_2}} \ , \ \lambda_2 = 2\pi \sqrt{\frac{G_2}{\rho_2}} \Big/ \omega . \tag{5.129}$$

Introducing these solutions under the boundary conditions [5.122] – [5.125], it follows:

$$D_1 = 0 , \tag{5.130}$$

$$R_1 C_1 \cos(k_1 L_1) = -R_2 C_2 , \tag{5.131}$$

$$\frac{G_1 I_1}{R_1} C_1 k_1 \sin(k_1 L_1) = \frac{G_2 I_2}{R_2} D_2 k_2 , \tag{5.132}$$

$$C_2 k_2 \sin(k_2 L_2) = D_2 k_2 \cos(k_2 L_2) . \tag{5.133}$$

To simplify let us take the case where the two shafts have identical characteristics:

$$\rho_1 = \rho_2 = \rho \ , \ G_1 = G_2 = G \ , \ I_1 = I_2 = I \ , \ L_1 = L_2 = L . \tag{5.134}$$

The equation with frequencies results from respecting equations [5.131] – [5.133]:

$$\begin{pmatrix} R_1 \cos(kL) & R_2 & 0 \\ \sin(kL)/R_1 & 0 & -1/R_2 \\ 0 & \sin(kL) & -\cos(kL) \end{pmatrix} \begin{pmatrix} C_1 \\ C_2 \\ D_2 \end{pmatrix} = \begin{pmatrix} 0 \\ 0 \\ 0 \end{pmatrix}. \qquad [5.135]$$

To obtain non-trivial solutions the determinant of the system must be nil:

$$\left(\frac{R_1}{R_2} + \frac{R_2}{R_1} \right) \sin(kL)\ \cos(kL) = 0. \qquad [5.136]$$

This equation is the characteristic equation for normal wave numbers; it admits two families of solutions:

$$k_n = \frac{n\pi}{L}, \quad n = 1, \dots, \infty, \qquad [5.137]$$

$$k_m = \frac{2m-1}{2}\frac{\pi}{L}, \quad m = 1, \dots, \infty. \qquad [5.138]$$

Let us examine the first family. The normal eigenfrequencies result from [5.128] and [5.137]:

$$\omega_n = \sqrt{\frac{G}{\rho}}\frac{n\pi}{L}, \quad n = 1, \dots, \infty. \qquad [5.139]$$

The system [5.135] becomes:

$$\begin{pmatrix} R_1(-1)^n & R_2 & 0 \\ 0 & 0 & -1/R_2 \\ 0 & 0 & (-1)^n \end{pmatrix} \begin{pmatrix} C_1 \\ C_2 \\ D_2 \end{pmatrix} = \begin{pmatrix} 0 \\ 0 \\ 0 \end{pmatrix}. \qquad [5.140]$$

It admits the solutions:

$$D_2 = 0 \ \text{ and } \ C_1 = (-1)^{n+1}\frac{R_2}{R_1}C_2 \cdot \qquad [5.141]$$

The mode shape associated to the mode n is obtained by introducing the particular values of C_1, D_1, C_2 and D_2 into the space form of the solutions [5.126] and [5.127], that is, after having normalized C_1A to one:

$$\left\{ f_n(x) \right\} = \begin{pmatrix} f_n^1(x) \\ f_n^2(x) \end{pmatrix} = \begin{pmatrix} \cos(n\pi/L) \\ (-1)^{n+1} (R_1/R_2) \cos(n\pi/L) \end{pmatrix}. \qquad [5.142]$$

Let us examine the second family. The normal angular frequencies result from [5.128] and [5.138]:

$$\omega_m = \sqrt{\frac{G}{\rho}} \frac{2m-1}{2} \frac{\pi}{L}, \quad m = 1, \ldots, \infty. \qquad [5.143]$$

The system [5.135] becomes:

$$\begin{pmatrix} 0 & R_2 & 0 \\ (-1)^m/R_1 & 0 & -1/R_2 \\ 0 & (-1)^m & 0 \end{pmatrix} \begin{pmatrix} C_1 \\ C_2 \\ D_2 \end{pmatrix} = \begin{pmatrix} 0 \\ 0 \\ 0 \end{pmatrix}. \qquad [5.144]$$

It creates the solutions:

$$C_2 = 0 \quad \text{and} \quad C_1 = (-1)^m \frac{R_1}{R_2} D_2, \qquad [5.145]$$

$$\left\{ f_m(x) \right\} = \begin{pmatrix} f_m^1(x) \\ f_m^2(x) \end{pmatrix} = \begin{pmatrix} (-1)^m (R_1/R_2) \cos\left(\frac{2m-1}{2} \frac{\pi}{L} x \right) \\ \sin\left(\frac{2m-1}{2} \frac{\pi}{L} x \right) \end{pmatrix}. \qquad [5.146]$$

To fix the ideas, let us trace the mode shapes of the two modes taking $R_2 = 2R_1$. The mode n = 1 is given in Figure 5.17, and the mode m = 1 is given in Figure 5.18.

Cutting into modes n and m is arbitrary and is introduced only for mathematical convenience. Physically we will find the traditional modal sequence by increasing the normal angular frequency: mode 1, that of lower normal frequency, is m = 1, mode 2 is that corresponding to n = 1, mode 3 to m = 2, etc.

Until now we have considered solutions resulting from [5.126] and [5.127] representing vibratory movement with deformation. The line of shafts being free at its ends, there will also be a mode of vibration without deformation (mode 0). The solutions are of the type:

$$\begin{cases} \alpha_1(x,t) = (A_0\, t + B_0)(C_{01}\, x + D_{01}) \\ \alpha_2(x,t) = (A_0\, t + B_0)(C_{02}\, x + D_{02}). \end{cases} \qquad [5.147]$$

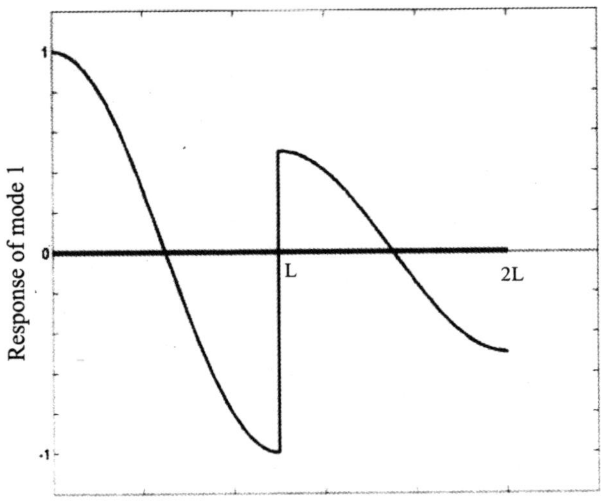

Figure 5.17. *Mode shape of the n=1 mode*

Respect the boundary conditions [5.122] to [5.125] leads to the solution:

$$\begin{cases} \alpha_1(x,t) = (A_0\, t + B_0) \\ \alpha_2(x,t) = (A_0\, t + B_0)\left(-R_2/R_1\right). \end{cases} \qquad [5.148]$$

In fact, this solution describes the uniform rotation of the two beams (contrary direction for the two beams with respect to the reduction R_2/R_1).

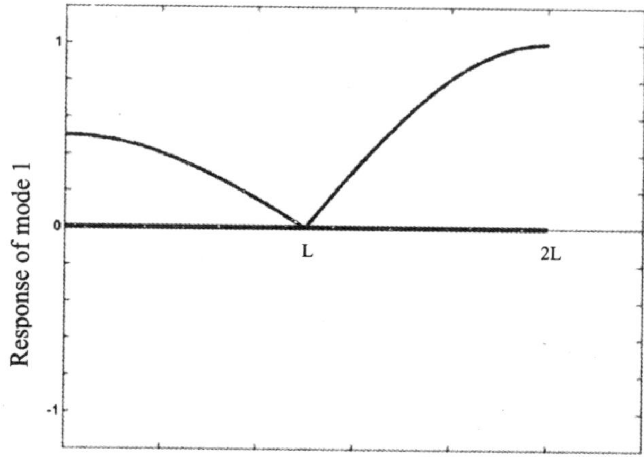

Figure 5.18. *Mode shape of the m=1 mode*

The free vibratory response is given by the accumulation of all the possibilities of movements:

$$\begin{pmatrix} \alpha_1(x,t) \\ \alpha_2(x,t) \end{pmatrix} = (A_0 t + B_0)\begin{pmatrix} 1 \\ -R_1/R_2 \end{pmatrix}$$

$$+\sum_{n=1}^{\infty}\left(A_n \cos(\omega_n t) + B_n \sin(\omega_n t)\right)\begin{pmatrix} (-1)^{n+1}(R_2/R_1)\cos(n\pi/L) \\ \cos(n\pi/L) \end{pmatrix}$$

$$+\sum_{m=1}^{\infty}\left(A_m \cos(\omega_m t) + B_m \sin(\omega_m t)\right) \begin{pmatrix} (-1)^{m+1}(R_1/R_2)\cos\left((2m-1)\pi/2L\right) \\ \sin\left((2m-1)\pi/2L\right) \end{pmatrix}. \qquad [5.149]$$

The values of the constants A_0, A_n, A_m, B_0, B_n and B_m are fixed by the initial conditions.

5.6. Conclusion

The wave equation that we have studied is directly applicable to the vibrations of beams in traction and torsion but also to cords and sound pipes. We have shown two methods to describe free vibration. The method of propagation which introduces the concept of image source when we take the finite aspect of the structures into account is well adapted to the calculation of free response of the large-sized systems. The method of separation of variables which leads to the key concept of normal mode introduces vibratory response as a superposition of the independent modal responses.

Two examples of calculations showed how to take initial conditions into account and to obtain vibratory displacements and stresses. A characteristic of modal behavior of the structures governed by the wave equation is the correspondence between antinodes of displacements and nodes of stresses and between nodes of displacements and antinodes of stresses.

In the case of coupled structures, we have extended the results obtained for isolated structures by connecting interfaces. This procedure, put into practice by a line of shafts composed of two beams and a reducer, can be generalized to the case of several coupled systems.

Chapter 6

Free Bending Vibration of Beams

6.1. Introduction

In this chapter we consider the vibratory movement of beams most commonly met in practice: bending vibrations.

This prevalence of the problems of bending results from the following aspects:

a) The normal angular frequency of bending is the first to appear when we describe the axis of frequencies increasing starting from zero. In other words, the first modes of a beam with all effects mixed together (bending, traction, torsion) are those of bending.

b) Transverse excitations on beams are the most current; they are the ones generating bending modes.

c) For a given value of dynamic stress and compared to the other vibratory movements, flexing movements generate very large displacements. Consequently, we may have non-dangerous constraints for the beam, which produce large transverse movements that can be uncomfortable.

d) Bending movements impact the surrounding air and lead to the generation of noise.

Various hypotheses have been proposed in Chapter 3 to describe the bending of beams; the most sophisticated theories introduce "secondary" effects (shearing, rotational inertia) which can prove to be important for high frequency or anisotropic materials. Our discussion will, nevertheless, be based on the strongest hypothesis (pure bending) because it allows an approach with fewer calculations. The influence of secondary effects will be discussed afterwards.

As a first development we solve the equation of motion by separation of variables. The solution is interpreted in term of traveling and vanishing waves, and the concepts of phase speed and group speed are introduced to characterize the propagation in infinite beams.

We then introduce boundary conditions and deduce from them the vibration modes. A summary table is given.

Examples of application are finally presented to describe the principal physical phenomena. Into the second part the secondary effects of rotational inertia and shearing are introduced separately and simultaneously. The results are then interpreted in terms of propagation velocity and of normal mode. We also establish the criteria that make it possible to determine *a priori* if the secondary effects are negligible.

6.2. The problem

We consider a straight beam oriented along axis 1 of an orthonormal system of reference, and we are interested in the movement generated by a transverse excitation. This vibratory state was analyzed in detail in Chapter 3, section 3.5. It was shown that several models can be proposed according to the level of simplification that we allow ourselves. The simplest theory, and thus the most restrictive one, will be used as a basis for this discourse. In section 6.7 we will demonstrate the influence of effects neglected in the traditional approach.

Thus, we consider the case of pure bending described in section 3.5.3 of Chapter 3. Displacements of the continuous medium are given by the expressions [6.1]:

$$\begin{aligned} &W_1(x_1, x_2, x_3, t) = -x_2 \frac{\partial W}{\partial x_1}(x_1, t), \\ &W_2(x_1, x_2, x_3, t) = W(x_1, t), \\ &W_3(x_1, x_2, x_3, t) = 0. \end{aligned} \qquad [6.1]$$

The function $W(x_1, t)$ represents the transverse displacement of the beam; it verifies the following equations:

Equation of motion:

$$\rho S \frac{\partial^2 W}{\partial t^2} + \frac{\partial^2}{\partial x_1^2}\left(EI \frac{\partial^2 W}{\partial x_1^2} \right) = 0 \quad \forall\, x_1 \in \left]0, L\right[,\ \forall\, t. \qquad [6.2]$$

In this expression, I is the inertia of the cross-section of the beam with respect to direction 3, E is Young's modulus of material which is here supposed to be isotropic, ρ is the density, and S is the cross-section of the beam. We no longer specify that the equations are valid in the time interval $\left]t_0, t_1\right[$, we simply note $\forall\, t$ since moments t_0 and t_1 are arbitrary.

Boundary conditions to verify in $x = 0$ and $x = L$, $\forall\, t$:

$$\text{either: } W(x_1, t) = 0, \qquad \text{or } : \frac{\partial}{\partial x_1}\left(EI\frac{\partial^2 W}{\partial x_1^2}(x_1, t)\right) = 0 \tag{6.3}$$

and:

$$\text{either: } \frac{\partial W}{\partial x_1}(x_1, t) = 0, \qquad \text{or } : EI\frac{\partial^2 W}{\partial x_1^2}(x_1, t) = 0. \tag{6.4}$$

Longitudinal stress is deduced from the value of transverse displacement by the relation [6.5]:

$$\sigma_{11}(x_1, x_2, x_3, t) = x_2\, E\frac{\partial^2 W}{\partial x_1^2}(x_1, t). \tag{6.5}$$

The solution of equation [6.2] is the basis for the calculation of vibratory response of beams in bending. It is only analytically possible in the particularly simple cases of variation of the mechanical characteristics along the beam length. We will base our discourse on the simplest case, that of the homogenous beam with a constant cross-section, which is easily solved. Under these conditions, ρS and EI are constants and equation [6.2] becomes:

$$\rho S\frac{\partial^2 W}{\partial t^2} + EI\frac{\partial^4 W}{\partial x_1^4} = 0 \quad \forall\, x_1 \in \left]0, L\right[,\ \forall\, t. \tag{6.6}$$

6.3. Solution of the equation of the homogenous beam with a constant cross-section

6.3.1. *Solution*

To solve equation [6.6] we employ the method of separation of variables. Thus, we pose:

$$W(x,t) = f(x)\ g(t)\,. \tag{6.7}$$

To simplify the notations from now on, we will write x instead of x_1.

Let us introduce the expression [6.7] into equation [6.6]; it follows:

$$\frac{d^2g}{dt^2}(t)\ f(x) + \frac{EI}{\rho S}\frac{d^4f}{dx^4}(x)\ g(t) = 0\,. \tag{6.8}$$

Let us separate the variables by grouping on the right the functions of time and on the left the functions of space:

$$\frac{d^2g}{dt^2}(t)\Big/g(t) = -\frac{EI}{\rho S}\frac{d^4f}{dx^4}(x)\Big/f(x) = a\,. \tag{6.9}$$

Following the traditional argumentation, we observe that “a” is a constant. Indeed, the first member of [6.9] is independent of x and the second member is independent of t: “a” is thus independent of time and of x; it is a constant.

Equation [6.9] separates into two equations:

$$\frac{d^2g}{dt^2}(t) - ag(t) = 0\,, \tag{6.10}$$

$$\frac{d^4f}{dx^4}(x) + a\frac{\rho S}{EI}f(x) = 0 \tag{6.11}$$

where the constant "a" can be positive, negative or nil. From that we deduce the three types of solutions:

1) $a = 0$, leading to the solutions:

$$g(t) = At + B \quad \text{and} \quad f(x) = C + Dx + Ex^2 + Fx^3. \tag{6.12}$$

I.e. with [6.7]:

$$W(x, t) = (At + B)(C + Dx + Ex^2 + Fx^3). \tag{6.13}$$

2) $a < 0$, we then pose $a = -\omega^2$.

The solutions of [6.10] and [6.11] are given by:

$$\begin{aligned} &g(t) = A\cos(\omega t) + B\sin(\omega t), \\ &f(x) = C\cos(kx) + D\sin(kx) + E\,\text{ch}(kx) + F\,\text{sh}(kx) \end{aligned} \tag{6.14}$$

$$\text{with} \ : \ \omega = \sqrt{\frac{EI}{\rho S}}\, k^2. \tag{6.15}$$

The quantity ω is the angular frequency of the vibratory movement; the quantity k which appears in the space solution is called a wave number. The relation [6.15] connecting angular frequency and the wave number is the relation of dispersion.

The solution is thus:

$$\begin{aligned} W(x, t) = \big(A\cos(\omega t) + B\sin(\omega t)\big)\big(&C\cos(kx) + D\sin(kx) \\ &+ E\,\text{ch}(kx) + F\,\text{sh}(kx)\big). \end{aligned} \tag{6.16}$$

3) $a > 0$

After all the calculations we obtain:

$$\begin{cases} g(t) = Ae^{\sqrt{a}\,t} + Be^{-\sqrt{a}\,t} \\ f(x) = Ce^{j\alpha x} + De^{-j\alpha x} + Ee^{\alpha x} + Fe^{-\alpha x}. \end{cases} \tag{6.17}$$

And finally the transverse displacement of the beam:

$$W(x,t) = \left(Ae^{\sqrt{a}\,t} + Be^{-\sqrt{a}\,t}\right)\left(Ce^{j\alpha x} + De^{-j\alpha x} + Ee^{\alpha x} + Fe^{-\alpha x}\right) \qquad [6.18]$$

with: $\alpha = \sqrt{j\sqrt{a\frac{\rho S}{EI}}}$.

As we have shown during the resolution of the wave equation using separation of variables (section 5.4 of Chapter 5), the case of the positive constant leads to the trivial solution $W(x, t) = 0$ if we respect the boundary conditions, so let us not exploit this any further before the solution [6.18].

The solution [6.13] is possible but marginal because it is to be considered only in the case of boundary conditions allowing movements without strain. Instead of vibratory movements in a strict sense, it induces uniform movements (translation or rotation of the whole beam).

Vibratory movement is introduced by the case of the negative constant and thus has the general form [6.16]. Let us stress that there are other equivalent forms of writing this solution down, in particular [6.19] and [6.20]:

$$W(x, t) = \left(A\cos(\omega t) + B\sin(\omega t)\right)\left(C\cos(kx) + D\sin(kx) + Ee^{kx} + Fe^{-kx}\right) \qquad [6.19]$$

and:

$$W(x, t) = \left(Ae^{j\omega t} + Be^{-j\omega t}\right)\left(Ce^{jkx} + De^{-jkx} + Ee^{kx} + Fe^{-kx}\right). \qquad [6.20]$$

In the continuation we will make use of the most adapted solution form depending on the case.

6.3.2. *Interpretation of the vibratory solution, traveling waves, vanishing waves*

We are particularly interested in the solution [6.16] or of course in its equivalent forms [6.19] and [6.20].

Let us take, for example, the form [6.19] and distribute the product. After suitable regrouping of the terms we obtain a new expression:

$$W(x,t) = a_1 \cos(\omega t - kx) + a_2 \sin(\omega t - kx) + a_3 \cos(\omega t + kx) + a_4 \sin(\omega t + kx) + a_5 \cos(\omega t)\, e^{kx} + a_6 \cos(\omega t)\, e^{-kx} + a_7 \sin(\omega t)\, e^{kx} + a_8 \sin(\omega t)\, e^{-kx}. \quad [6.21]$$

Each term of [6.21] is interpreted as a wave. The first four terms are traveling waves; the four following are vanishing waves. We need to develop this further in order to understand clearly the physical significance associated with these concepts.

Let us take the first traveling wave of equation [6.21], that is [6.22] and represent the state of displacement of the beam at the moments t_0 and $t_0 + \Delta t$ in Figure 6.1:

$$a_1 \cos(\omega t - kx). \quad [6.22]$$

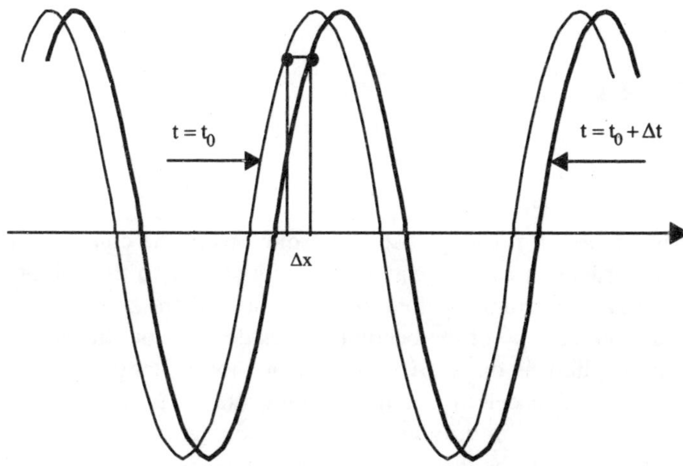

Figure 6.1. *Propagation of the traveling bending wave*

We see that at two consecutive moments the displacements of the beam progresses towards positive x. To determine the speed with which this progression occurs, it is enough to note that the two points x and $x + \Delta x$ of the same vibratory level at the moments t_0 and $t_0 + \Delta t$ must verify:

$$a_1 \cos(\omega t_0 - kx) = a_1 \cos\big(\omega (t_0 + \Delta t) - k(x + \Delta x)\big) \quad [6.23]$$

and, more exactly, to select two points with the same amplitude in the same phase congruence (there is an infinite number of point with the same amplitude):

$$\omega (t_0 + \Delta t) - k (x + \Delta x) = \omega (t_0) - k (x) .$$

From that we draw:

$$\frac{\Delta x}{\Delta t} = \frac{\omega}{k} . \qquad [6.24]$$

Passing to the limit, we observe that the propagation velocity of the waves of bending (or celerity of bending), c_F, is given by:

$$\frac{dx}{dt} = \frac{\omega}{k} = c_F .$$

But the variables k and ω are not independent, they must verify the relation of dispersion [6.15]. Consequently, we can express the propagation velocity of the bending waves according as a function of k with [6.25] or as a function of ω with [6.26]:

$$c_F = \sqrt{EI/\rho S}\, k , \qquad [6.25]$$

$$c_F = \sqrt[4]{EI/\rho S}\, \sqrt{\omega} . \qquad [6.26]$$

This result is important: it shows that if the solutions of the equation of bending are interpreted in terms of traveling waves in a similar way to the solutions of the equation of longitudinal or torsion vibrations of beams, there exists a fundamental difference because the celerity of the bending waves depends on the frequency (it is said then that the medium is dispersive) whereas torsion or longitudinal vibrations are independent of it. The celerity of the bending waves is nil for zero angular frequency and tends towards infinity together with the angular frequency. A similar calculation would show that the three other traveling waves of [6.21] have the same celerity, and the wave $a_2 \sin(\omega t - kx)$ is also propagated towards growing x, whereas the waves $a_3 \cos(\omega t + kx)$ and $a_4 \sin(\omega t + kx)$ are propagated towards decreasing x.

A second type of wave is present in the solution [6.21]; let us take, for example, $a_7 \sin(\omega t)\, e^{kx}$ and display in Figure 6.2 the displacements of the beam at various consecutive moments t_0, t_1, t_2, t_3.

The wave is not propagated: for a given observation moment, it has an exponential variation with x. As time passes, the space form of the vibratory movement is preserved and only its amplitude is modified. This movement is characterized by a very strong space variation which is comparable to a phenomenon of disappearance of the signal with distance, since the amplitude quickly becomes undetectable in experiments when we move away from the wave origin of the wave (from which the name of the vanishing wave originates).

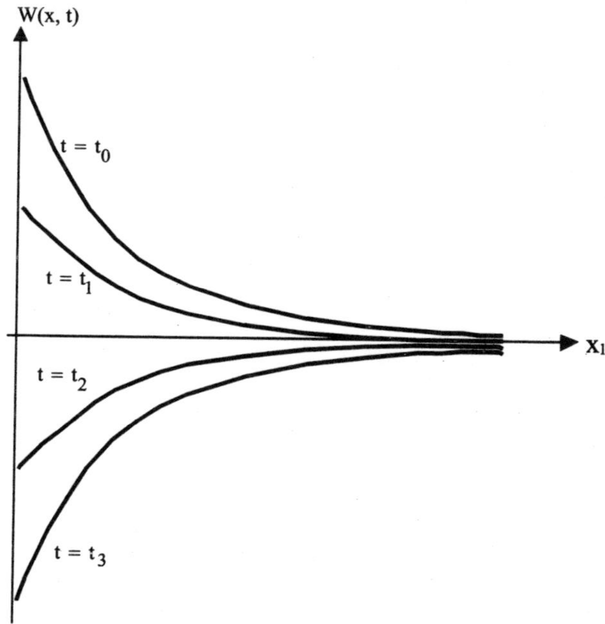

Figure 6.2. *Vibratory movement of a vanishing wave*

6.4. Propagation in infinite beams

6.4.1. *Introduction*

We consider an infinite beam, which is of course unrealistic, but in certain cases constitutes a correct approximation of "sufficiently long" finite beams. After all, introducing boundary conditions, despite appearing more satisfactory at first sight, often results in very imperfect modeling.

A first comment has to be made: it concerns the unrealism of the movement produced by vanishing waves during the response of an infinite beam, because they introduce infinite displacements when x tends towards infinity.

The vibratory field of an infinite physically acceptable beam will thus include only traveling waves, that is to say by taking the form [6.20] thereof and cumulating all the possible movements:

$$W(x,t) = \int_{-\infty}^{\infty} \left(A(k)e^{j\omega t}\ e^{jkx} + B(k)\, e^{-j\omega t}\ e^{jkx} \right) dk\,. \qquad [6.27]$$

The integral comes due to the fact that the most general solution is obtained by the summation of all the traveling waves.

A transformation of [6.27] provides the equivalent form which we are going to use:

$$W(x,t) = \int_{-\infty}^{\infty} g(k,t)\, e^{jkx}\, dk \qquad [6.28]$$

with: $g\,(k,t) = \Gamma\,(k)\ \cos\,(\omega\, t + \varphi(k))$ and $\omega = \sqrt{\dfrac{EI}{\rho S}} k^2$. [6.29]

The function $g(k,t)$ contains constants $\Gamma(k)$ and $\varphi(k)$ that have to be determined. Of course, their values can be fixed by the conditions of initial displacements and velocity of the beam.

The general form of the solution given in [6.28] can be found by solving the equation of motion [6.6] using the Fourier space transform. Let $\widetilde{W}(k,t)$ be the Fourier space transform of $W(x,t)$:

$$\widetilde{W}(k,t) = \int_{-\infty}^{\infty} W(x,t)\, e^{-jkx}\, dx\,.$$

We will determine $\widetilde{W}(k,t)$ by taking the Fourier transform of equation [6.6], that is after all the calculations:

$$-\rho S \frac{d^2\widetilde{W}}{dt^2}(k,t) - EIk^4\, \widetilde{W}(k,t) = 0\,.$$

For each value of k, this differential in time equation is integrated without difficulty to give:

$$\widetilde{W}(k,t) = \Theta(k)\ \cos(\omega\, t + \psi(k)) \ \text{ and } \ \omega = \sqrt{\frac{EI}{\rho S}}k^2 \cdot$$

The space-time solution is obtained by an inverse Fourier transformation $\widetilde{W}(k,t)$:

$$W(x,t) = \frac{1}{2\pi}\int_{-\infty}^{\infty}\Theta(k)\ \cos(\omega\, t + \psi(k))\, e^{jkx}dk\,.$$

This solution clearly coincides with [6.28] and [6.29] where we posed:

$$\Gamma(k) = \frac{\Theta(k)}{2\pi} \ \text{ and } \ \varphi(k) = \Psi(k) \cdot$$

6.4.2. *Propagation of a group of waves*

To explain the phenomenon of propagation of a group of waves, we will consider the particular case of an initial displacement of the type [6.30], which we display in Figure 6.3, and with zero initial speed.

$$W(x,0) = 2\cos(kx)\frac{\sin(\Delta x)}{x}\ , \quad \text{with : } \Delta << k \qquad [6.30a]$$

$$\frac{\partial W}{\partial t}(x,0) = 0\,. \qquad [6.30b]$$

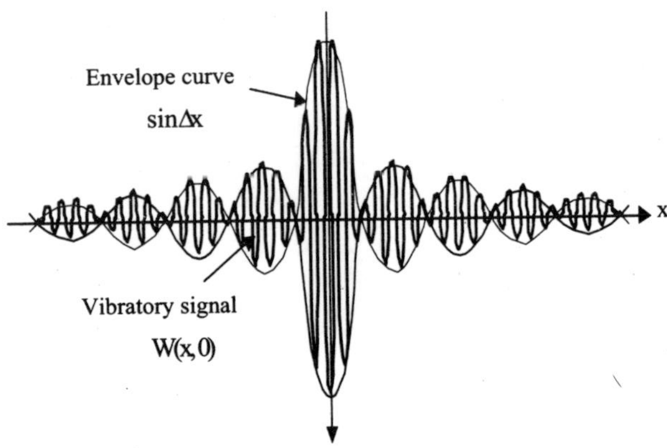

Figure 6.3. *Initial displacements of the beam*

We can calculate the function $g(k, t)$ contained in [6.28] in the following manner. At $t = 0$, the general form [6.28] becomes:

$$W(x, 0) = \int_{-\infty}^{\infty} g(x, 0)\, e^{jkx}\, dx\,. \qquad [6.31]$$

This expression indicates that $g(k, 0)$ is the Fourier transform of $W(x, 0)$:

$$g(k, 0) = \int_{-\infty}^{\infty} W(x, 0)\, e^{-jkx}\, dx \qquad [6.32]$$

To observe the initial condition of imposed displacement [6.30a] we may use the known result on the Fourier transformation of the gate function. Initial displacement [6.30a] was selected to coincide with the Fourier transform of the following "gate" function $g(k, 0)$:

$$g(k,0) = \begin{cases} 1 \text{ if } k \in [-k-\Delta, -k+\Delta] \\ 1 \text{ if } k \in [k-\Delta, k+\Delta] \\ 0 \text{ if } k \notin [-k-\Delta, -k+\Delta] \cup [k-\Delta, k+\Delta]. \end{cases} \qquad [6.33]$$

By using this result and the solution [6.29] we obtain the two relations:

$$k \in [-k-\Delta,\ -k+\Delta] \cup [k-\Delta,\ k+\Delta] \Rightarrow \Gamma(k)\cos\varphi = 1$$

$$k \notin [-k-\Delta,\ -k+\Delta] \cup [k-\Delta,\ k+\Delta] \Rightarrow \Gamma(k)\cos\varphi = 0.$$

To introduce the initial condition of zero speed [6.30b], we calculate dg/dt (k, t) with [6.29] and then its value at $t = 0$. We then use the equality with the Fourier transform of the initial speed.

$$\frac{dg}{dt}(k,0) = \int_{-\infty}^{\infty} \frac{\partial W}{\partial t}(x,0)\, e^{-jkx}\, dx\,. \qquad [6.34]$$

In our case initial speed is nil and:

$$\frac{dg}{dt}(k,0) = 0\,,$$

that is:

$$\Gamma(k)\,\omega\ \sin\varphi = 0\,.$$

We deduce:

$$\varphi = 0\,,$$

$$k \in [-k-\Delta,\ -k+\Delta] \cup [k-\Delta,\ k+\Delta] \Rightarrow \Gamma(k) = 1\,,$$

$$k \notin [-k-\Delta,\ -k+\Delta] \cup [k-\Delta,\ k+\Delta] \Rightarrow \Gamma(k) = 0\,.$$

From a physical point of view this initial condition amounts to exciting a group of waves with wave numbers close to k.

Transferring this to [6.28], we obtain finally:

$$W(x,t) = \int_{k-\Delta}^{k+\Delta} \cos(\omega\, t)\, e^{jkx}\, dk + \int_{-k-\Delta}^{-k+\Delta} \cos(\omega\, t)\, e^{jkx}\, dk\,. \qquad [6.35]$$

The relation [6.15] between ω and k must be clearly taken into account in this expression:

$$\omega = \sqrt{\frac{EI}{\rho S}} k^2 .$$

Since we have supposed $\Delta << k$, we may approach the function $\omega(k)$ using its Taylor development truncated in the first order:

$$\begin{cases} \omega(k+\varepsilon) \approx \omega(k) + \varepsilon \dfrac{\partial \omega}{\partial k}(k) \\ \omega(-k+\varepsilon) \approx \omega(-k) + \varepsilon \dfrac{\partial \omega}{\partial k}(-k). \end{cases} \qquad [6.36]$$

After change of variable and use of [6.36], the integral [6.35] becomes:

$$\begin{aligned} W(x,t) = & \int_{-\Delta}^{+\Delta} \cos\left(\left(\omega(k) + \varepsilon \frac{\partial \omega}{\partial k}(k)\right) t\right) e^{j(k+\varepsilon)x} \, d\varepsilon \\ & + \int_{-\Delta}^{+\Delta} \cos\left(\left(\omega(-k) + \varepsilon \frac{\partial \omega}{\partial k}(-k)\right) t\right) e^{j(-k+\varepsilon)x} \, d\varepsilon . \end{aligned} \qquad [6.37]$$

After a long but not difficult calculation we obtain the result [6.38]:

$$\begin{aligned} W(x,t) = & \cos\left(\omega(k)t + kx\right) \frac{\sin\left(\Delta\left(\dfrac{\partial \omega}{\partial k}(k)t + x\right)\right)}{\dfrac{\partial \omega}{\partial k}(k)t + x} \\ & + \cos\left(\omega(k)t - kx\right) \frac{\sin\left(\Delta\left(\dfrac{\partial \omega}{\partial k}(k)t - x\right)\right)}{\dfrac{\partial \omega}{\partial k}(k)t - x} \end{aligned} \qquad [6.38]$$

To obtain this result we have supposed:

$$\omega(k) = \omega(-k) \text{ and } \frac{\partial \omega}{\partial k}(k) = -\frac{\partial \omega}{\partial k}(-k). \qquad [6.39]$$

As the relation of dispersion [6.15] indicates, this hypothesis is valid in the case studied. Of course, if the relation of dispersion does not allow verifying [6.39], the solution [6.38] is not valid and, consequently, must be modified.

If we pose t = 0, the expression [6.38] of vibratory displacement coincides with the initial condition [6.30]. Two signals are created at successive moments; they move in opposite directions. Let us consider the one moving towards negative x $W_{\_}(x,t)$:

$$W_{\_}(x,t) = \cos(\omega(k)t + kx)\frac{\sin\left(\Delta\left(\frac{\partial\omega}{\partial k}(k)t + x\right)\right)}{\frac{\partial\omega}{\partial k}(k)t + x}. \quad [6.40]$$

Figure 6.4 illustrates the propagation movement of the group of waves.

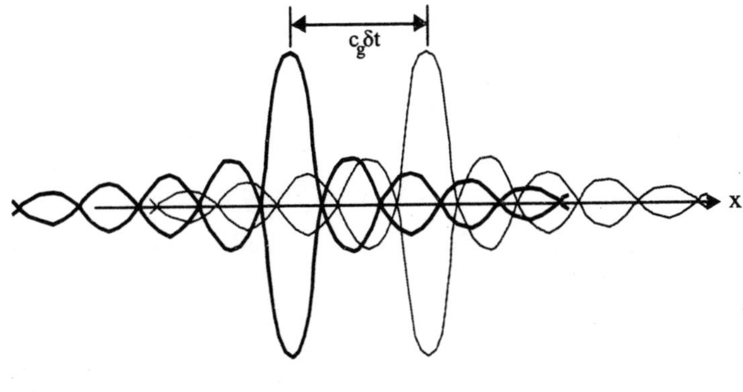

position at the moment t + δt ——— *position at the moment* t

Figure 6.4. *Propagation of the group of waves towards negative x-coordinates; only envelope curves are represented*

The signal is composed of the product of two terms with very different frequencies: a classical wave with a short wavelength and a pseudo-wave with a large wavelength.

The classical wave $\cos(\omega(k)t + kx)$ describes a vibratory movement with the angular frequency ω, which propagates at the bending wave speed c_F with a wave number k [6.25] or [6.26]:

$$\cos(\omega(k)t + kx) \quad [6.41]$$

The pseudo-wave, the second term of the product in [6.40], characterizes the envelope curve of the signal and depends on the size of the group of waves by its angular frequency equal to $\Delta\,\partial\omega/\partial k$ and at the same time its maximum amplitude, equal to Δ. The maximum is taken at the moment t at the point x given by:

$$x = -t\frac{\partial\omega}{\partial k}(k). \qquad [6.42]$$

The displacement of this point of maximum amplitude over time is characteristic of the pseudo-wave propagation. Thus, we obtain the propagation velocity c_g of this signal with:

$$c_g = \left|\frac{dx}{dt}\right| = \frac{d\omega}{dk}(k). \qquad [6.43]$$

In the case of the equation of dispersion of beams in pure bending [6.15], it follows:

$$c_g = 2\sqrt{\frac{EI}{\rho S}}\,k. \qquad [6.44]$$

Let us recall that the propagation velocity of the bending waves was given by [6.25]:

$$c_F = \sqrt{\frac{EI}{\rho S}}\,k. \qquad [6.45]$$

The celerity c_g of the pseudo-wave is called group speed; it is different from celerity c_F, known as phase speed, associated with the traditional wave. In the case considered, we observe that group speed is twice higher than phase speed. It follows that the combined movement of the product of the two signals, which are propagated at different speeds, does not keep the same form over time; it is a fundamental difference with respect to the case of the wave equation that we have examined in Chapter 5.

Note: in the case of the wave equation, the relation of dispersion is of the $\omega = kc$ type; from it stems the equality of phase and group speed, since $\partial\omega/\partial k = \omega/k$. The medium is known to be not dispersive because the propagation of a package of waves occurs without modification of the space form of the signal over time.

6.5. Introduction of boundary conditions: vibration modes

6.5.1. *Introduction*

The solutions of the equation of motion were obtained in section 6.3 using separation of variables. We will now introduce boundary conditions, which will highlight the set of vibratory movements that the beam can undergo. Naturally, the vibrations are different when the boundary conditions change and it is out of the question to consider all the possibilities here (there are 16 types of boundary conditions, since there are 4 possibilities at each end and there are two ends). We will consider only certain types of boundary conditions to illustrate this point.

The resolution by separation of variables revealed three types of solutions: [6.13], [6.16] and [6.17]. While studying longitudinal vibrations we saw that the type [6.17] yielded only the zero solution since it had to respect the boundary conditions. The same applies here and we will not consider this solution further. The [6.13] type of solution is characteristic of rigid beam movements and is only possible if boundary conditions allow them; this is particularly the case for free ends, but there are other possibilities which we will examine below. The type [6.16] will give an infinite number of solutions for all the cases of boundary conditions.

6.5.2. *The case of the supported-supported beam*

The boundary conditions are as follows:

$$\begin{cases} W(0,t) = 0 \\ EI\dfrac{\partial^2 W}{\partial x^2}(0,t) = 0 \end{cases} \qquad [6.46]$$

and:

$$\begin{cases} W(L,t) = 0 \\ EI\dfrac{\partial^2 W}{\partial x^2}(L,t) = 0. \end{cases} \qquad [6.47]$$

The relations [6.46] and [6.47] mean that transverse displacement is nil at each end and that longitudinal movement linked to the rotation of cross-sections is free, which imposes the nullity of the torque. This modeling of boundaries is well adapted to the description of a beam supported by ball bearings that block transverse movements but allow the rotation of cross-sections and, thus, the longitudinal movements.

This type of boundary conditions does not allow rigid movements, and the solution [6.13] of the equation of motion leads only to the zero solution. Vibratory movements thus all result from the solution [6.16]:

$$W(x,t) = \big(A\cos(\omega t) + B\sin(\omega t)\big)\big(C\cos(kx) + D\sin(kx) + E\,\mathrm{ch}(kx) + F\,\mathrm{sh}(kx)\big) \qquad [6.48]$$

with: $\omega = \sqrt{\dfrac{EI}{\rho S}}\, k^2 .$ $\qquad$ [6.49]

Let us impose that [6.48] satisfy the boundary conditions [6.46] and [6.47]. After calculations it follows:

$$\begin{cases} C + E = 0 \\ C - E = 0 \\ C\cos(kL) + D\sin(kL) + E\,\mathrm{ch}(kL) + F\,\mathrm{sh}(kL) = 0 \\ -C\cos(kL) - D\sin(kL) + E\,\mathrm{ch}(kL) + F\,\mathrm{sh}(kL) = 0 . \end{cases} \qquad [6.50]$$

The first two relations lead to: $C = E = 0$ from which using the last two relations we deduce:

$$\begin{pmatrix} \sin(kL) & \mathrm{sh}(kL) \\ -\sin(kL) & \mathrm{sh}(kL) \end{pmatrix} \begin{pmatrix} D \\ F \end{pmatrix} = \begin{pmatrix} 0 \\ 0 \end{pmatrix} . \qquad [6.51]$$

If the linear system [6.51] has a non-nil determinant, we infer a single solution:

$$\begin{pmatrix} D \\ F \end{pmatrix} = \begin{pmatrix} 0 \\ 0 \end{pmatrix} . \qquad [6.52]$$

Taking into account the fact that the constants C and E are nil, we deduce from the general form [6.48] that:

$$W(x,t) = 0 . \qquad [6.53]$$

To obtain a non-zero solution, it is thus necessary that the determinant of [6.51] be nil. That gives:

$$\sin(kL)\,\text{sh}(kL) = 0\,, \tag{6.54}$$

that is:

$$\sin(kL) = 0\,. \tag{6.55}$$

This characteristic equation shows that there is an infinite number of values of the wave number k_n that verify it:

$$k_n = \frac{n\pi}{L},\ n = 1,\ \ldots,\ \infty\,. \tag{6.56}$$

Each wave number solution k_n is associated with a normal angular frequency due to the equation of dispersion [6.49]:

$$\omega_n = \sqrt{\frac{EI}{\rho S}}\left(\frac{n\pi}{L}\right)^2,\ n = 1,\ \ldots,\ \infty\,. \tag{6.57}$$

To fully characterize the solutions it remains to solve the linear system [6.51] for the values k_n given by [6.56], which cancel its determinant. We obtain:

$$\begin{pmatrix} 0 & \text{sh}(n\pi) \\ 0 & \text{sh}(n\pi) \end{pmatrix}\begin{pmatrix} D \\ F \end{pmatrix} = \begin{pmatrix} 0 \\ 0 \end{pmatrix}, \tag{6.58}$$

that is $F = 0$ and any D.

Taking into account all these results in the [6.48] form of vibratory displacement there follows for each modal index n a modal movement $W_n(x, t)$ given by [6.59]:

$$W_n(x, t) = \left(A_n \cos \omega_n t + B_n \sin \omega_n t\right) \sin\left(\frac{n\pi}{L} x\right). \tag{6.59}$$

This expression characterizes the movement of the n mode of vibration. It occurs with the normal angular frequency ω_n (equation [6.57]) and with the mode shape $f_n(x)$ given by [6.60] (the arbitrary constant D has been posed as equal to 1):

$$f_n(x) = \sin\left(\frac{n\pi}{L}x\right). \qquad [6.60]$$

Figure 6.5 illustrates the mode shape of the first 3 modes.

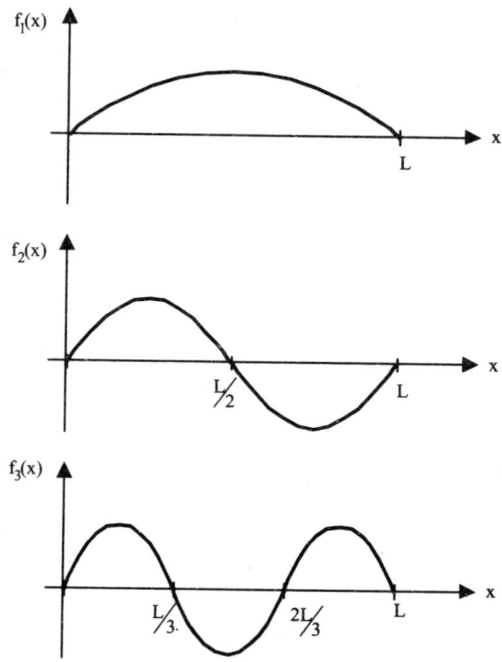

Figure 6.5. *Mode shapes of the first three modes of vibration of bending of a supported-supported beam*

General vibratory movement is the sum of all the modal movements, that is:

$$W(x,t) = \sum_{n=1}^{\infty}\left(A_n \cos(\omega_n t) + B_n \sin(\omega_n t)\right)\sin\left(\frac{n\pi}{L}x\right). \qquad [6.61]$$

The constants A_n and B_n will be fixed by the initial conditions at the origin of the free vibrations. Their calculation requires the use of the properties of orthogonality of mode shapes. We will not proceed further with this calculation.

6.5.3. *The case of the supported-clamped beam*

The boundary conditions are now:

$$\begin{cases} W(0,t) = 0 \\ EI\dfrac{\partial^2 W}{\partial x^2}(0,t) = 0 \end{cases} \qquad [6.62]$$

and:

$$\begin{cases} W(L,t) = 0 \\ \dfrac{\partial W}{\partial x}(L,t) = 0. \end{cases} \qquad [6.63]$$

Clamping in L blocks the two displacements, transverse and longitudinal, which amounts to setting to zero the slope since it is equal to the rotation of the cross-sections, which is in turn connected to longitudinal displacement. With respect to the preceding case, only the fourth relation of [6.50] is modified and becomes:

$$-C\sin(kL) + D\cos(kL) + E\,\mathrm{sh}(kL) + F\,\mathrm{ch}(kL) = 0. \qquad [6.64]$$

The solution, which is completely similar to that of the preceding section, leads to:

$$C = E = 0. \qquad [6.65]$$

Two other constants being subjected to verification of [6.66]:

$$\begin{pmatrix} \sin(kL) & \mathrm{sh}(kL) \\ \cos(kL) & \mathrm{sh}(kL) \end{pmatrix}\begin{pmatrix} D \\ F \end{pmatrix} = \begin{pmatrix} 0 \\ 0 \end{pmatrix}. \qquad [6.66]$$

Setting to zero the determinant of [6.66] leads to the characteristic equation:

$$\mathrm{tg}(kL) = \mathrm{th}(kL). \qquad [6.67]$$

This characteristic equation is not as simple to solve as in the case of the supported-supported beam and requires computerized processing. We propose here a graphic method, which is not very precise but makes it possible to clearly locate the solutions.

In Figure 6.6 we have plotted the two curves tg(kL) and th(kL) so that each intersection provides a root. First of all, it should be noticed that there is an infinite number of solutions if we take into account the periodicity of the tangent function. It should also be noted that as the hyperbolic tangent very quickly tends towards one, we have an approximation of the roots by approximating the characteristic equation [6.67] by:

$$\mathrm{tg}(kL) = 1. \qquad [6.68]$$

That is, values:

$$k_n = \frac{4n-1}{4}\frac{\pi}{L} \quad \text{for } n = 1,\ldots,\infty. \qquad [6.69]$$

The very first modes require computerized processing if we wish to be very precise, but the approximation [6.69] is already quite good. We can deduce from it the values of normal angular frequencies using the equation of dispersion [6.49]:

$$\omega_n = \sqrt{\frac{EI}{\rho S}}\left(\frac{4n-1}{4}\frac{\pi}{L}\right)^2. \qquad [6.70]$$

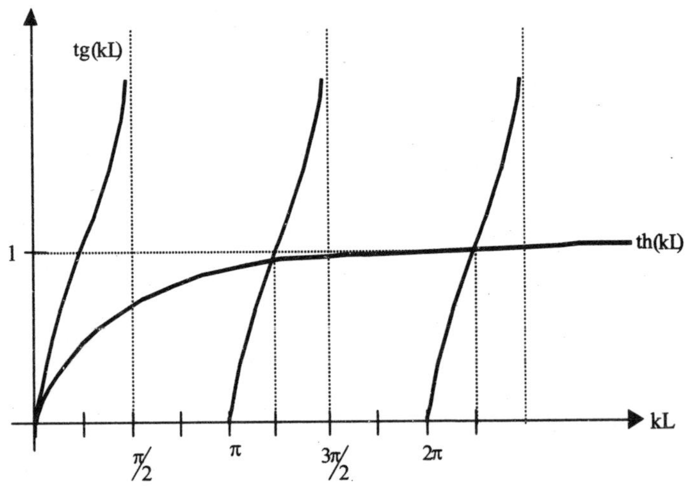

Figure 6.6. *Roots of the characteristic equation of the supported-clamped beam*

The calculation of mode shapes is carried out using the wave numbers solution [6.69] in the linear system [6.66]. After all the calculations it follows:

$$D = -\sqrt{2}\,\mathrm{sh}\left(\frac{4n-1}{4}\pi\right)F. \qquad [6.71]$$

That is, posing $D = 1$ and replacing the various quantities by their respective values in the general mode shape [6.48]:

$$W(x,t) = \sum_{n=1}^{\infty}\left(A_n \cos(\omega_n t) + B_n \sin(\omega_n t)\right)\cdot f_n(x) \qquad [6.72]$$

with:

$$f_n(x) = \left(\sin\left(\frac{4n-1}{4}\frac{\pi}{L}x\right) + F_n \mathrm{sh}\left(\frac{4n-1}{4}\frac{\pi}{L}x\right)\right)$$

and [6.73]

$$F_n = -\frac{1}{\sqrt{2}\,\mathrm{sh}\left(\frac{4n-1}{4}\pi\right)}$$

And for large enough n: $-\sqrt{2}e^{-\left(\frac{4n-1}{4}\pi\right)}$. [6.74]

The approximation of F_n [6.74] is better the higher the rank of the mode.

Let us consider the normal strains of modes; they are given by the expression [6.73]. For high ranked modes, by expressing the sine hyperbolic by form into an exponential we obtain:

$$f_n(x) = \sin\left(\frac{4n-1}{4}\frac{\pi}{L}x\right) - \frac{1}{\sqrt{2}}e^{\frac{4n-1}{4}\frac{\pi}{L}(x-L)} + \frac{1}{\sqrt{2}}e^{-\frac{4n-1}{4}\frac{\pi}{L}(x+L)}. \qquad [6.75]$$

The third term of the second member is small for all the values of x and can be neglected; the second, on the other hand, is not negligible when x is close to L and must be preserved. Consequently, mode shape consists of two dominating terms:

$$f_n(x) = \sin\left(\frac{4n-1}{4}\frac{\pi}{L}x\right) - \frac{1}{\sqrt{2}}e^{\frac{4n-1}{4}\frac{\pi}{L}(x-L)}. \qquad [6.76]$$

Figure 6.7 illustrates the variations of the two terms depending on x: the first term is important everywhere in the beam and characterizes the internal solution, while the second decreases very quickly when we move away from the $x = L$ end; it introduces the edge effect. When the total displacement is traced, it is obviously very close to the internal solution as long as we are far from the clamped end. Towards the L end of the beam, the edge effect is of the same order of magnitude as the internal solution.

This interpretation calls for several observations:

1. The tendency described is general for problems of bending. The presence of edge effects is characteristic of the influence of vanishing waves present in the solution of the equation of motion.

2. Edge effects appear in the vicinities of beams singularities, naturally, with boundary conditions, but also in the case of beams with a variable section at the level of each inertia variation; see Figure 6.8.

3. The boundary condition of support does not introduce edge effects (the same is true for the guided condition).

4. The edge effect has real influence only at a distance of Δ from the singularity, which is lower than the quarter wavelength λ of the internal solution; that is:

$$\Delta < \frac{\lambda}{4} = \frac{\pi}{2}\frac{c_F}{\omega} = \frac{\pi}{2}\frac{\sqrt[4]{\dfrac{EI}{\rho S}}}{\sqrt{\omega}}. \qquad [6.77]$$

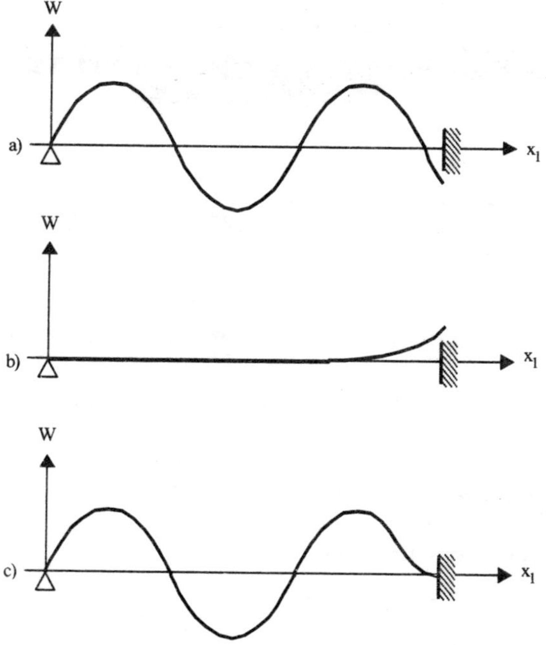

Figure 6.7. *Mode shapes of the 3rd mode of a supported-clamped beam a) Internal solution, b) Edge effect, c) Mode shape*

For mode n we can introduce into [6.77] the normal angular frequency ω_n, the distance of influence of the edge effect results from it:

$$\Delta < \frac{\pi}{2} \frac{\sqrt[4]{\frac{EI}{\rho S}}}{\sqrt{\omega_n}} = \frac{\pi}{2k_n}. \qquad [6.78]$$

In the case considered, k_n is provided in [6.69]. From that we draw:

$$\Delta < \frac{2L}{4n-1}. \qquad [6.79]$$

We note with [6.79] that the distance characteristic of the zone of influence of the edge effect strongly decreases when the order of the mode grows.

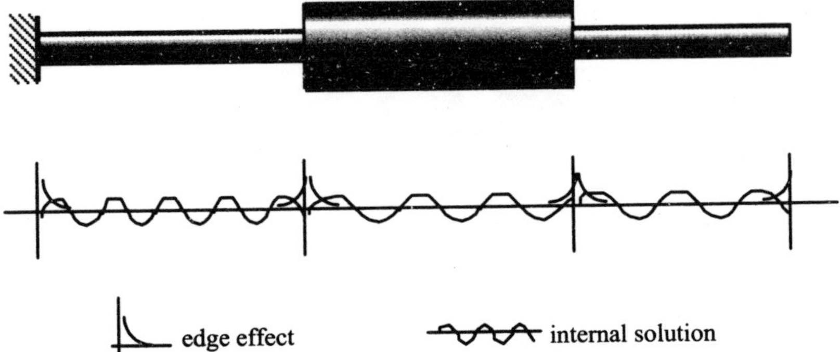

Figure 6.8. *Localization of the edge effects on a clamped-free beam with variable inertia*

6.5.4. *The free-free beam*

The boundary conditions are in this case:

$$\begin{cases} EI\dfrac{\partial^2 W}{\partial x^2}(0,t)=0 \\ EI\dfrac{\partial^3 W}{\partial x^3}(0,t)=0 \end{cases} \qquad [6.80]$$

and:

$$\begin{cases} EI\dfrac{\partial^2 W}{\partial x^2}(L,t)=0 \\ EI\dfrac{\partial^3 W}{\partial x^3}(L,t)=0. \end{cases} \qquad [6.81]$$

The bending moment and shearing force are nil at both ends. These boundary conditions that leave the ends free for transverse and rotation displacement allow movements without strain (or of rigid solid). Thus, it is necessary to consider solutions of the [6.13] type in addition to the solutions of the [6.16] type.

Let us consider the solution of the equation of motion of the [6.13] type. Applying boundary conditions [6.80] and [6.81] leads to:

$$E = F = 0.$$

That is, to solutions of the form:

$$W(x, t) = (At + B)(C + Dx) \tag{6.82}$$

where C and D are unspecified.

We can separate [6.82] into two solutions, representative of the rigid modes of translation and rotation of the beam so that the two mode shapes are orthogonal:

– translation mode:

$$W_T(x, t) = (A_0 t + B_0) \; ; \tag{6.83}$$

– rotation mode:

$$W_R(x, t) = (A'_0 t + B'_0)\left(x - \frac{L}{2}\right). \tag{6.84}$$

We often call these movements "zero modes" because they represent uniform movements, which have one infinite period and, therefore, a zero frequency.

Let us consider the solution of the equation of motion of the [6.16] type; the application of boundary conditions [6.81] leads to:

$$C = E,$$

$$D = F$$

and:

$$\begin{pmatrix} (\cos(kL) - \text{ch}(kL)) & (\sin(kL) - \text{sh}(kL)) \\ (\sin(kL) - \text{sh}(kL)) & (\text{ch}(kL) - \cos(kL)) \end{pmatrix} \begin{pmatrix} C \\ D \end{pmatrix} = \begin{pmatrix} 0 \\ 0 \end{pmatrix}. \tag{6.85}$$

To obtain non-trivial solutions it is necessary that the determinant of [6.85] be nil, that is:

$$\cos(kL)\,\text{ch}(kL) = 1. \tag{6.86}$$

The solutions of [6.86] must be approximated using a computer; we will find the values of the roots in the summary table of the following section. For the first modes there is no obvious approximation, but for the higher modes $(kL >> 1)$ we may approximate the characteristic equation by:

$$\cos(kL) = 0, \tag{6.87}$$

that is:

$$kL = (2n+1)\frac{\pi}{2}. \qquad [6.88]$$

An infinite number of solutions is thus obtained:

– normal wave numbers:

$$k_n = (2n+1)\frac{\pi}{2L}; \qquad [6.89]$$

– normal angular frequencies:

$$\omega_n = \sqrt{\frac{EI}{\rho S}}\, k_n^2; \qquad [6.90]$$

– mode shapes:

$$f_n(x) = \left(\cos(k_n x) + ch(k_n x)\right) + D_n\left(\sin(k_n x) + sh(k_n x)\right) \qquad [6.91]$$

with $D_n = -\dfrac{\cos(k_n L) - ch(k_n L)}{\sin(k_n L) - sh(k_n L)}$. [6.92]

For the high rank modes we note that $D_n \approx -1$. Mode shapes are then approximated by:

$$f_n(x) = \cos(k_n x) - \sin(k_n x) + ch(k_n x) - sh(k_n x). \qquad [6.93]$$

The first two terms of the second member of [6.93] are characteristic of the internal solution, while the two last ones apply to the edge effects at the ends of the beam ($x = 0$ and $x = L$).

The most general free vibration movement is obtained by an accumulation of all the modal movements including zero modes:

$$W(x,t) = (A_0 t + B_0) + (A'_0 t + B'_0)\left(x - \frac{L}{2}\right) + \sum_{n=1}^{\infty}(A_n \cos\omega_n t + B_n \sin\omega_n t)\, f_n(x). \qquad [6.94]$$

6.5.5. *Summary table*

Boundary conditions	Characteristic equations	$(k_n L)^2$	Mode shapes
Supported-supported	$\sin(kL) = 0$	9.87, 39.50, 88.9,… $(n\pi)^2$	$\sin\left(\frac{n\pi}{L}x\right)$
Clamped-clamped	$ch(kL)\cos(kL) = 1$	22.4, 61.7, 121,… $\left(\frac{2n+1}{2}\pi\right)^2$	$\frac{ch(k_n x) - \cos(k_n x)}{ch(k_n L) - \cos(k_n L)} - \frac{sh(k_n x) - \sin(k_n x)}{sh(k_n L) - \sin(k_n L)}$
Clamped-supported	$th(kL) = tg(kL)$	15.4, 50, 104,… $\left(\frac{4n+1}{4}\pi\right)^2$	$\frac{ch(k_n x) - \cos(k_n x)}{ch(k_n L) - \cos(k_n L)} - \frac{sh(k_n x) - \sin(k_n x)}{sh(k_n L) - \sin(k_n L)}$
Clamped-free	$ch(kL)\cos(kL) = -1$	3.52, 22.4, 61.7, 121,… $\left(\frac{2n+1}{2}\pi\right)^2$	$\frac{ch(k_n x) - \cos(k_n x)}{ch(k_n L) + \cos(k_n L)} - \frac{sh(k_n x) - \sin(k_n x)}{sh(k_n L) - \sin(k_n L)}$
Free-free	$ch(kL)\cos(kL) = 1$	22.4, 61.7, 121,… $\left(\frac{2n+1}{2}\pi\right)^2$	$\frac{ch(k_n x) + \cos(k_n x)}{ch(k_n L) - \cos(k_n L)} - \frac{sh(k_n x) + \sin(k_n x)}{sh(k_n L) - \sin(k_n L)}$
Supported-free	$th(kL) = tg(kL)$	15.4, 50, 104,… $\left(\frac{4n+1}{4}\pi\right)^2$	$\frac{\sin(k_n x)}{\sin(k_n L)} + \frac{sh(k_n x)}{sh(k_n L)}$

Table 6.1. *Table giving the vibration modes of beams in bending for various boundary conditions. We provide the first numerical values of* $(k_n L)^2$ *and then an asymptotic form for large n*

Calculations completely similar to those of the preceding sections can be made in all the cases of boundary conditions and provide the normal modes of beams in bending. We have drawn a table which recapitulates the results in several cases of boundary conditions. The calculation of normal angular frequencies is performed using the equation $\omega_n = \sqrt{\frac{EI}{\rho S L^4}} \cdot (k_n L)^2$ where the value of $(k_n L)^2$ is provided in the table.

For the free-free case, there are two rigid modes in addition to the vibration modes; for the free-supported case, there is one rigid mode.

6.6. Stress-displacement connection

During the study of longitudinal or torsion vibrations we have seen that modal stresses varied inversely to vibratory displacement: a node of displacement corresponding to an antinode of stress and vice versa. What happens in the case of bending? To clarify this point we will take the case of the supported-clamped beam analyzed previously.

Longitudinal stress is calculated on the basis of transverse displacement by the relation [6.5]. Replacing $W(x_1, t)$ with the expression [6.75], it follows (we reintroduce the notation x_1 instead of x in order to avoid any ambiguity):

$$\sigma_{11}(x_1, x_2, x_3, t) = x_2 \, E \sum_{n=1}^{\infty} k_n^2 \, (A_n \cos \omega_n t + B_n \sin \omega_n t) \; h_n(x_1) \qquad [6.95a]$$

with: $h_n(x_1) = -\sin(k_n x_1) + F_n \, \mathrm{sh}(k_n x_1)$, [6.95b]

where k_n is given by [6.69], ω_n by [6.70] and F_n by [6.73].

Expression [6.95a] shows that the bending stress is nil for the neutral fiber $(x_2 = 0)$ and maximum for the upper and lower surfaces $(x_2 = \pm h/2)$. We also observe that bending stress breaks up into a modal series whose normal functions are $h_n(x_1)$ [6.95b].

Figure 6.9 illustrates the variation of the normal stress functions $h_n(x_1)$. The results are to be compared with those in Figure 6.7 which represented mode shapes. We note that in the part of the beam dominated by the internal solution, a node (or an antinode) of displacement corresponds to a node (or an antinode in opposing phase) of constraint. This situation is the reverse of that of longitudinal and torsion vibrations. In the part of beam close to the clamped end dominated by the edge effect the situation is different since a zero displacement corresponds to maximum stress. In the case of a free end, we would note that for a maximum displacement at the end we record zero stress. The correspondence between modal displacements and modal stresses is thus differentiated. In the part of beam dominated by the internal solution an antinode (or a node) of displacement corresponds an antinode (or a node) of stress. When the edge effect is greater, the tendency is reversed since an antinode (or a node) of displacement corresponds to a node (or an antinode) of stress. It should be noted that for a condition of support, there is no edge effect and the internal solution dominates until the end.

The calculation of the response by modal decomposition poses the problem of the number of terms to be considered in the calculation of the series. It is not possible to give a general rule since amplitudes A_n and B_n depend on the initial conditions. We can, on the other hand, affirm that convergence in stress would be more difficult than in displacement taking into account the multiplicative term k_n^2 which appears in [6.95]. Indeed, the generic term of the modal stress series [6.95] will always decrease slower than the term of the modal displacement series [6.72].

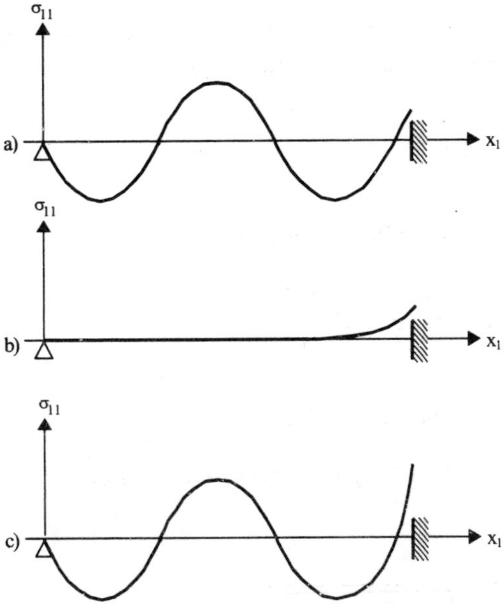

Figure 6.9. *Normal stress of the 3rd mode of a supported-clamped beam: a) internal solution, b) edge effect, c) normal constraint*

6.7. Influence of secondary effects

Equation [6.2], which is at the foundation of our discourse, represents the effect of pure bending and results from simplifying hypotheses which we have examined in Chapter 3. The validity of the simplified approach poses the problem of the influence of secondary effects that have been neglected. There are two secondary effects: rotational inertia and transverse shearing. Taking these effects into account raises various technical difficulties; the calculation which is simple enough for rotational inertia is more difficult for shearing. We will consider the two cases successively.

6.7.1. *Influence of rotational inertia*

The equation representative of vibratory movement was provided in Chapter 3, equation [3.66].

$$\rho S \frac{\partial^2 W}{\partial t^2} - \rho I \frac{\partial^4 W}{\partial x^2 \partial t^2} + EI \frac{\partial^4 W}{\partial x^4} = 0 . \qquad [6.96]$$

Here we have, on the one hand, adopted a simplified notation and, on the other hand, supposed a homogenous beam. Let us seek the solutions of the vibrations problem in the form:

$$W(x, t) = f(x)\, e^{j\omega t} . \qquad [6.97]$$

Replacing it in equation [6.96], it follows:

$$-\rho S \omega^2 f(x) + \rho I \omega^2 \frac{d^2 f}{dx^2} + EI \frac{d^4 f}{dx^4} = 0 . \qquad [6.98]$$

The solution of this equation is of the type:

$$f(x) = C\, sh(\alpha x) + D\, ch(\alpha x) + E \sin(kx) + F \cos(kx) \qquad [6.99]$$

with:

$$\alpha = \sqrt{-\frac{\rho}{2E}\omega^2 + \sqrt{\left(\frac{\rho}{E}\right)^2 \frac{\omega^4}{4} + \frac{\rho S}{EI}\omega^2}} \qquad [6.100]$$

and:

$$k = \sqrt{\frac{\rho}{2E}\omega^2 + \sqrt{\left(\frac{\rho}{E}\right)^2 \frac{\omega^4}{4} + \frac{\rho S}{EI}\omega^2}} . \qquad [6.101]$$

The general solution consists of four terms: the first two characterized by a wave number α are vanishing waves, while the last two with a wave number k are traveling waves.

We can calculate the phase speed c_F of the traveling waves with pulsation ω in a traditional way using the ratio:

$$c_F = \frac{\omega}{k} = \frac{\omega}{\sqrt{\frac{\rho}{2E}\omega^2 + \sqrt{\left(\frac{\rho}{E}\right)^2 \frac{\omega^4}{4} + \frac{\rho S}{EI}\omega^2}}}. \qquad [6.102]$$

At low frequencies we can approximate this expression by:

$$c_F = \sqrt[4]{\frac{EI}{\rho S}}\sqrt{\omega}. \qquad [6.103]$$

The propagation velocity given by [6.103] corresponds to the phase speed of beams in bending without secondary effects that we have calculated (equation [6.26]).

At high frequencies the approximation of [6.102] is:

$$c_F = \sqrt{\frac{E}{\rho}}. \qquad [6.104]$$

The speed of traveling waves with a high frequency becomes independent of ω, which shows that at a high frequency, a beam in bending with rotational inertia is a non-dispersive medium. This celerity is equal to that of longitudinal waves.

Rotational inertia is thus an important effect at high frequencies, since it modifies the propagation velocity of waves; at low frequencies, however, it is negligible.

On the basis of the solution [6.99] it is easy to take into account the boundary conditions to determine the vibration modes of a beam in bending. Here we will consider the case leading to the simplest solution: that of a beam supported at both ends. The application of boundary conditions leads to the characteristic equation:

$$\sin(kL) = 0 \qquad [6.105]$$

with : $C = D = F = 0$ and any E.

The solutions of [6.105] are given by the sequence of values:

$$k_n = \frac{n\pi}{L}, \ n = 1, \ldots, \infty \qquad [6.106]$$

from which, with [6.101], we can draw the values of normal angular frequencies:

$$\omega_n = \sqrt{\frac{EI}{\rho S + \rho I \cdot (n\pi/L)^2}} \left(\frac{n\pi}{L}\right)^2. \qquad [6.107]$$

The mode shapes are given by:

$$f_n(x) = \sin\left(\frac{n\pi}{L} x\right). \qquad [6.108]$$

The comparison with solutions obtained without the effect of rotational inertia reveals that normal pulsations are modified but the mode shape stays the same. This second property is not general; it appears for the conditions of supported ends, but mode shapes would be modified in the case of clamped or of a free end. Nonetheless, the tendency concerning angular frequencies is general, since rotational inertia makes normal pulsation decrease. If we introduce the relationship ε_n between normal angular frequencies of the n mode calculated, taking rotational inertia into account and omitting it, we obtain:

$$\varepsilon_n = \sqrt{\frac{1}{1 + \frac{I}{S}\left(\frac{n\pi}{L}\right)^2}}. \qquad [6.109]$$

This factor ε_n is characteristic of the influence of rotational inertia: when it tends towards 1, the effect is negligible; the weaker it is in front of 1, the more influence rotational inertia has. Thus, we may state that the effect of rotational inertia is increasing with the order of the mode and that the characteristics of the beam are also important. The non-dimensional value (I/SL^2) is characteristic of the influence of rotational inertia; it reflects the geometry of the beam, but at the same time is independent of the material. To reinforce these ideas, let us consider a one meter-long beam with a circular cross-section of two centimeters in diameter: we then have: $(I/SL^2) = 4.10^{-6}$ and the error over the normal angular frequencies does not exceed 10% for the modes of the order n smaller than 25.

6.7.2. *Influence of transverse shearing*

The equations representative of the bending of beams with transverse shearing and rotational inertia were provided and interpreted in Chapter 3, equations [3.53] and [3.54]. We recall them here using a simplified notation in order to be concise: $W(x,t)$ is the vector of the beam (previously noted $W_2^0(x,t)$) and $\beta(x,t)$ is the rotation of the cross-section (previously noted $W_1^2(x,t)$).

$$\rho I\frac{\partial^2\beta}{\partial t^2}-\frac{S}{4S_{1212}}\left(\beta+\frac{\partial W}{\partial x}\right)+I\frac{\partial}{\partial x}\left(\frac{1}{S_{1111}}\frac{\partial\beta}{\partial x}\right)=0 \qquad [6.110]$$

$$-\rho S\frac{\partial^2 W}{\partial t^2}+\frac{\partial}{\partial x}\left(\frac{S}{4S_{1212}}\left(\beta+\frac{\partial W}{\partial x}\right)\right)=0\,. \qquad [6.111]$$

Let us note that the effect of rotational inertia is introduced by the first term of the left-hand part of equation [6.110] and that it suffices to remove it to take into account nothing but the shearing effect.

We will consider harmonic movements of the type:

$$W(x,t)=f(x)\,e^{j\omega t}, \qquad [6.112]$$

$$\beta(x,t)=h(x)\,e^{j\omega t}. \qquad [6.113]$$

Inserting these expressions into equations [6.110] and [6.111], we obtain a system with a differential equation [6.114] where rotational inertia is neglected:

$$\begin{pmatrix} -SG\alpha+EI\dfrac{d^2}{dx^2} & -SG\alpha\dfrac{d}{dx} \\ SG\alpha\dfrac{d}{dx} & \rho S\omega^2+SG\dfrac{d^2}{dx^2}\end{pmatrix}\begin{pmatrix}h(x)\\ f(x)\end{pmatrix}=\begin{pmatrix}0\\0\end{pmatrix}. \qquad [6.114]$$

In order to be even more concise, we note:

$$E=1/S_{1111} \text{ and } \alpha G=1/4S_{1212}\,.$$

For an isotropic material, these quantities correspond to the Coulomb and Young moduli. For an orthotropic material E is the longitudinal module of the beam, G is the shearing modulus with respect to the axes (1,2) and α is the shearing correction (see Chapter 3, section 3.5.2).

The resolution of this system is classical; it suffices to seek the solutions in the form:

$$\begin{pmatrix} h(x) \\ f(x) \end{pmatrix} = \begin{pmatrix} M \\ N \end{pmatrix} e^{kx}. \qquad [6.115]$$

Replacing in the system [6.114]:

$$\begin{pmatrix} -SG\alpha + EIk^2 & -SG\alpha G \\ SG\alpha G\alpha k & \rho S\omega^2 + SGk^2 \end{pmatrix} \begin{pmatrix} M \\ N \end{pmatrix} = \begin{pmatrix} 0 \\ 0 \end{pmatrix}. \qquad [6.116]$$

To obtain non-trivial solutions, the determinant of the system [6.116] must be nil, that is:

$$Ak^4 + Bk^2 + C = 0 \qquad [6.117a]$$

with:

$$A = ISEG\alpha S, B = \rho S\omega^2 IE, C = -SG\alpha G\alpha\rho^2. \qquad [6.117b]$$

The solutions of equation [6.117a] are obtained easily since they are those of a polynomial of the second degree in k^2. There are four $\pm k_1$ and $\pm jk_2$ solutions with:

$$k_1 = \sqrt{\frac{-B + \sqrt{\Delta}}{2A}}, \qquad [6.118]$$

$$k_2 = \sqrt{\frac{+B + \sqrt{\Delta}}{2A}} \qquad [6.119]$$

and:

$$\Delta = B^2 - 4AC.$$

To characterize the solutions completely, it is also necessary to determine the unknowns M and N of the linear system [6.116]. This is done by calculating the terms of the matrix for the values of k canceling the determinant. Obviously there is no unique pair (M, N), and we choose to fix N equal to one and to calculate M consequently. With the first line from the system it follows:

$$\text{if } \ k = \pm k_1 \ , \ M = \pm M_1 = SG\alpha \frac{k_1}{\rho I\omega^2 - SG\alpha + IEk_1^2} \ ; \qquad [6.120]$$

$$\text{if } \ k = \pm k_2 \ , \ M = \pm M_2 = jSG\,\alpha \frac{k_2}{\rho I\omega^2 - SG\alpha + IEk_2^2} \ . \qquad [6.121]$$

In short, there are four elementary solutions; the general solution is the linear combination.

$$\begin{pmatrix} h(x) \\ f(x) \end{pmatrix} = C\begin{pmatrix} M_1 \\ 1 \end{pmatrix} e^{k_1 x} + D\begin{pmatrix} -M_1 \\ 1 \end{pmatrix} e^{-k_1 x} + E\begin{pmatrix} M_2 \\ 1 \end{pmatrix} e^{jk_2 x} + F\begin{pmatrix} -M_2 \\ 1 \end{pmatrix} e^{-jk_2 x} . \qquad [6.122]$$

Vibratory movement results from the superposition of two vanishing waves with a wave number k_1 and two traveling waves with a wave number k_2.

Let us consider the celerity of the traveling waves:

$$c_F = \frac{\omega}{k_2} \qquad [6.123]$$

with k_2 given by [6.119], that is:

$$k_2 = \sqrt{\frac{\rho SEI\,\omega^2 + \sqrt{(\rho SE\,\omega^2)^2 + 4\rho^3 EI\,\omega^2 G^2 \alpha^2}}{2EISG\alpha}} \ . \qquad [6.124]$$

This celerity depends on the angular frequency; therefore, the medium is dispersive. We can calculate it in the extreme cases of low and high frequencies.

At a low frequency we have the approximation:

$$k_2 = \sqrt[4]{\frac{\rho S}{EI}}\,\sqrt{\omega} \ . \qquad [6.125]$$

The use of [6.125] in [6.123] gives the celerity found with the theory of bending without secondary effects (equation [6.26]). We may thus conclude that transverse shearing is negligible at low frequencies. The boundary frequency of validity of the theory without shearing is rather difficult to obtain; it would result from the comparison of the values of k_2 provided by [6.124] and [6.125] for increasing frequencies; as long as the two wave numbers are close, shearing does not have an influence.

In the extreme case of high frequencies we have the approximation:

$$k_2 = \sqrt{\frac{G\alpha}{\rho}}\omega \qquad [6.126]$$

and thus:

$$c_F = \sqrt{\frac{G\alpha}{\rho}}. \qquad [6.127]$$

This propagation velocity corresponds to that of the waves of shearing, already highlighted in the problem of torsion in Chapter 5. The value is in fact a little different, taking into account the correction of the shearing modulus.

We can also observe the asymptotic value of M_2 thanks to [6.121] and to the low and high frequency values of k_2. For low frequencies it follows:

$$M_2 = -jk_2.$$

With [6.118] we deduce that the propagation part of the solution is given by:

$$\begin{pmatrix} h(x) \\ f(x) \end{pmatrix} = E\begin{pmatrix} -jk_2 \\ 1 \end{pmatrix} e^{jk_2x} + F\begin{pmatrix} jk_2 \\ 1 \end{pmatrix} e^{-jk_2x}. \qquad [6.128]$$

We note that:

$$h(x) = -\frac{df}{dx}(x),$$

that is:

$$\beta(x,t) = -\frac{\partial W}{\partial x}(x,t).$$

This corresponds to the hypothesis of pure bending where the rotation of the cross-sections is equal to the slope and confirms the fact that transverse shearing is negligible at a low frequency.

At high frequencies, M_2 tends towards zero $\left(M_2 = O(1/\omega)\right)$. This means that transverse displacement occurs without rotation of cross-sections and, therefore, that the longitudinal constraint is nil. The movement is a pure shearing wave, as the propagation velocity would lead us to believe [6.127].

Figure 6.10 illustrates the type of movements of the beam at a low frequency.

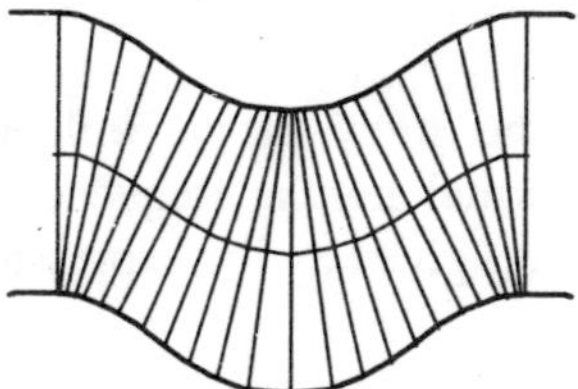

Figure 6.10. *Low frequency movement: standard bending, transverse movement is accompanied by a rotation of the cross-sections*

Figure 6.11 illustrates the type of movements of the beam at a high frequency:

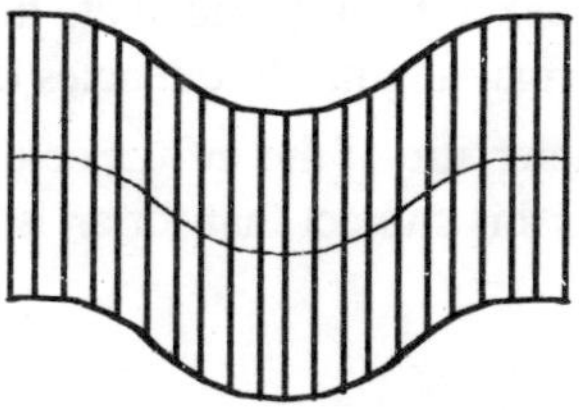

Figure 6.11. *High frequency movement: pure shearing wave*

The calculation of the modal system is performed in a traditional fashion by requiring the solution [6.122] to respect the boundary conditions. We will consider the simplest case, that of the supported-supported beam without, however, developing calculations. The mode shape of the index n is given by the vector:

$$\begin{pmatrix} h_n(x) \\ f_n(x) \end{pmatrix} = \begin{pmatrix} M_n \cos\left(\dfrac{n\pi}{L} x\right) \\ \sin\left(\dfrac{n\pi}{L} x\right) \end{pmatrix} \qquad [6.129]$$

with: $$M_n = -\frac{n\pi}{L}\frac{1}{1+\frac{IE}{SG\alpha}\left(\frac{n\pi}{L}\right)^2}\ . \qquad [6.130]$$

The corresponding normal angular frequency is given by the expression:

$$\omega_n = \sqrt{\frac{EI}{\rho S}\left(\frac{n\pi}{L}\right)^2}\sqrt{\frac{1}{1+\frac{IE}{SG\alpha}\left(\frac{n\pi}{L}\right)^2}}\ . \qquad [6.131]$$

Let us introduce, as for the study of the influence of rotational inertia, the relationship ε_n between the normal angular frequencies calculated with and without secondary effects (here shearing) given by the expressions [6.131] and [6.57]:

$$\varepsilon_n = \sqrt{\frac{1}{1+\frac{EI}{SG\alpha}\left(\frac{n\pi}{L}\right)^2}}\ . \qquad [6.132]$$

We observe that, as for rotational inertia:

– the factor is always weaker than 1 and that, therefore, taking shearing into account always makes the normal angular frequencies decrease;

– the value $1/SL^2$ is characteristic of the influence of shearing but is amplified by the ratio $E/G\alpha$. Let us add on this subject that for an isotropic material:

$$\frac{E}{G\alpha} = 2(1+\nu)/\alpha$$

where ν is the Poisson's ratio of material and α is the shearing correction which depends on the form of the cross-section (see Chapter 3).

Thus, for a steel beam with a circular cross-section, this amplifying factor is equal to 2.88. For an orthotropic material the amplification is much stronger because these materials are characterized by a weak shearing modulus. For a composite material composed of bidirectional fiberglass immersed in resin, the $E/G\alpha$ ratio normally reaches 7. The shearing effect thus acquires a considerable importance for these materials. With [6.132] and [6.109] we also note that if the shearing and rotational inertia effects are of the same order of magnitude, shearing dominates.

The shearing effect is greater the higher the rank n of the mode.

The free vibratory response is obtained by cumulating modal vibratory movements:

$$\begin{pmatrix} \beta(x,t) \\ W(x,t) \end{pmatrix} = \sum_{n=1}^{\infty} \left(A_n \cos(\omega_n t) + B_n \sin(\omega_n t)\right) \begin{pmatrix} M_n \cos\left(\frac{n\pi}{L} x\right) \\ \sin\left(\frac{n\pi}{L} x\right) \end{pmatrix} \qquad [6.133]$$

where the constants A_n and B_n are set by the initial conditions. We will not proceed further with these calculations.

6.7.3. *Taking into account shearing and rotational inertia*

6.7.3.1. *Propagation of waves*

In the two preceding sections, we treated the two secondary effects separately. Here we will consider the case where the two effects are considered simultaneously. We will limit ourselves to the analysis using only the W vector of the beam, that is, on the basis of equation [3.57b] from Chapter 3:

$$EI\frac{\partial^4 W}{\partial x^4} + \rho S\frac{\partial^2 W}{\partial t^2} + \frac{\rho^2 I}{G}\frac{\partial^4 W}{\partial t^4} - \rho I\left(1+\frac{E}{G}\right)\frac{\partial^4 W}{\partial x^2 \partial t^2} = 0. \qquad [6.134]$$

The case without shearing is obtained in this form where only the W vector appears while removing the terms which depend on G in [6.134] (in fact, it is the limit when G tends towards infinity, which demonstrates that this hypothesis consists of rigidifying the shearing modulus in an artificial way, thus blocking this movement). The case without rotational inertia amounts to ignoring the fourth term of the first member of [6.134]; the standard equation consists of ignoring all these terms.

For a harmonic movement, we pose:

$$W(x,t) = f(x)\,e^{j\omega t}.$$

Equation [6.134] becomes:

$$EI\frac{d^4 f}{dx^4}(x) + \left(\frac{\rho^2 I}{G\alpha}\omega^4 - \rho S\omega^2\right)f(x) + \rho I\omega^2\left(1+\frac{E}{G\alpha}\right)\frac{d^2 f}{dx^2} = 0. \qquad [6.135]$$

The solution of the differential equation [6.135] is standard; we seek the solution in the form:

$$f(x) = e^{kx} \qquad [6.136]$$

where k is determined by the characteristic equation:

$$EIk^4 + \rho I\omega^2\left(1+\frac{E}{G\alpha}\right)k^2 + \left(\frac{\rho^2 I}{G\alpha}\omega^2 - \rho S\right)\omega^2 = 0. \qquad [6.137]$$

For each frequency ω we associate the solutions:

$$k_1^2 = \frac{-\rho I\omega^2\left(1+\frac{E}{G\alpha}\right)+\sqrt{\Delta}}{2EI}, \qquad [6.138]$$

$$k_2^2 = \frac{-\rho I\omega^2\left(1+\frac{E}{G\alpha}\right)-\sqrt{\Delta}}{2EI} \qquad [6.139]$$

where Δ is the positive or zero quantity defined by:

$$\Delta = \rho^2 I^2 \omega^4\left(1-\frac{E}{G\alpha}\right)^2 + 4\,EI\,\rho\,S\,\omega^2. \qquad [6.140]$$

Let us introduce the angular frequency Ω_{lim} which satisfies $k_1 = 0$ and delimits two different vibratory behaviors:

$$\Omega_{lim} = \sqrt{\frac{SG\alpha}{\rho I}}. \qquad [6.141]$$

For $\omega < \Omega_{lim}$, there are two pure imaginary wave numbers $\pm jk_p$ resulting from [6.139] and two real wave numbers $\pm k_e$ resulting from [6.138]. The solution of the problem is:

$$f(x) = C\sin(k_p x) + D\cos(k_p x) + E\,\text{sh}(k_e x) + F\,\text{ch}(k_e x). \qquad [6.142]$$

The expressions of k_p characteristic of the propagation part of the vibrations and k_e characteristic of the vanishing part of the vibrations are provided hereunder:

$$k_p = \sqrt{\frac{\rho I\omega^2\left(1+\frac{E}{G\alpha}\right)+\sqrt{\Delta}}{2EI}}, \qquad [6.143]$$

$$k_e = \sqrt{\frac{-\rho I\omega^2\left(1+\frac{E}{G\alpha}\right)-\sqrt{\Delta}}{2EI}}. \qquad [6.144]$$

For $\omega > \Omega_{lim}$, the four solutions of [6.138] and [6.139] become imaginary and vibratory movement is composed of four traveling waves:

$$f(x) = C\sin(k_p x) + D\cos(k_p x) + E\sin(k'_e x) + F\cos(k'_e x). \qquad [6.145]$$

Wave numbers k_p and k'_e are given respectively by [6.143] and [6.146]:

$$k'_e = \sqrt{\frac{\rho I\omega^2\left(1+\frac{E}{G\alpha}\right)-\sqrt{\Delta}}{2EI}}. \qquad [6.146]$$

We can associate a physical significance to the angular frequency Ω_{lim}. Let us take the case of a rectangular section to settle the ideas. Equation [6.141] becomes:

$$\Omega_{lim} = \sqrt{\frac{G\alpha}{\rho}}\frac{2\sqrt{3}}{h} \qquad [6.147]$$

where $\sqrt{\frac{G\alpha}{\rho}} = c_T$ is the celerity of transverse waves and h is the thickness of the beam.

By introducing the wavelength λ_T of transverse waves (defined by [6.148]) into equation [6.147] we obtain the relation [6.149]:

$$\lambda_T = 2\pi\sqrt{\frac{G\alpha}{\rho}}, \qquad [6.148]$$

$$h = \lambda_T\frac{\sqrt{3}}{\pi}. \qquad [6.149]$$

The angular frequency Ω_{lim} is thus characteristic of a shearing wavelength close to the thickness of the beam. We can thus conclude that at higher frequencies, the beam hypothesis is no longer well adapted to describe the phenomena.

The calculation of propagation velocities clarifies the physical phenomena; it is necessary to distinguish the behavior at frequencies higher and lower than Ω_{lim}.

For $\omega < \Omega_{lim}$, the celerity of the traveling waves is calculated easily thanks to [6.143]:

$$c_p = \sqrt{\frac{2EI}{\rho I\left(1+\frac{E}{G\alpha}\right)+\sqrt{\frac{\Delta}{\omega^4}}}}. \quad [6.150]$$

When ω tends towards Ω_{lim}, Δ tends towards zero and celerity tends towards a constant value: c_{lim}.

$$c_{lim} = \sqrt{\frac{E}{\rho}}\sqrt{\frac{2}{(1+E/G\alpha)}} = \sqrt{\frac{G\alpha}{\rho}}\sqrt{\frac{2}{(1+G\alpha/E)}}. \quad [6.151]$$

This celerity lies between the two values $c_L = \sqrt{E/\rho}$ and $c_T = \sqrt{G\alpha/\rho}$ corresponding to the velocities of longitudinal and transversal waves which characterized the bending waves speeds with either rotational inertia or shearing alone.

For $\omega > \Omega_{lim}$, there are two speeds of propagation associated with the two types of traveling waves appearing in the solution:

$$c_p = \sqrt{\frac{2EI}{\rho I\left(1+\frac{E}{G\alpha}\right)+\sqrt{\frac{\Delta}{\omega^4}}}} \quad [6.152]$$

and:

$$c_e = \sqrt{\frac{2EI}{\rho I\left(1+\frac{E}{G\alpha}\right)-\sqrt{\frac{\Delta}{\omega^4}}}}. \quad [6.153]$$

The analysis at the extremes of these expressions is interesting: if ω tends towards Ω_{lim} by a higher value, we note that $c_p = c_e = c_{lim}$; if ω tends towards infinity, we have c_p tending towards c_L and c_e tending towards c_T.

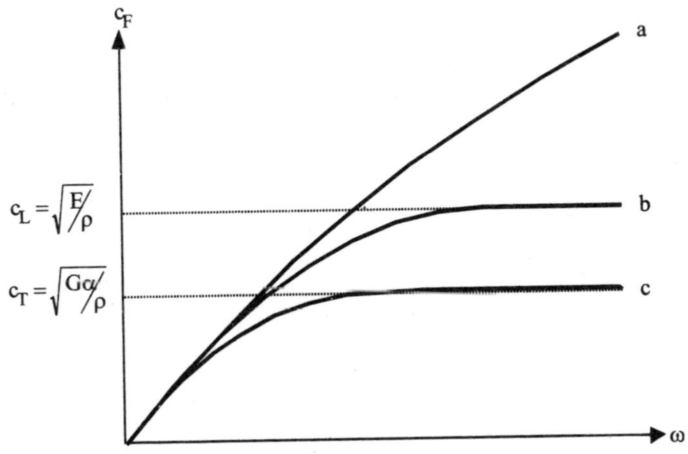

Figure 6.12. *Velocity of the bending waves: a) pure bending, b) with rotational inertia, c) with shearing*

Thus, above "omegalim", two different types of traveling waves take part in the vibratory movement. In high frequency approximation, the first are propagated with the velocity of longitudinal waves, and the second with the velocity of transverse waves. In fact, they coincide with the high frequency approximations noted in sections 6.7.1 and 6.7.2 for waves of bending with rotational inertia and waves of bending with shearing respectively. Thus, at high frequencies, the two secondary effects are uncoupled.

Figure 6.12 shows the celerity of the waves of bending in the three cases of the bending equation: fundamental case, with rotational inertia, with shearing; Figure 6.13 has the celerity of the waves of bending when the two secondary effects are considered. We note that at low frequencies all the theories coincide but that they deviate from one another at high frequencies.

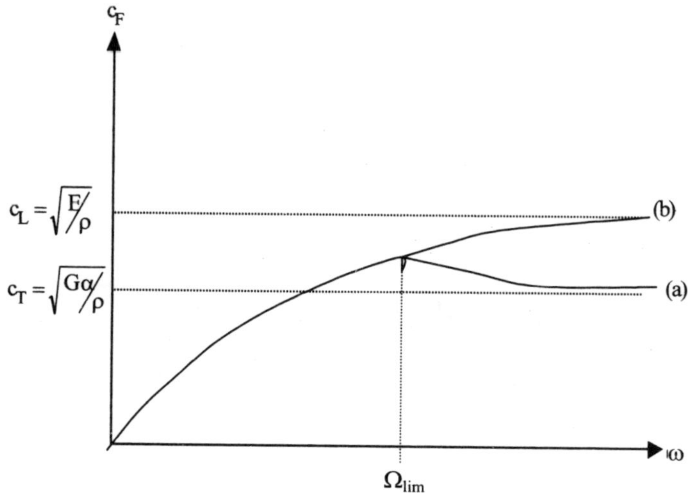

Figure 6.13. *Velocity of bending waves with shearing and rotational inertia (a), with rotational inertia only (b)*

6.7.3.2 *Vibration modes*

The vibration modes of finite beams is traditionally outlined by seeking solutions of the type [6.142] and [6.145] verifying the boundary conditions. This calculation is rather long and we will not give it here. We directly give the result obtained in the simple case of the supported-supported beam.

The normal strains remain identical to the simple case of pure flexing, that is:

$$f_n(x) = \sin\left(\frac{n\pi}{L}x\right).$$

For a given mode shape $f_n(x)$ we associate two normal angular frequencies ω_n and ω'_n. The ω_n angular frequencies are much lower than them ω'_n.

$$\omega_n = \sqrt{\frac{\left(\rho I\left(1+\frac{E}{G\alpha}\right)\left(\frac{n\pi}{L}\right)^2 + \rho S - \sqrt{\Delta_n}\right)G\alpha}{2I\rho^2}},$$

$$\omega'_n = \sqrt{\frac{\left(\rho I\left(1+\frac{E}{G\alpha}\right)\left(\frac{n\pi}{L}\right)^2 + \rho S + \sqrt{\Delta_n}\right)G\alpha}{2I\rho^2}}$$

with:

$$\Delta_n = \rho^2 I\left(1-\frac{E}{G\alpha}\right)^2\left(\frac{n\pi}{L}\right)^4 + (\rho S)^2 + 2\rho^2 SI\left(1-\frac{E}{G\alpha}\right)^2\left(\frac{n\pi}{L}\right)^2.$$

This unfolding of the normal angular frequencies comes from the existence of a hidden variable $\beta(x,t)$ which we excluded to obtain the simple form of the equation of motion [6.134] which only depends on $W(x,t)$. The apparent simplification of this procedure, in fact, pays off in the calculation of the response, since the modes of angular frequency ω_n and ω'_n are not orthogonal (they have the same mode shape) which presents a problem for the introduction of boundary conditions. This problem of non-orthogonality of modes would not appear had we kept equations [6.110] and [6.111] to solve the problem in a similar fashion to the solution in section 6.7.2.

6.8. Conclusion

This chapter described the vibrations of bending of beams according to the various hypotheses used to model this problem. We showed, in particular, that the influence of the secondary effects of shearing and rotational inertia could be important for high frequencies and for the thicker beams. The anisotropy of material considerably amplifies the effect of shearing which cannot be neglected even for the first modes.

The analysis of vibratory phenomena was performed for the beam in pure bending because the relative simplicity of the equation of motion allows easier exploitation. Two types of solutions appear during the resolution of the equation of motion: waves traveling as for longitudinal vibrations so for those of torsion, but also vanishing waves which acquire their importance at the singularities of the beam (boundary conditions, excitation, variation of inertia). Traveling waves are characterized by a propagation velocity which varies with frequency; we then say that the medium is dispersive. This property marks an important difference with the vibrating mediums described by the wave equation where the velocity of waves is constant. The dispersive nature of the solution modifies the space form of a disturbance during its propagation and prohibits the use of the images method presented in Chapter 4.

The propagation of a package of waves with very close frequencies reveals the group speed characterizing the overall displacement of the disturbance.

The vibrations modes of finite beams were presented in various cases and a summary table was drawn up. Mode shapes were characterized by a different behavior according to whether the point of observation is localized in the vicinity of the boundaries or not. With relation to this, we introduced the concepts of internal solution and edge effect: the internal solution is generated by traveling waves, vanishing waves effectively appear only in the edge effect. Finally the relation between stress and transverse displacement was studied; we may derive the following tendencies: when the edge effect is present (near the singularities) an antinode (or a node) of modal displacement corresponds to a node (or an antinode) of stress; when the internal solution dominates a node (or an antinode) of displacement corresponds to a node (or an antinode) of stress.

Chapter 7

Bending Vibration of Plates

7.1. Introduction

Vibrations of thin plates constitute a problem that is difficult to solve analytically; in fact, only a small number of cases make it possible to find analytical expressions of vibration modes. A first limitation is due to the shape of the plate, which must be rectangular or circular; in this chapter we mainly study rectangular plates and give a short example for circular plates. A second limitation comes from the type of boundary conditions of the plate, which must be particular. Let us underline that the *a priori* simple case of the rectangular plate with all of its four edges clamped does not form part of the cases where an analytical solution of the vibration modes can be found.

A difference with the vibrations of beams appears when plates are studied; it is the complexity of calculations, if only at the level of the presentation of problems and more particularly of the writing of boundary conditions. The first part of this chapter will consist of a systematic recording of the various boundary conditions that can be applied to rectangular plates in order to familiarize the reader with the subject.

We will then determine the modal system in the cases that are treated analytically and which we will interpret physically. Finally, we develop the edge effect method which provides the approximated vibration modes for high rank modes in analytical form.

At the end of the chapter we propose an example of vibrations of a circular plate.

There are a large number of publications on this area; let us point out that there are several works which have tabulated the first normal angular frequency and mode shapes of plates with various boundary conditions and which constitute a complement to this discussion. We give a non-exhaustive list of these works in the bibliography.

7.2. Posing the problem: writing down boundary conditions

Our discourse is built on the equation of pure bending which neglects rotational inertia, that is, on the simplest possible approach. This approach is linked to our desire to limit as mush as possible the cumbersomeness of calculations, so as to emphasize the physical aspects.

Let us recall the equation of bending of plates [7.1] stated in Chapter 4 (equation [4.56]):

$$\rho h \frac{\partial^2 w}{\partial t^2} + D\left(\frac{\partial^4 w}{\partial x_1^4} + 2\frac{\partial^4 w}{\partial x_1^2 \partial x_2^2} + \frac{\partial^4 w}{\partial x_2^4}\right) = 0. \qquad [7.1]$$

ρ is the density of material, h is the thickness of the plate, D is the bending rigidity $D = Eh^3 / 12(1-\nu^2)$ where E is the Young modulus and ν is the Poisson's ratio.

Let us consider a rectangular plate with sides a and b as shown in Figure 7.1. To write the boundary conditions for each edge, it is necessary to define the external normal $\vec{n}$ and the tangent $\vec{s}$:

$$\vec{n} = \begin{pmatrix} n_1 \\ n_2 \end{pmatrix}, \quad \vec{s} = \begin{pmatrix} -n_2 \\ n_1 \end{pmatrix}.$$

n_1 and n_2 are the direction cosines of the external normal vector. On the edge $x = 0$, for example, we have:

$$\vec{n} = \begin{pmatrix} -1 \\ 0 \end{pmatrix}, \quad \vec{s} = \begin{pmatrix} 0 \\ -1 \end{pmatrix}.$$

All of the normal vectors and tangents are clarified in Figure 7.1.

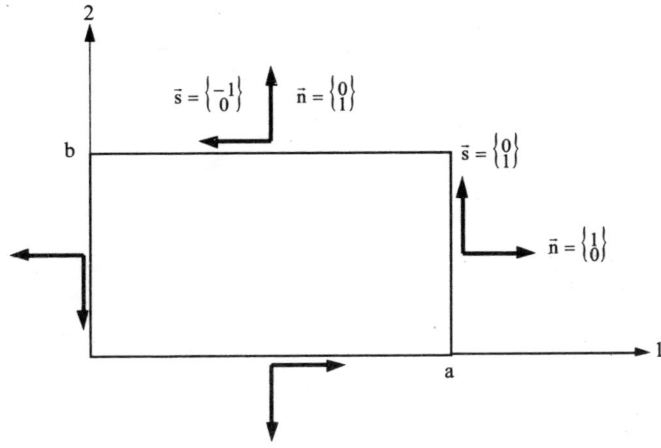

Figure 7.1. *Rectangular plate, normal and tangent vectors at the edges*

To consolidate our understanding we will explicitly write the four possible boundary conditions: clamped, supported, guided and free.

The boundary conditions associated with the equation of motion [7.1] are provided by the two alternatives [7.2] and [7.3].

$$\begin{cases} \text{either} : W(x_1, x_2, t) = 0, \\ \\ \text{or} \quad : T(x_1, x_2, t) = 0 \end{cases} \qquad [7.2]$$

and:

$$\begin{cases} \text{either} : \partial W/\partial n(x_1, x_2, t) = 0, \\ \\ \text{or} \quad : M(x_1, x_2, t) = 0. \end{cases} \qquad [7.3]$$

In these equations, T is the shearing force and M is the normal bending moment at the edge of the plate. We recall their general expressions provided in Chapter 4, equations [4.58] and [4.57], concerning the set-up of the equation of thin plates:

$$T = \frac{\partial}{\partial n}\{M\} + D\frac{\partial}{\partial s}\left\{2\left(\frac{\partial^2 W}{\partial x_1^2} - \frac{\partial^2 W}{\partial x_2^2}\right)(1+\nu)\, n_1 n_2 \right.$$
$$\left. + 2(1-\nu)\frac{\partial^2 W}{\partial x_1 \partial x_2}(n_1^2 - n_2^2)\right\}, \qquad [7.4]$$

$$M = D\left\{\left(\frac{\partial^2 W}{\partial x_1^2} + \nu\frac{\partial^2 W}{\partial x_2^2}\right) n_1^2 + 2(1-\nu)\frac{\partial^2 W}{\partial x_1 \partial x_2} n_1 n_2 \right.$$
$$\left. + \left(\nu\frac{\partial^2 W}{\partial x_1^2} + \frac{\partial^2 W}{\partial x_2^2}\right) n_2^2\right\}. \qquad [7.5]$$

The normal and tangent derivatives are by definition equal to:

$$\frac{\partial}{\partial n} = \vec{n}\cdot\vec{\nabla}\,, \quad \frac{\partial}{\partial s} = \vec{s}\cdot\vec{\nabla}\,,$$

that is:

$$\frac{\partial}{\partial n} = n_1\frac{\partial}{\partial x_1} + n_2\frac{\partial}{\partial x_2}\,, \quad \frac{\partial}{\partial s} = n_1\frac{\partial}{\partial x_2} - n_2\frac{\partial}{\partial x_1}. \qquad [7.6]$$

Using the expressions of these derivatives and those of the direction cosines of the normal vectors at each edge of the plate we may write down the corresponding boundary conditions a) at the clamped edge, b) at the supported edge, c) at the guided edge and d) at the free edge.

Clamped edge conditions, written for the edge $x_1 = 0$:

$$\begin{cases} W(0, x_2, t) = 0\,, \\ \dfrac{\partial W}{\partial x_1}(0, x_2, t) = 0\,. \end{cases} \qquad [7.7]$$

Condition of support, written for the edge $x_2 = 0$:

$$\begin{cases} W(x_1, 0, t) = 0, \\ D\left(\dfrac{\partial^2 W(x_1, 0, t)}{\partial x_2^2} + \nu \dfrac{\partial^2 W(x_1, 0, t)}{\partial x_1^2}\right) = 0. \end{cases} \qquad [7.8]$$

This boundary condition is simplified a little further because the first condition $W(x_1, 0, t) = 0$ implies $\dfrac{\partial^2 W}{\partial x_1^2}(x_1, 0, t) = 0$; we thus have for the supported edge:

$$\begin{cases} W(x_1, 0, t) = 0, \\ D\left(\dfrac{\partial^2 W(x_1, 0, t)}{\partial x_2^2}\right) = 0. \end{cases} \qquad [7.9]$$

Condition of guidance, written for the edge $x_1 = a$:

$$\begin{cases} \dfrac{\partial W}{\partial x_1}(a, x_2, t) = 0, \\ D\left(\dfrac{\partial^3 W(a, x_2, t)}{\partial x_1^3} + (2-\nu)\dfrac{\partial^3 W(a, x_2, t)}{\partial x_1 \partial x_2^2}\right) = 0. \end{cases} \qquad [7.10]$$

Free condition, written for the edge $x_2 = b$:

$$\begin{cases} D\left(\dfrac{\partial^2 W(x_1, b, t)}{\partial x_2^2} + \nu \dfrac{\partial^2 W(x_1, b, t)}{\partial x_1^2}\right) = 0, \\ D\left(\dfrac{\partial^3 W(x_1, b, t)}{\partial x_2^3} + (2-\nu)\dfrac{\partial^3 W(x_1, b, t)}{\partial x_2 \partial x_1^2}\right) = 0. \end{cases} \qquad [7.11]$$

7.3. Solution of the equation of motion by separation of variables

7.3.1. *Separation of the space and time variables*

Let us pose:

$$W(x_1, x_2, t) = f(x_1, x_2)\ g(t) . \qquad [7.12]$$

Let us inject the form [7.12] in the equation of motion [7.1]. It follows:

$$\rho h f(x_1, x_2) \frac{d^2 g}{dt^2}(t) + D\left(\frac{\partial^4 f}{\partial x_1^4}(x_1, x_2) + \frac{\partial^4 f}{\partial x_2^4}(x_1, x_2) + 2\frac{\partial^4 f}{\partial x_1^2 \partial x_2^2}(x_1, x_2) \right) g(t) = 0 . \qquad [7.13]$$

Let us perform the separation of the variables of space and time in [7.13]:

$$\frac{d^2 g}{dt^2}(t) \Big/ g(t) = -D\left(\frac{\partial^4 f}{\partial x_1^4} + \frac{\partial^4 f}{\partial x_2^4} + 2\frac{\partial^4 f}{\partial x_1^2 \partial x_2^2} \right)(x_1, x_2) \Big/ \rho h\ f(x_1, x_2) = \text{constant} . \qquad [7.14]$$

The constant can be negative, nil or positive, thus leading to different solutions. However, as we saw for beams, the positive constant leads to the trivial solution when boundary conditions are observed; thus, we exploit only the two cases of zero and negative constants.

a) Zero constant. In this case, equation [7.14] is reduced to:

$$\frac{d^2 g}{dt^2}(t) = 0 \qquad [7.15]$$

and:

$$D\left(\frac{\partial^4 f}{\partial x_1^4} + \frac{\partial^4 f}{\partial x_2^4} + 2\frac{\partial^4 f}{\partial x_1^2 \partial x_2^2} \right)(x_1, x_2) = 0 . \qquad [7.16]$$

The solution of these equations does not represent the vibratory movements themselves but uniform displacements which characterize the rigid movements that can appear for certain boundary conditions (free plate conditions, in particular). The general form of rigid movement of the plate is composed of the translation and of two rotations with respect to axes 1 and 2, i.e.:

$$W(x_1, x_2, t) = (At + B)(C + Dx_1 + Ex_2). \qquad [7.17]$$

b) Negative constant

By posing the constant equal to $-\omega^2$ we obtain:

$$\frac{d^2g}{dt^2}(t) + \omega^2 g(t) = 0 \qquad [7.18]$$

and:

$$-\omega^2 \rho h\, f(x_1, x_2) + D\left(\frac{\partial^4 f}{\partial x_1^4} + \frac{\partial^4 f}{\partial x_2^4} + 2\frac{\partial^4 f}{\partial x_1^2 \partial x_2^2}\right)(x_1, x_2) = 0. \qquad [7.19]$$

The temporal equation [7.18] is solved very easily. We obtain:

$$g(t) = A\cos(\omega t) + B\sin(\omega t) \qquad [7.20]$$

The equation of space is, however, difficult to solve since it remains a partial derivative equation. The search for solutions in the form of separate variables is fruitless in many cases of boundary conditions. In spite of this restrictive aspect, it is interesting because the only cases where the modal system has an analytical form are those where the separation of variables is applicable.

7.3.2. *Solution of the equation of motion by separation of space variables*

Let us suppose that the solution of space can be broken up as follows:

$$f(x_1, x_2) = \varphi_1(x_1)\,\varphi_2(x_2). \qquad [7.21]$$

Using relation [7.21] in equation [7.19] it follows:

$$-\omega^2 \rho h \, \varphi_1 \, \varphi_2 + D\left(\frac{d^4\varphi_1}{dx_1^4} \varphi_2 + \frac{d^4\varphi_2}{dx_2^4} \varphi_1 + 2 \frac{d^2\varphi_1}{dx_1^2} \frac{d^2\varphi_2}{dx_2^2} \right) = 0 . \qquad [7.22]$$

Let us seek the solutions of equation [7.22] on a basis of exponentials:

$$\varphi_1(x_1) = A_1 \, e^{k_1 x_1} \;, \; \varphi_2(x_2) = A_2 \, e^{k_2 x_2} . \qquad [7.23]$$

Making a replacement in [7.22] it follows:

$$\left(-\omega^2 \rho h + D \, (k_1^2 + k_2^2)^2 \right) A_1 \, e^{k_1 x_1} A_2 \, e^{k_2 x_2} = 0 . \qquad [7.24]$$

That is, to have a non-trivial solution:

$$(k_1^2 + k_2^2) = \pm \, \omega \Big/ \sqrt{\frac{D}{\rho h}} . \qquad [7.25]$$

To each particular value γ_1^2 of k_1^2 there correspond two particular values γ_2^2 and δ_2^2 of k_2^2 verifying [7.25]:

$$(\gamma_1^2 + \gamma_2^2) = \omega \Big/ \sqrt{\frac{D}{\rho h}} \qquad [7.26]$$

and:

$$(\gamma_1^2 + \delta_2^2) = -\,\omega \Big/ \sqrt{\frac{D}{\rho h}} . \qquad [7.27]$$

The solutions for $\varphi_1(x_1)$ and $\varphi_2(x_2)$ are then respectively:

$$\varphi_1(x_1) = a_1 \, e^{\gamma_1 x_1} + b_1 \, e^{-\gamma_1 x_1} , \qquad [7.28]$$

$$\varphi_2(x_2) = a_2 \, e^{\gamma_2 x_2} + b_2 \, e^{-\gamma_2 x_2} + c_2 \, e^{\delta_2 x_2} + d_2 \, e^{-\delta_2 x_2} . \qquad [7.29]$$

We can of course find a symmetrical form by inverting the indices 1 and 2.

It should immediately be noted that the solutions [7.28] and [7.29] will not in general make it possible to satisfy the boundary conditions for the $x_1 = 0$ and $x_1 = a$ edges of the plate, since we have two constants of integration in [7.28] and four boundary conditions to impose. In general, the application of boundary conditions to the solution obtained in the form of separate variables will lead to the trivial solution $f(x_1, x_2) = 0$. There are, however, certain boundary conditions for which we obtain non-trivial solutions.

7.3.3. *Solution of the equation of motion (second method)*

A second method of solving equation [7.19] is possible; it is based on a rewriting of the following equation in factorized form:

$$\left(\left(\frac{\partial^2}{\partial x_1^2}+\frac{\partial^2}{\partial x_2^2}\right)+\omega\sqrt{\frac{\rho h}{D}}\right)\left(\left(\frac{\partial^2}{\partial x_1^2}+\frac{\partial^2}{\partial x_2^2}\right)-\omega\sqrt{\frac{\rho h}{D}}\right) f(x_1, x_2) = 0 . \qquad [7.30]$$

Let us consider the solutions $f^+(x_1, x_2)$ and $f^-(x_1, x_2)$ of the two equations built with the differential operators appearing in the product:

$$\left(\left(\frac{\partial^2}{\partial x_1^2}+\frac{\partial^2}{\partial x_2^2}\right)+\omega\sqrt{\frac{\rho h}{D}}\right) f^+(x_1, x_2) = 0 \qquad [7.31]$$

and:

$$\left(\left(\frac{\partial^2}{\partial x_1^2}+\frac{\partial^2}{\partial x_2^2}\right)-\omega\sqrt{\frac{\rho h}{D}}\right) f^-(x_1, x_2) = 0 . \qquad [7.32]$$

We can easily demonstrate that the sum of the two solutions $f^+(x_1, x_2)+f^-(x_1, x_2)$ is the solution of equation [7.19]. Indeed, to simplify this, let us note:

$$L^+ = \left(\left(\frac{\partial^2}{\partial x_1^2}+\frac{\partial^2}{\partial x_2^2}\right)+\omega\sqrt{\frac{\rho h}{D}}\right)$$

and:

$$L^- = \left(\left(\frac{\partial^2}{\partial x_1^2} + \frac{\partial^2}{\partial x_2^2}\right) - \omega\sqrt{\frac{\rho h}{D}}\right)$$

In its factorized form, equation [7.30] is written:

$$L^+ (L^- \{ f \}) = 0 .$$

Replacing $f(x_1, x_2)$ by the decomposition $f^+(x_1, x_2) + f^-(x_1, x_2)$ it follows:

$$L^+\left(L^-\left\{f^+ + f^-\right\}\right) = L^+\left(L^-\left\{f^+\right\}\right) + L^+\left(L^-\left\{f^-\right\}\right) = 0 . \qquad [7.33]$$

Taking into account [7.32] we have the relation $L^-\left\{f^-\right\} = 0$, from which we draw:

$$L^+\left(L^-\left\{f^+\right\}\right) = 0 .$$

Inverting the order of derivations we can also write:

$$L^+\left(L^-\left\{f^+\right\}\right) = L^-\left(L^+\left\{f^+\right\}\right).$$

Finally, taking into account [7.31] $L^+\left\{f^+\right\} = 0$, and we state by grouping all these results in [7.33]:

$$L^+\left(L^-\left\{f^+ + f^-\right\}\right) = 0 .$$

Thus, the sum of $f^+(x_1, x_2)$ and $f^-(x_1, x_2)$ is the solution of the equation of the vibrations of plates [7.30].

Let us seek the solution $f^+(x_1, x_2)$ in the traditional form of an exponential product:

$$f^+(x_1, x_2) = e^{k_1 x_1} e^{k_2 x_2}.$$

Introducing this form into equation [7.31] leads to the result:

$$f^+(x_1, x_2) = \left(a_1\, e^{\gamma_1 x_1} + b_1\, e^{-\gamma_1 x_1}\right) \left(a_2\, e^{\gamma_2 x_2} + b_2\, e^{-\gamma_2 x_2}\right)$$

with: $\gamma_2^2 = \omega \sqrt{\dfrac{\rho h}{D}} - \gamma_1^2$.

In the same manner we obtain for equation [7.32]:

$$f^-(x_1, x_2) = \left(a_1\, e^{\gamma_1 x_1} + b_1\, e^{-\gamma_1 x_1}\right) \left(a_2\, e^{\delta_2 x_2} + b_2\, e^{-\delta_2 x_2}\right)$$

with: $\delta_2^2 = -\omega \sqrt{\dfrac{\rho h}{D}} - \gamma_1^2$.

This method of resolution leads to the same result as the method used in section 7.2.2. Indeed, the superposition of the two solutions $f^+(x_1, x_2)$ and $f^-(x_1, x_2)$ clearly coincides with the forms [7.28] and [7.29]:

$$f(x_1, x_2) = f^+(x_1, x_2) + f^-(x_1, x_2).$$

The decomposition of the equation of plates vibration into a product equation is particularly interesting in the case of circular plates, which we consider briefly at the end of the chapter.

7.4. Vibration modes of plates supported at two opposite edges

7.4.1. *General case*

This case of boundary conditions makes it possible to find an analytical solution for the modal system.

We suppose that the two supported edges are the edges $x_1 = 0$ and $x_1 = a$; the solutions [7.28] and [7.29] are the convenient forms. If these are the edges $x_2 = 0$

and $x_2 = b$, it is necessary to take the symmetrical form of [7.28] and [7.29] by inverting the indices 1 and 2.

Let us require the boundary conditions in $x_1 = 0$ and $x_1 = a$ to be respected; it follows:

$$f(0, x_2) = 0 \Rightarrow \varphi_1(0) = 0,$$

$$f(a, x_2) = 0 \Rightarrow \varphi_1(a) = 0,$$

$$\frac{\partial^2 f}{\partial x_1^2}(0, x_2) = 0 \Rightarrow \frac{d^2\varphi_1}{dx_1^2}(0) = 0,$$

$$\frac{\partial^2 f}{\partial x_1^2}(a, x_2) = 0 \Rightarrow \frac{d^2\varphi_1}{dx_1^2}(a) = 0.$$

The application of the first two conditions to [7.28] is redundant and leads to:

$$a_1 = -b_1. \qquad [7.34]$$

The third and fourth conditions are also redundant and give:

$$a_1 e^{\gamma_1 a} + b_1 e^{-\gamma_1 a} = 0.$$

By combining the two preceding equations we obtain:

$$\mathrm{sh}(\gamma_1 a) = 0.$$

I.e. there is an infinity of wave numbers γ_{1n} solution:

$$\gamma_{1n} = j\frac{n\pi}{a}. \qquad [7.35]$$

Using equations [7.34] and [7.35] in the general form [7.28], we associate the function $\varphi_{1n}(x_1)$ to each wave number:

$$\varphi_{1n} = a_{1n} \sin\left(\frac{n\pi}{a} x_1\right). \qquad [7.36]$$

Note that it is the two by two redundancy of the boundary conditions that makes it possible to have non-trivial solutions for $\varphi_1(x_1)$ verifying four boundary conditions with two integration constants. We conceive that it is an exceptional situation and is only valid for particular boundary conditions.

For the two other edges $x_2 = 0$ and $x_2 = b$, we can impose any type of boundary conditions: supported, clamped, free or guided edge. As an example we will take the case of two supported edges, which has the advantage of leading to rather short calculations.

7.4.2. *Plate supported at its four edges*

The function $\varphi_2(x_2)$ must verify the other boundary conditions:

$$\varphi_2(0) = 0 \ , \ \varphi_2(0) = 0 \ , \ \frac{d^2\varphi_2}{dx_2^2}(0) = 0 \ , \ \frac{d^2\varphi_2}{dx_2^2}(b) = 0 \, .$$

With the form [7.29] of $\varphi_2(x_2)$ it follows for the conditions in $x_2 = 0$:

$$a_2 + b_2 + c_2 + d_2 = 0 \, ,$$

$$\gamma_2^2 \, (a_2 + b_2) + \delta_2^2 \, (c_2 + d_2) = 0 \, .$$

We can note with [7.26] and [7.27] that:

$$\delta_2^2 - \gamma_2^2 = -2 \, \omega \sqrt{\frac{\rho h}{D}} \neq 0 \, .$$

The linear system has a non-zero determinant and its solution is:

$$a_2 = - b_2 \quad \text{and} \, c_2 = - d_2 \, . \qquad [7.37]$$

Taking into account the preceding relations [7.37], the conditions in b give:

$$\begin{cases} a_2 \, \text{sh}(\gamma_2 b) + c_2 \, \text{sh}(\delta_2 b) = 0 \, , \\ a_2 \, \gamma_2^2 \, \text{sh}(\gamma_2 b) + c_2 \, \delta_2^2 \, \text{sh}(\delta_2 b) = 0 \, . \end{cases} \qquad [7.38]$$

Non-trivial solutions are obtained if the determinant of the linear system [7.38] is nil:

$$(\delta_2^2 - \gamma_2^2)\ \mathrm{sh}(\gamma_2 b)\ \mathrm{sh}(\delta_2 b) = 0\,. \qquad [7.39]$$

The quantity $\delta_2^2 - \gamma_2^2$ being non-zero equation [7.39] is satisfied if:

$$\mathrm{sh}(\gamma_2 b) = 0 \quad \text{or} \quad \mathrm{sh}(\delta_2 b) = 0\,.$$

Let us consider these two possibilities:

a) The equation $\mathrm{sh}(\gamma_2 b) = 0$ is verified for an infinite number of γ_{2m} values with:

$$\gamma_{2m} = j\frac{m\pi}{b}\,. \qquad [7.40]$$

Using this result in [7.38], we note that if $\mathrm{sh}(\delta_2 b) \neq 0$, then $c_2 = 0$ and a_2 is unspecified. Bringing together all the results, the form $\varphi_{2m}(x_2)$ solution is given by [7.41] for each value γ_{2m}:

$$\varphi_{2m}(x_2) = a_{2m} \sin\left(\frac{m\pi}{b} x_2\right), \quad m = 1, \ldots, \infty\,. \qquad [7.41]$$

The amplitude a_{2m} is arbitrary.

b) The equation $\mathrm{sh}(\delta_2 b) = 0$ is verified for an infinite number of values δ_{2p}, with:

$$\delta_{2p} = j\frac{p\pi}{b}\,. \qquad [7.42]$$

Using this result in [7.38], we note that if $\mathrm{sh}(\gamma_2 b) \neq 0$, then $a_2 = 0$ and c_2 is unspecified. Bringing all the results together, the form $\varphi_{2p}(x_2)$ solution is given by [7.43] for each value γ_{2p}:

$$\varphi_{2p}(x_2) = a_{2p} \sin\left(\frac{p\pi}{b} x_2\right), \quad p = 1, \ldots, \infty\,. \qquad [7.43]$$

The amplitude a_{2p} is arbitrary.

It should be noted with [7.41] and [7.43] that the two possibilities lead to the same solutions; it is thus necessary to also recognize here an effect of the redundancy of the boundary conditions of support. It is enough to consider one of the two sets of solutions. We take the one with the index m.

The vibration modes of the plate supported at its four edges now stems from the set of the results obtained. To each pair of indices (n, m) we associate the normal angular frequency ω_{nm} deduced from equations [7.26], [7.35] and [7.42]:

$$\omega_{nm} = \sqrt{\frac{D}{\rho h}}\left(\left(\frac{n\pi}{a}\right)^2 + \left(\frac{m\pi}{b}\right)^2\right). \quad [7.44]$$

The mode shape $f_{nm}(x_1, x_2)$ is calculated using [7.21], [7.36] and [7.43], that is:

$$f_{nm}(x_1, x_2) = \sin\left(\frac{n\pi}{a}x_1\right)\sin\left(\frac{m\pi}{b}x_2\right). \quad [7.45]$$

The constants have been normalized to one due to a preoccupation with simplification. The modal vibratory movement is obtained by multiplying the solutions of space and of time:

$$W_{nm}(x_1, x_2, t) = \left(A_{nm}\cos(\omega_{nm}t) + B_{nm}\sin(\omega_{nm}t)\right) \sin\left(\frac{n\pi}{a}x_1\right)\sin\left(\frac{m\pi}{b}x_2\right). \quad [7.46]$$

Finally, the most general movement is produced by the superposition of modal movements:

$$W(x_1, x_2, t) = \sum_{n=1}^{\infty}\sum_{m=1}^{\infty} W_{nm}(x_1, x_2, t). \quad [7.47]$$

The constants of integration A_{nm} and B_{nm} are fixed by the initial conditions.

The use of the alternative solutions [7.40] and [7.41] instead of [7.42] and [7.43] naturally leads to the same modal vibratory movements. However, there appears a difference in calculation, because the normal angular frequencies are negative in this case. It is of course just a sleight of hand, because at the level of [7.46], introducing negative angular frequencies amounts to changing the sign of the constant B_{np}, which in any case is arbitrary at this stage of calculation.

7.4.3. *Physical interpretation of the vibration modes*

The vibration modes of the rectangular plate supported at its four edges was provided by [7.44] and [7.45]. We can note that the modes are defined by a double index with respect to the directions 1 and 2. The mode (1,1) is that of the lower normal angular frequency ω_{11} and of the mode shape $f_{11}(x_1, x_2)$ whose values are given below:

$$\omega_{11} = \sqrt{\frac{D}{\rho h}}\left(\left(\frac{\pi}{a}\right)^2 + \left(\frac{\pi}{b}\right)^2\right)$$

and $f_{11}(x_1, x_2) = \sin\left(\frac{\pi}{a}x_1\right)\sin\left(\frac{\pi}{b}x_2\right)$.

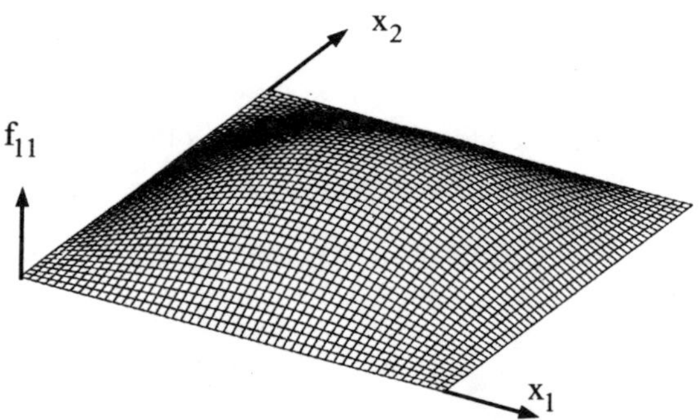

Figure 7.2a. *Mode shape of the mode (1,1) of a supported rectangular plate. Three-dimensional image*

Figure 7.2 presents the mode shape. We note that for this mode of lower normal angular frequencies, all thc points of the plate vibrate in phase. A second representation is used traditionally, the layout of nodal lines ($\{x_1, x_2\}$ being such that $f_{nm}(x_1, x_2) = 0$).

Figure 7.2b. *Mode shape of the mode (1,1) of a supported rectangular plate. Nodal lines (none for this mode)*

Let us consider the mode (2,1); its modal characteristics are given by:

$$\omega_{21} = \sqrt{\frac{D}{\rho h}}\left(\left(\frac{2\pi}{a}\right)^2 + \left(\frac{\pi}{b}\right)^2\right)$$

and $f_{21}(x_1, x_2) = \sin\left(\frac{2\pi}{a}x_1\right)\ \sin\left(\frac{\pi}{b}x_2\right).$

This mode is the second in frequency if $a > b$. It presents a nodal line for $x_1 = a/2$ as Figure 7.3b indicates. On both sides of the nodal line, the vibrations occur in opposition of phase.

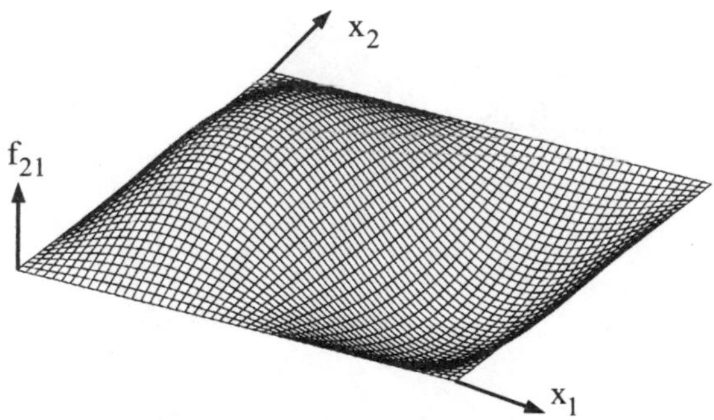

Figure 7.3a. *Mode shape of the mode (2,1) of a supported rectangular plate. Three-dimensional image*

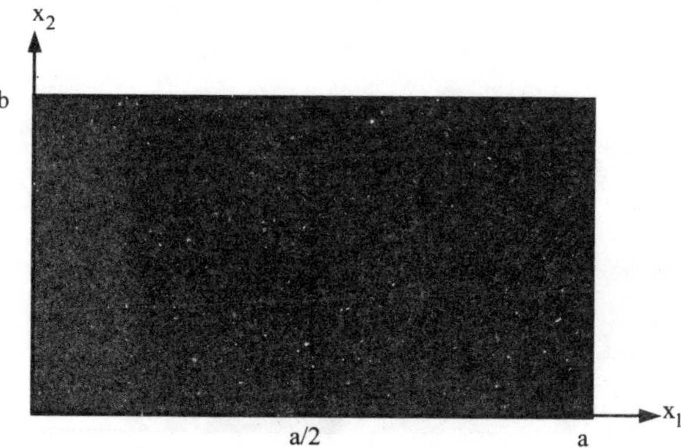

Figure 7.3b. *Mode shape of the mode (2,1) of a supported rectangular plate. Nodal line*

Finally, we take the example of a higher order mode, the mode (3,4):

$$\omega_{34} = \sqrt{\frac{D}{\rho h}}\left(\left(\frac{3\pi}{a}\right)^2 + \left(\frac{4\pi}{b}\right)^2\right)$$

and $f_{34}(x_1, x_2) = \sin\left(\frac{3\pi}{a}x_1\right)\,\sin\left(\frac{4\pi}{b}x_2\right)$.

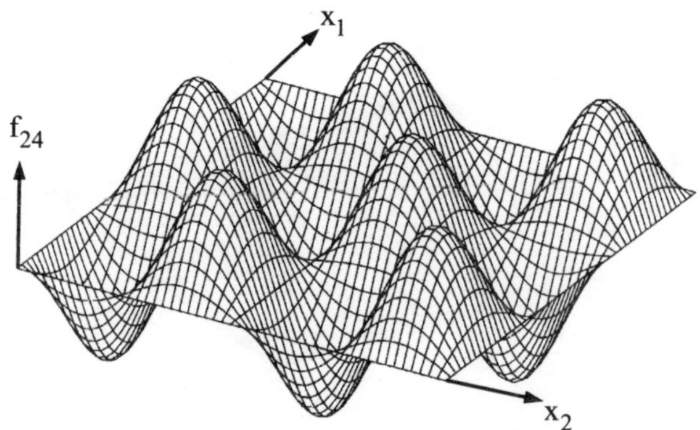

Figure 7.4a. *Mode shape of the mode (3,4) of a supported rectangular plate. Three-dimensional image*

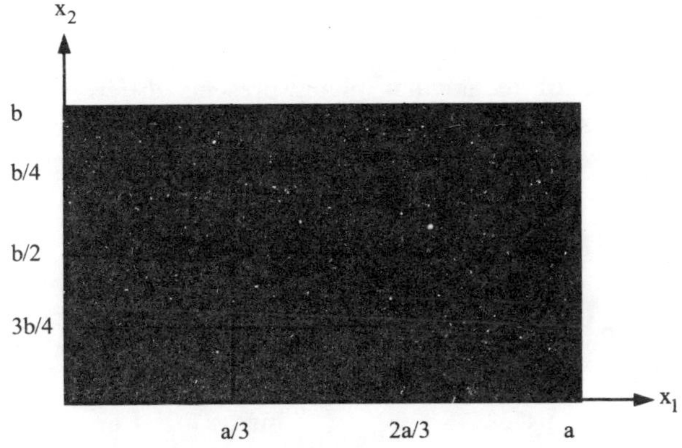

Figure 7.4b. *Mode shape of the mode (3,4) of a supported rectangular plate. Nodal lines*

This mode presents two nodal lines in the x_1 direction (in $a/3$ and $2a/3$) and three nodal lines in the x_2 direction (in $b/4$, $b/2$ and $3b/4$). On both sides of a nodal line the vibrations are in opposition of phase.

These mode shapes present nodal lines parallel to axes 1 and 2; it is an obligatory characteristic of the solutions obtained in the form of separate space variables. On the contrary, we may affirm that the modes of vibration of rectangular plates that do not have nodal lines parallel to the axes cannot be obtained by separation of space variables. Figure 7.5 shows the mode shape of a free rectangular plate: the nodal lines are not parallel to the axes; for this case of boundary conditions the technique of separation of space variables used in this chapter does not yield a result.

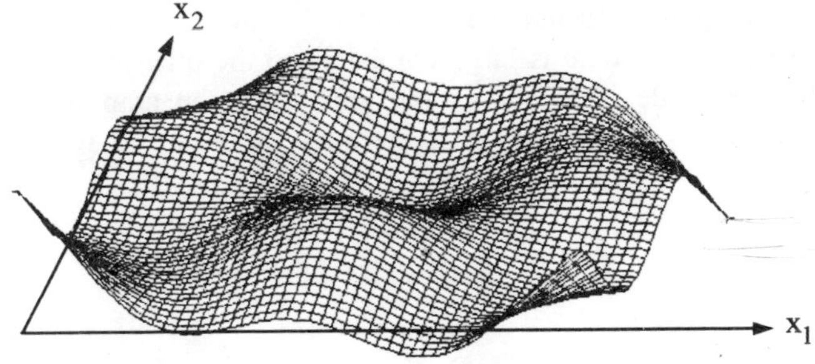

Figure 7.5. *Mode shape of the 17th mode of a rectangular plate free at its edges*

7.4.4. *The particular case of square plates*

This particular case of rectangular plates presents different pairs of normal modes, which have the same normal angular frequencies. These are the modes (n, m) and (m, n). The response of a pair of modes follows the general formula [7.46]:

$$W(x_1, x_2, t) = \left(A_{nm} \cos(\omega_{nm} t) + B_{nm} \sin(\omega_{nm} t)\right)$$

$$\sin\left(\frac{n\pi}{a} x_1\right) \sin\left(\frac{m\pi}{a} x_2\right) + \left(A_{mn} \cos(\omega_{mn} t) + B_{mn} \sin(\omega_{mn} t)\right)$$

$$\sin\left(\frac{m\pi}{a} x_1\right) \sin\left(\frac{n\pi}{a} x_2\right).$$

To simplify the analysis, let us suppose that the constants B_{nm} and B_{mn} are nil. Two modes with the same normal angular frequency can be grouped:

$$W(x_1, x_2, t) = \cos(\omega_{nm} t) \left(A_{nm} \sin\left(\frac{n\pi}{a} x_1\right) \sin\left(\frac{m\pi}{a} x_2\right)\right.$$

$$\left. + A_{mn} \sin\left(\frac{m\pi}{a} x_1\right) \sin\left(\frac{n\pi}{a} x_2\right)\right).$$

Everything occurs as if there was only one combined mode whose shape is the linear combination of the mode shapes of the 2 modes (n, m) and (m, n) with the same normal angular frequency. The effect obtained is rather spectacular when we visualize the nodal lines of the combined modes. Of course the amplitudes A_{nm} and A_{mn} greatly influence the result; they depend on the initial conditions of the vibration initiating and can thus vary greatly. Figure 7.6 gives several examples of results; when one amplitude is large compared to the other, we observe a situation where only one mode is barely visible, but when the amplitudes are close, the resulting nodal lines have forms that are very far removed from primitive nodal lines.

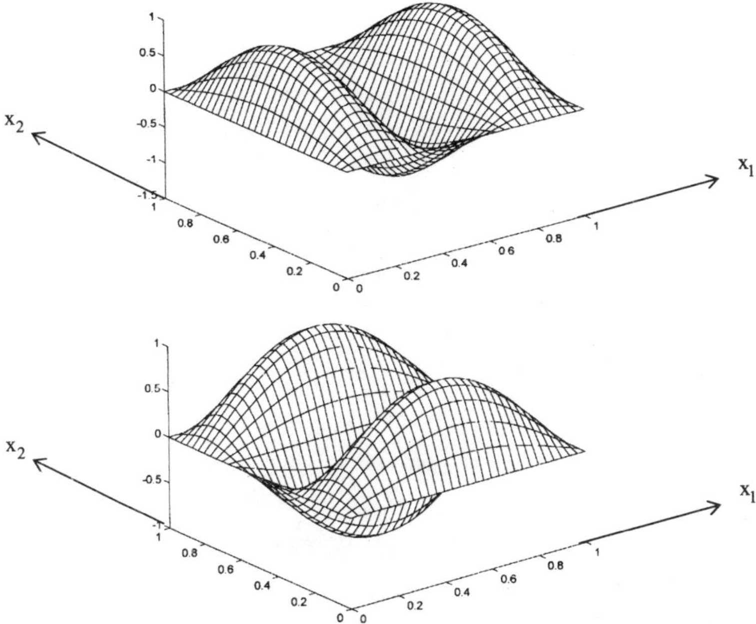

Figure 7.6a. *The two basic mode shapes* f_{13} *and* f_{31} *for a square plate*

If the constants B_{nm} and B_{mn} are not nil, the combination of the two modes produces two different combined modes, one is associated with $\cos(\omega_{nm}t)$ and the other to $\sin(\omega_{nm}t)$. Strictly speaking, there are no nodal lines, since the lines of zero displacement over time move away from the nodal lines of a mode combined to another.

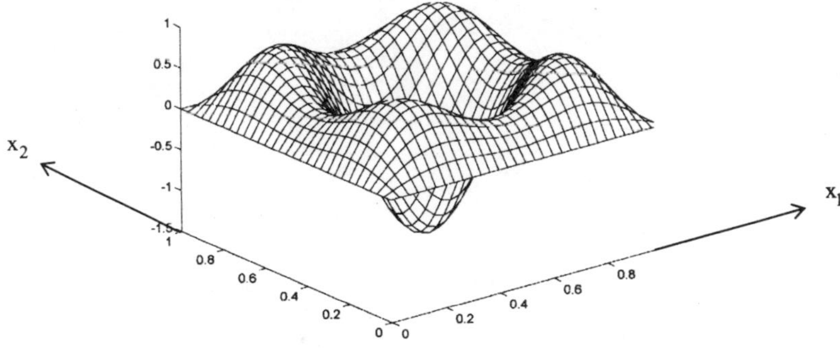

Amplitudes of the mode shapes: A13 = 0.703, A31 = 0.707

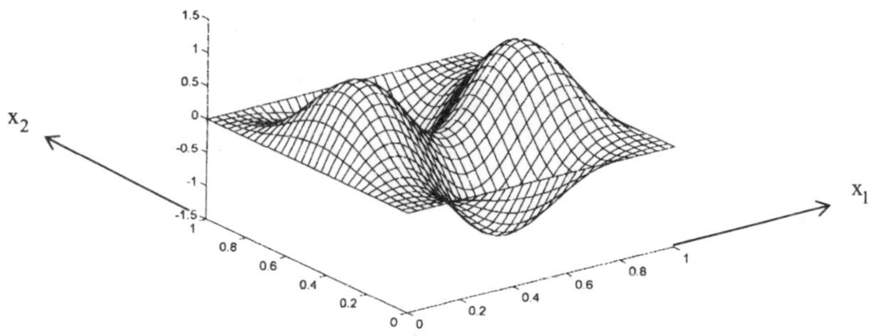

Amplitudes of the mode shapes: A13 = –0.707, A31 = 0.707

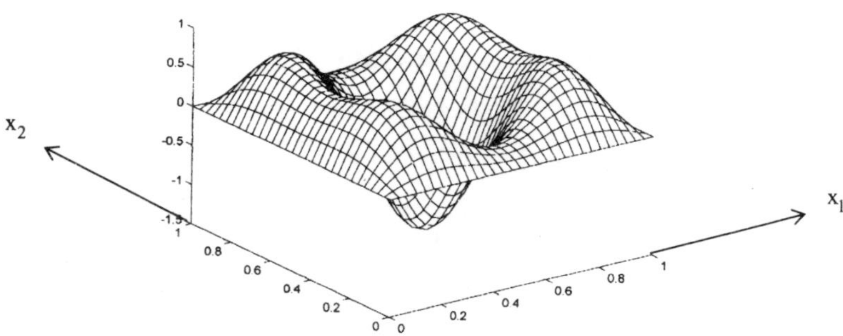

Amplitudes of the mode shapes: A13 = 0.5, A31 = 0.86

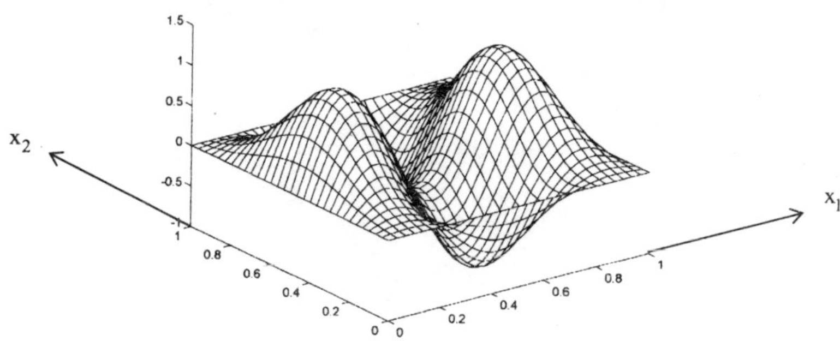

Amplitudes of the mode shapes: A13 = –0.5, A31 = 0.86

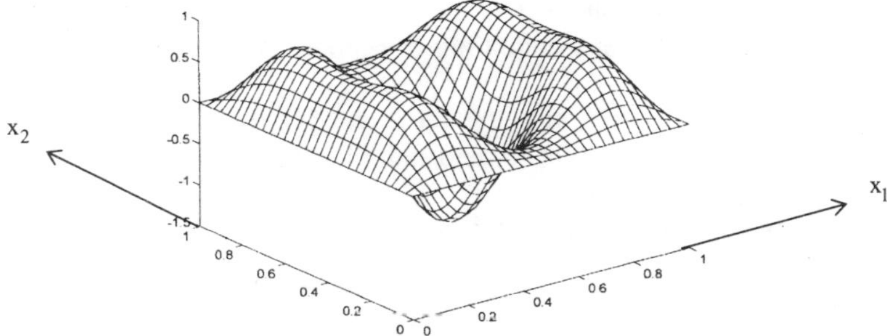

Amplitudes of the mode shapes: A13 = 0.38, A31 = 0.92

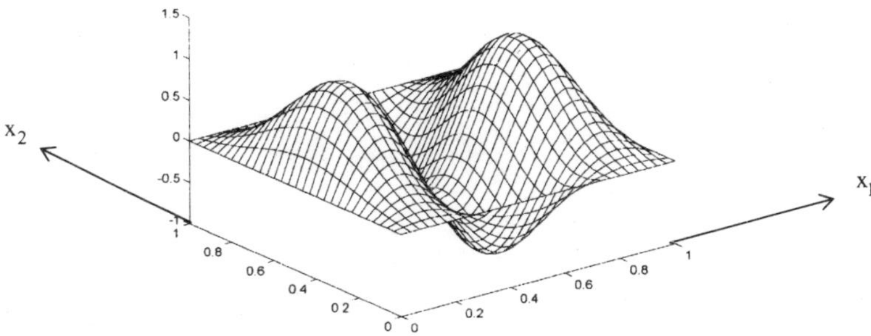

Amplitudes of the mode shapes: A13 = 0.38, A31 = 0.92

Figure 7.6b. *Combined modes stemming from the combination of the modes (1.3) and (3.1) for various amplitudes of the two modes*

These particular phenomena, associated with simply supported square plates, remain true when the boundary conditions are identical for the four edges, because the symmetry of the problem necessarily involves the existence of normal angular frequency doublets.

7.4.5. *Second method of calculation*

A second method of calculation of the modal system of plates supported at two opposite edges can be employed; it is more direct and poses the space solution of the problem *a priori* in the form:

$$f(x_1, x_2) = \sum_{n=1}^{\infty} \sin\left(\frac{n\pi}{a} x_1\right) h_n(x_2). \qquad [7.48]$$

The decomposition into sine offers the double property of defining a functional base and verifying the boundary conditions of support for $x_1 = 0$ and $x_1 = a$.

Now it remains to verify, firstly, the equation of space [7.19] and, secondly, the remaining boundary conditions for $x_2 = 0$ and $x_2 = b$.

Let us start with verifying equation [7.19]; injecting [7.48] in [7.19], we obtain:

$$\sum_{n=1}^{\infty} \sin\left(\frac{n\pi}{a} x_1\right)\left(\frac{d^4 h_n}{dx_2^4} - 2\left(\frac{n\pi}{a}\right)^2 \frac{d^2 h_n}{dx_2^2} + \left(\left(\frac{n\pi}{a}\right)^4 - \omega^2 \frac{\rho h}{D}\right) h_n\right) = 0. \quad [7.49]$$

To uncouple the equations, it is sufficient to use the orthogonality of the sine functions:

$$\int_0^a \sin\left(\frac{n\pi}{a} x_1\right) \sin\left(\frac{m\pi}{b} x_1\right) dx_1 = \begin{cases} 0 \text{ if } n \neq m \\ \frac{a}{2} \text{ if } n = m \end{cases}.$$

Multiplying equation [7.49] by $\sin\left(\frac{m\pi}{a} x_1\right)$, then integrating it from 0 to a, it follows:

$$\frac{d^4 h_m}{dx_2^4} - 2\left(\frac{m\pi}{a}\right)^2 \frac{d^2 h_m}{dx_2^2} + \left(\left(\frac{m\pi}{a}\right)^4 - \omega^2 \frac{\rho h}{D}\right) h_m = 0.$$

It is a differential equation which is easily integrated. Noting:

$$r_{1m} = \sqrt{\left(\frac{m\pi}{a}\right)^2 + \omega\sqrt{\frac{\rho h}{D}}}$$

$$\text{and } r_{2m} = j\sqrt{\omega\sqrt{\frac{\rho h}{D}} - \left(\frac{m\pi}{a}\right)^2}, \quad [7.50]$$

we obtain:

$$h_m(x_2) = C_m \sin(r_{2m}x_2) + D_m \cos(r_{2m}x_2) + E_m \,\mathrm{sh}(r_{1m}x_2) + F_m \,\mathrm{ch}(r_{1m}x_2). \qquad [7.51]$$

Let us introduce the boundary conditions for the edges $x_2 = 0$ and $x_2 = b$. All the boundary conditions for an edge are possible; however, to consolidate our concepts we consider the clamped edges:

$$\begin{aligned} &f(x_1, 0) = 0 \Rightarrow h_m(0) = 0, \\ &f(x_1, b) = 0 \Rightarrow h_m(b) = 0, \\ &\frac{df}{dx_2}(x_1, 0) = 0 \Rightarrow \frac{dh_m}{dx_2}(0) = 0, \\ &\frac{df}{dx_2}(x_1, b) = 0 \Rightarrow \frac{dh_m}{dx_2}(b) = 0. \end{aligned} \qquad [7.52]$$

Using the form [7.51] under the four conditions [7.52], we obtain the homogenous linear system [7.53]:

$$\begin{pmatrix} 0 & 1 & 0 & 1 \\ r_{2m} & 0 & r_{1m} & 0 \\ \sin(r_{2m}b) & \cos(r_{2m}b) & \mathrm{sh}(r_{1m}b) & \mathrm{ch}(r_{1m}b) \\ r_{2m}\cos(r_{2m}b) & -r_{2m}\sin(r_{2m}b) & r_{1m}\mathrm{sh}(r_{1m}b) & r_{1m}\mathrm{sh}(r_{1m}b) \end{pmatrix} \begin{pmatrix} C_m \\ D_m \\ E_m \\ F_m \end{pmatrix} = \begin{pmatrix} 0 \\ 0 \\ 0 \\ 0 \end{pmatrix}. \qquad [7.53]$$

To obtain non-trivial solutions, it is necessary that the determinant of the system be nil. Upon some calculation this leads to the characteristic equation:

$$2r_{1m}\, r_{2m}\left(\cos(r_{2m}b)\ \mathrm{ch}(r_{1m}b) - 1\right) = (r_{2m}^2 - r_{1m}^2)\ \sin(r_{2m}b)\ \mathrm{sh}(r_{1m}b). \qquad [7.54]$$

The solution of this characteristic equation is not trivial and requires computerized treatment. Let us note, however that r_{1m} and r_{2m} are not independent variables but, on the contrary, are connected to the angular frequency ω by equations [7.50]. Thus, we seek to determine the angular frequencies ω verifying [7.54]. For each index m we find an infinite number of solutions $m = 1, \ldots, \infty$. Each solution is consequently identified by a double index ω_{pq}. For a square plate we may draw up Table 7.1 indicating the first six normal angular frequencies.

mode	1.1	2.1	1.2	2.2	3.1	1.3	
$\omega_{pq}\, a^2 \sqrt{\frac{\rho h}{D}}$	28,946	54,743	69,320	94,584	102,213	129,086	A-E-A-E
	19,739	49,348	49,348	78,956	98,696	98,696	A-A-A-A

Table 7.1. *Adimensional normal angular frequencies of the first 6 modes of a square plate, with a side a. (boundary conditions: all supported (A-A-A-A) or supported-clamped (A-E-A-E)*

We may note thanks to Table 7.1 that clamping two edges produces an effect of stiffness compared to the condition of support, which increases the normal angular frequencies. This effect is all the more pronounced when the mode is of a low rank.

The calculation of mode shapes requires the resolution of the linear system [7.53] for each root of the equation. Computerized processing also generally proves necessary here. In certain particular cases we will be able to find tabulations of mode shapes in [LEI 93] and [CORN 84]. We may, however, state the technique of calculation of mode shapes rather simply. Let us consider the mode (n, m) with an angular frequency ω_{nm} ; thanks to equations [7.50] we associate to it the two wave numbers r_{1nm} and r_{2nm}, which of course give a determinant equal to zero of the system [7.53]. Consequently, the solution vector of [7.53] is not unique. We may, nonetheless, choose one of them by normalizing the solution vector.

The mode shape is then provided by:

$$f_{nm}(x_1, x_2) = \sin\left(\frac{n\pi}{a} x_1\right)\Big(C_{nm} \sin(r_{2nm}x_2) + D_{nm} \cos(r_{2nm}x_2)$$

$$+ E_{nm} \operatorname{sh}(r_{1nm}x_2) + F_{nm} \operatorname{ch}(r_{1nm}x_2)\Big).$$

7.5. Vibration modes of rectangular plates: approximation by the edge effect method

7.5.1. *General issues*

As we saw previously, the calculation of the vibration modes in analytical form is impossible in the majority of the cases of boundary conditions. Several techniques of approximation are possible, most commonly based on the Rayleigh-Ritz method. A possible choice of the test functions used in this method consists of approximating the mode shape (n, m) of the rectangular plate by the product of the mode shapes of beam of the orders n and m in the directions 1 and 2 respectively. We treat this type of calculation in Chapters 11 and 12 which deal with the approximation of the modal system by the Rayleigh-Ritz method.

A different technique suggested by Bolotin is based on a physical property of the modes which stipulates that the high order modes of homogenous rectangular plates present different behaviors away from their boundaries and near them. During the study of beam vibrations we observed this phenomenon linked to vanishing waves which have real influence only in the vicinity of the boundaries and are characteristic of the edge effect. Far from the boundaries, the solution stems entirely from traveling waves; it is the internal solution.

7.5.2. *Formulation of the method*

We seek an approximate solution of equation [7.19] of the space function $f(x_1, x_2)$. The idea of the method consists of using different approximations of the solution far and near the boundaries, these solutions being based on the one obtained by separation of space variables [7.28] and [7.29]. The form presumed valid far from the edges $x_1 = 0$ and $x_1 = a$ is:

$$f(x_1, x_2) = \sin(k_1 x_1 + \varphi_1)\ h_2(x_2). \qquad [7.55]$$

This solution supposes that in this part of the plate the solution in direction 1 contains only the internal solution $\sin(k_1 x_1 + \varphi_1)$.

Symmetrically, the form presumed valid far from the edges $x_2 = 0$ and $x_2 = b$ is:

$$f(x_1, x_2) = \sin(k_2 x_2 + \varphi_2)\ h_1(x_1). \qquad [7.56]$$

These forms of solution [7.55] and [7.56] must, on the one hand, verify the equation of motion [7.19] and, on the other hand, coincide when the point of observation is far from the four edges.

Introducing, to begin with, the respect of equation [7.19] by the form [7.56], it follows:

$$\frac{d^4 h_1}{dx_1^4} - 2k_2^2 \frac{d^2 h_1}{dx_1^2} + \left(k_2^4 + \frac{\rho h}{D} \omega^2 \right) h_1 = 0. \qquad [7.57]$$

Equation [7.57] has as a solution:

$$h_1(x_1) = C_1 \sin(s_2 x_1 + \psi_1) + D_1 e^{-s_1(a_1 - x_1)} + E_1 e^{-s_1 x_1} \quad [7.58]$$

$$\text{with : } s_2 = \sqrt{\sqrt{\frac{\rho h}{D}}\,\omega - k_2^2} \quad [7.59]$$

and:

$$s_1 = \sqrt{\sqrt{\frac{\rho h}{D}}\,\omega + k_2^2}\,. \quad [7.60]$$

Performing the same operation with the form [7.55] we obtain:

$$h_2(x_2) = C_2 \sin(r_2 x_2 + \psi_2) + D_2 e^{-r_1(b_2 - x_2)} + E_2 e^{-r_1 x_2} \quad [7.61]$$

$$\text{with: } r_2 = \sqrt{\sqrt{\frac{\rho h}{D}}\,\omega - k_1^2} \quad [7.62]$$

and:

$$r_1 = \sqrt{\sqrt{\frac{\rho h}{D}}\,\omega + k_1^2}\,. \quad [7.63]$$

The solutions [7.58] and [7.61] are formed using the same model: they contain an internal sinusoidal solution and edge effects characterized by decreasing exponentials when we move away from the boundaries. This conforms well with the initial hypothesis: far from the boundaries only the internal solution remains. This hypotheses will be verified all the better the greater the value of the wave number s_1 (or r_1).

In the case of beams in bending vibrations we observed the same structure of solutions. The zone where the edge effect is important is about a quarter of the wavelength of natural waves and, therefore, reduces sharply with frequency.

Let us use the fact that the two forms must coincide far from the boundaries and that the internal solutions provided by the two forms of solutions must, therefore, be equal:

$$C_1 \sin(r_2\, x_2 + \psi_2)\ \sin(k_1\, x_1 + \varphi_1) = C_2 \sin(s_2\, x_1 + \psi_1)\ \sin(k_2\, x_2 + \varphi_2).$$

This lead to the identification:

$$\begin{cases} s_2 = k_1 \\ \Psi_1 = \varphi_1 \\ r_2 = k_2 \\ \Psi_2 = \varphi_2 \\ C_1 = C_2 = C \end{cases} \qquad [7.64]$$

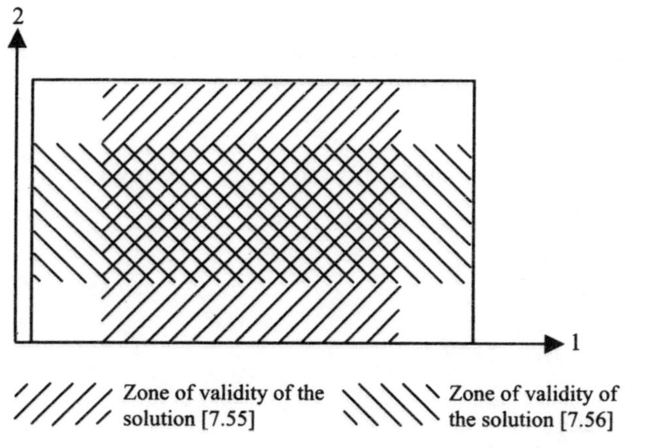

Figure 7.7. *Zones of validity of the solutions of the edge effect method*

The internal solution is thus provided by:

$$C \sin(k_1\, x_1 + \varphi_1)\ \sin(k_2\, x_2 + \varphi_2). \qquad [7.65]$$

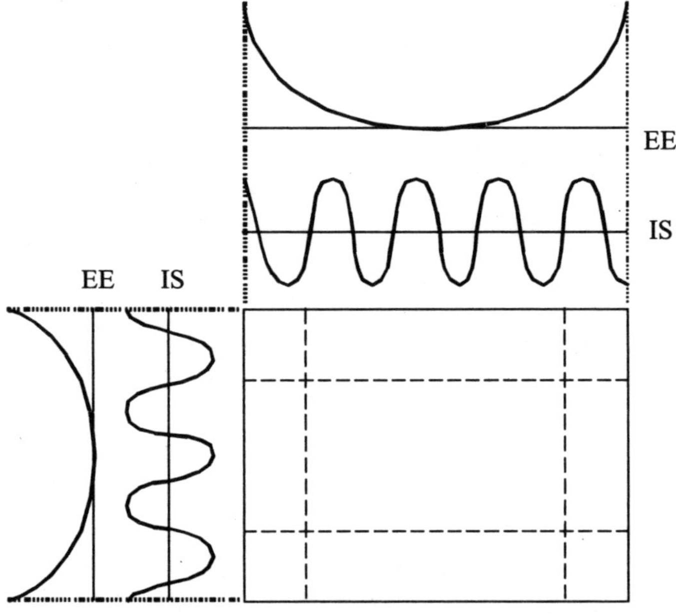

Figure 7.8. *Representation of the solutions of the edge effect method IS: internal solution, EE: edge effect*

Moreover, taking into account [7.59] and [7.62], it follows:

$$\omega = \sqrt{\frac{D}{\rho h}} \, (k_1^2 + k_2^2) \, . \tag{7.66}$$

That is, with [7.60] and [7.63]:

$$s_1 = k_1^2 + 2k_2^2 \, , \tag{7.67}$$

$$r_1 = k_2^2 + 2k_1^2 \, . \tag{7.68}$$

The two forms of solutions are thus, finally, provided by:

– for x_1 far from the edges $x_1 = 0$ and $x_1 = a$:

$$\begin{aligned} f(x_1, x_2) &= C \sin(k_1 x_1 + \varphi_1) \sin(k_2 x_2 + \varphi_2) \\ &+ D_2 \sin(k_1 x_1 + \varphi_1) \, e^{-r_1(b-x_2)} + E_2 \sin(k_1 x_1 + \varphi_1) \, e^{-r_1 x_2} ; \end{aligned} \tag{7.69}$$

– for x_2 far from the edges $x_2 = 0$ and $x_2 = b$:

$$f(x_1, x_2) = C\sin(k_1 x_1 + \varphi_1)\ \sin(k_2 x_2 + \varphi_2) \\ + D_1 \sin(k_2 x_2 + \varphi_2)\ e^{-s_1(a-x_1)} + E_1 \sin(k_2 x_2 + \varphi_2)\ e^{-s_1 x_1}. \qquad [7.70]$$

7.5.3. *The plate clamped at its four edges*

It is a question of imposing the respect of the boundary conditions on the four edges considering the form [7.69] or [7.70] adapted to the selected edge. Let us consider, to begin with, the edge $x_1 = 0$. It is of course the form [7.70] which is adapted to it. However, it is simplified because the second term of [7.70] represents the effect edge for $x_1 = a$, which is negligible for $x_1 = 0$. The approximation of the solution in the vicinity of the edge $x_1 = 0$ is, therefore, reduced to two terms:

$$f(x_1, x_2) = C\sin(k_1 x_1 + \varphi_1)\ \sin(k_2 x_2 + \varphi_2) + E_1 \sin(k_2 x_2 + \varphi_2)\ e^{-s_1 x_1}.$$

This form of solution is applicable at the edge, far from the corners because the edge effect in the second direction becomes important there. On the assumption that this edge effect decreases sharply when we move away from the boundary, it will only have influence very locally, very close to the corners, and will not have a very significant effect on the result.

The application of the boundary conditions of clamped edge provides two relations:

$$f(0, x_2) = 0 \Rightarrow C\sin\varphi_1 + E_1 = 0, \qquad [7.71]$$

$$\frac{\partial f}{\partial x_1}(0, x_2) = 0 \Rightarrow C k_1 \cos\varphi_1 - E_1 s_1 = 0. \qquad [7.72]$$

Proceeding in a similar way for the edge $x_1 = a$ we obtain:

$$f(a, x_2) = 0 \Rightarrow C\sin(k_1 a + \varphi_1) + D_1 = 0, \qquad [7.73]$$

$$\frac{\partial f}{\partial x_1}(a, x_2) = 0 \Rightarrow C k_1 \cos(k_1 a + \varphi_1) + D_1 s_1 = 0. \qquad [7.74]$$

Using the linear system formed by equations [7.71] and [7.72] and supposing that the determinant is nil, we obtain the first line from [7.75]. Proceeding in the same way with equations [7.73] and [7.74] we draw the second line from [7.75] in order to obtaining a non-trivial solution:

$$\begin{cases} k_1 \cos \varphi_1 + s_1 \sin \varphi_1 = 0 \\ k_1 \cos (k_1 a_1 + \varphi_1) - s_1 \sin (k_1 a_1 + \varphi_1) = 0 \end{cases} \quad [7.75]$$

The system of equations [7.75] has the solutions:

$$\varphi_1 = -k_1 a/2 , \quad [7.76]$$

$$\frac{k_1}{s_1} = tg(k_1 a/2) . \quad [7.77]$$

Symmetrically for the edges $x_2 = 0$ and $x_2 = b$ after all the calculations, we obtain:

$$\varphi_2 = -k_2 b/2 , \quad [7.78]$$

$$\frac{k_2}{r_1} = tg(k_2 b/2) . \quad [7.79]$$

It now remains to determine the angular frequencies which make it possible to simultaneously verify [7.77] and [7.79]. These angular frequencies are provided by [7.66] from the moment when the wave numbers k_1 and k_2 verify the two equations:

$$\frac{k_1}{k_1^2 + 2k_2^2} = tg(k_2 \ b/2)$$

$$\text{and} \quad \frac{k_2}{k_2^2 + 2k_1^2} = tg(k_1 \ a/2) , \quad [7.80]$$

where we have made use of equations [7.67] and [7.68].

The solution of equations [7.80] requires computerized processing, which we will not perform here. However, to show the quality of the prediction, by way of an example, in Table 7.2 we give the value obtained for the first mode of a square plate.

	Edge effect method	Rayleigh-Ritz method
$\omega_{11}\, a^2 \sqrt{\dfrac{\rho h}{D}}$	35.09	35.99

Table 7.2. *Comparison of the values of the first adimensional normal angular frequency of a clamped square plate, calculated by the edge effect method and by the Rayleigh-Ritz method (according to [KIN 74])*

The result of the table shows an already satisfactory prediction while the edge effect method converges all the better the higher the wave numbers k_1 and k_2 are. It is this characteristic of better convergence for the higher modes which marks the specificity of this approach.

7.5.4. *Another type of boundary conditions*

A similar calculation can be carried out when the plate has other boundary conditions. For a square plate clamped at two adjacent edges and supported at the other two edges the first normal angular frequency is given in Table 7.3. In this case the approximation is still better, because the support does not generate the edge effect (error of 0.7% instead of 2.6% in the case where all the edges are clamped).

	Edge effect method	Ritz method
$\omega_{11}\, a^2 \sqrt{\dfrac{\rho h}{D}}$	26.87	27.06

Table 7.3. *Comparison of the values of the first adimensional normal angular frequency, calculated by the edge effect method and by the Ritz method (according to [KIN 74])*

The edge effect method is thus applicable to obtain an approximation of the normal angular frequencies of rectangular plates, the approximation being better the higher the ranks of the modes are. There is, however, a limitation to its use when the boundary conditions leave the transverse displacement of the plate free. This fact is highlighted in Table 7.4.

We may note that the edge effect method does not predict all the frequencies of resonance when two opposite edges are free: it is then dangerous to use it even if it does provide a correct approximation of the normal angular frequencies, which it is able to predict. This phenomenon is explained by the fact that the hypothesis of an edge effect localized at the boundaries is not acceptable when two opposite edges are free because the plate then presents modes of the beam type.

In the case of boundary conditions where transverse displacement is blocked at the edges (clamping and supports), the edge effect method gives good results and more so when the modes have a high rank, as opposed to the methods of traditional discretization.

Boundary conditions	**Edge effect method**	**Results from written works**	
clamped in $x_1 = 0$ and $x_1 = a$ free in $x_2 = 0$ and $x_2 = a$	26.73 44.56 67.29 80.60 88.17	22.17 26.40 43.6 61.2 67.2 79.8 87.5	
clamped in $x_1 = 0$ free in $x_1 = a$, $x_2 = 0$ and $x_2 = a$	7.78 26.27 30.06 53.21	Lower limit 3.43 7.26 20.87 26.50 28.55 51.50 60.25	Upper limit 3.473 8.54 21.30 27.29 31.17 54.26 61.28
clamped in $x_1 = 0$ and $x_2 = 0$ free in $x_1 = a$ and $x_2 = a$	5.866 25.05 25.05 47.13 63.87	Lower limit 6.958 24.80 26.80 48.05 63.14	

Table 7.4. *Adimensional normal angular frequencies of square plates presenting free edges (according to [KIN 74])*

7.5.5. *Approximation of the mode shapes*

The approximation of mode shape is simple to carry out: let us take the example of the clamped plate. Each normal mode (n, m) is characterized by wave numbers k_{1nm} and k_{2nm}, which are the solutions of the characteristic equation [7.80]. From them with [7.76] and [7.78] we deduce the values of φ_{1nm} and φ_{2nm}:

$$\varphi_{1nm} = - k_{1nm}\, a/2 \quad \text{and} \quad \varphi_{2nm} = - k_{2nm}\, b/2 .$$

Finally, thanks to [7.71] and [7.73], we can express the constants E_{1nm} and D_{1nm}:

$$E_{1nm} = - C \sin\varphi_{1nm} ,$$

$$D_{1nm} = - C \sin(k_{1nm}\, a + \varphi_{1nm}) .$$

By symmetry we also draw E_{2nm} and D_{2nm}:

$$E_{2nm} = - C \sin\varphi_{2nm} ,$$

$$D_{2nm} = - C \sin(k_{2nm}\, a + \varphi_{2nm}) .$$

It is enough to report these values in the general form [7.69] – [7.70] to obtain the mode shape $f_{nm}(x_1, x_2)$. The constant C is not fixed but traditionally one can make it equal to the unit.

7.6. Calculation of the free vibratory response following the application of initial conditions

To reduce the calculations as much as possible, we consider a plate supported at all its edges. The general form of the vibratory response is then given by [7.47], that is:

$$W(x_1, x_2, t) = \sum_{n=1}^{\infty} \sum_{m=1}^{\infty} (A_{nm} \cos \omega_{nm} t + B_{nm} \sin \omega_{nm} t) \sin \frac{n\pi}{a} x_1 \sin \frac{m\pi}{b} x_2 . \qquad [7.81]$$

The initial conditions are of the type:

$$W(x_1, x_2, 0) = d_0(x_1, x_2), \tag{7.82}$$

$$\frac{\partial W}{\partial t}(x_1, x_2, 0) = v_0(x_1, x_2). \tag{7.83}$$

With [7.81] we draw from this the two equations:

$$\sum_{n=1}^{\infty}\sum_{m=1}^{\infty} A_{nm} \sin\frac{n\pi}{a}x_1 \sin\frac{m\pi}{b}x_2 = d_0(x_1, x_2) \tag{7.84}$$

and:

$$\sum_{n=1}^{\infty}\sum_{m=1}^{\infty} B_{nm}\,\omega_{nm} \sin\frac{n\pi}{a}x_1 \sin\frac{m\pi}{b}x_2 = v_0(x_1, x_2). \tag{7.85}$$

To calculate the modal amplitudes A_{nm} and B_{nm}, it is necessary to use the orthogonality of mode shapes:

$$\int_0^a\int_0^b \sin\frac{n\pi}{a}x_1 \sin\frac{m\pi}{b}x_2 \sin\frac{p\pi}{a}x_1 \sin\frac{q\pi}{b}x_2\, dx_1 dx_2 = \begin{cases} 0 \text{ if } (n,m) \neq (p,q) \\ \dfrac{ab}{4} \text{ if } (n,m) = (p,q) \end{cases}.$$

Following the classical procedure we obtain:

$$A_{pq} = \frac{4}{ab}\int_0^a\int_0^b d_0(x_1, x_2) \sin\frac{p\pi}{a}x_1 \sin\frac{q\pi}{b}x_2\, dx_1 dx_2, \tag{7.86}$$

$$B_{pq} = \frac{1}{\omega_{pq}}\frac{4}{ab}\int_0^a\int_0^b v_0(x_1, x_2) \sin\frac{p\pi}{a}x_1 \sin\frac{q\pi}{b}x_2\, dx_1 dx_2. \tag{7.87}$$

By way of an example, an excitation through impact at the point (X_1, X_2) produces initial conditions of the type:

$$\begin{cases} d_0(x_1, x_2) = 0 \\ v_0(x_1, x_2) = V_0\, \delta(x_1 - X_1)\ \delta(x_2 - X_2) \end{cases}$$

where V_0 is speed at the point of impact.

Using of these expressions in [7.86] and [7.87] leads to the result:

$$A_{pq} = 0$$

$$B_{pq} = \frac{4V_0}{ab} \frac{\sin \frac{p\pi}{a} X_1 \sin \frac{q\pi}{b} X_2}{\omega_{pq}}.$$

The vibratory movement following the shock is thus:

$$W(x_1, x_2, t) =$$

$$\sum_{n=1}^{\infty} \sum_{m=1}^{\infty} \frac{4V_0}{ab} \sin(\omega_{nm} t) \frac{\sin \frac{n\pi}{a} X_1 \sin \frac{m\pi}{b} X_2 \sin \frac{n\pi}{a} x_1 \sin \frac{m\pi}{b} x_2}{\omega_{nm}} \qquad [7.88]$$

We have already studied the calculation of the free response for beams; the case of the plates is the same and does not introduce any fundamental differences. In particular, we find the classical tendency: excitation through point impact shock does not produce a response of a mode if the point of impact coincides with a nodal line of this mode.

7.7. Circular plates

7.7.1. *Equation of motion and solution by separation of variables*

This case is not treated in detail like that of the rectangular plates; we present the method and some sufficiently explicit cases of application so that the reader can generalize the approach.

Expressing the equation of motion of circular plates in polar co-ordinates has an obvious interest. We gave the equations within the framework of various hypotheses in Chapter 4. Hereafter we will consider the case of the Love-Kirchhoff hypotheses, and equations [4.79] and [4.76] which we recall below.

$$\rho h \frac{\partial^2 W}{\partial t^2} + D\Delta^2 \{W\} = 0 \qquad [7.89]$$

$$\text{with: } \Delta^2 = \left(\frac{\partial^2}{\partial r^2} + \frac{1}{r}\frac{\partial}{\partial r} + \frac{1}{r^2}\frac{\partial^2}{\partial \theta^2}\right)\left(\frac{\partial^2}{\partial r^2} + \frac{1}{r}\frac{\partial}{\partial r} + \frac{1}{r^2}\frac{\partial^2}{\partial \theta^2}\right). \qquad [7.90]$$

To solve this equation we will use separation of variables. First of all, at the level of the time and variables of space (r, θ):

$$W(r, \theta, t) = g(t)\ f(r, \theta). \qquad [7.91]$$

The introduction of [7.91] in [7.89] with subsequent separation of variables yields:

$$\frac{d^2}{dt^2} g(t) + \omega^2 g(t) = 0 \qquad [7.92]$$

and:

$$-\omega^2 \rho h\, f(r, \theta) + D\Delta^2 \{ f(r, \theta) \} = 0. \qquad [7.93]$$

Equation [7.92] admits the traditional solution of the vibratory problems:

$$g(t) = E \cos \omega t + F \sin \omega t. \qquad [7.94]$$

To solve equation [7.93] let us once again apply the separation of variables; we seek a solution in the form of [7.95].

$$f(r, \theta) = h(r)\ s(\theta). \qquad [7.95]$$

Let us note, moreover, that taking into account the periodicity of the function $s(\theta)$, we can break it up into a Fourier series and obtain the general form:

$$s(\theta) = A_0 + \sum_{n=1}^{\infty} A_n \cos n\theta + B_n \sin n\theta. \qquad [7.96]$$

Let us use the technique of resolution from section 7.3.3, introducing the two, operators L^+ and L^- :

$$L^+ = \left(\frac{\partial^2}{\partial r^2} + \frac{1}{r}\frac{\partial}{\partial r} + \frac{1}{r^2}\frac{\partial^2}{\partial \theta^2}\right) + \omega\sqrt{\frac{\rho h}{D}}, \quad [7.97]$$

$$L^- = \left(\frac{\partial^2}{\partial r^2} + \frac{1}{r}\frac{\partial}{\partial r} + \frac{1}{r^2}\frac{\partial^2}{\partial \theta^2}\right) - \omega\sqrt{\frac{\rho h}{D}}. \quad [7.98]$$

The solution of [7.93] is the sum of the solutions of [7.97] and [7.98]. Let us first consider [7.97]. Introducing the form [7.96] for $s(\theta)$ we obtain:

$$B_0\left(\frac{d^2}{dr^2} + \frac{1}{r}\frac{d}{dr} + \omega\sqrt{\frac{\rho h}{D}}\right)\{h(r)\} + \sum_{n=1}^{\infty}(B_n \cos n\theta + A_n \sin n\theta) \left(\frac{d^2}{dr^2} + \frac{1}{r}\frac{d}{dr} + \omega\sqrt{\frac{\rho h}{D}} - \frac{n2}{r2}\right)\{h(r)\} = 0. \quad [7.99]$$

Using the orthogonality of the functions $\sin n\theta$ and $\cos n\theta$, we determine a set of functions:

$$h_n(r)\cos n\theta, \quad [7.100a]$$

$$h_n(r)\sin n\theta, \quad [7.100b]$$

$$h_0(r), \quad [7.100c]$$

which are solutions of [7.99] if they verify [7.101]:

$$\left(\frac{d^2}{dr^2} + \frac{1}{r}\frac{d}{dr} + \omega\sqrt{\frac{\rho h}{D}} - \frac{n^2}{r^2}\right)\{h_n(r)\} = 0. \quad [7.101]$$

Expressions [7.100] define shapes which will characterize the modes of vibration when boundary conditions are applied.

We can also write equation [7.101] in a different form by multiplying it by r^2 :

$$\left(r^2 \frac{d^2}{dr^2} + r\frac{d}{dr} + \omega r^2 \sqrt{\frac{\rho h}{D}} - n^2\right)\left\{h_n(r)\right\} = 0 . \qquad [7.102]$$

In order to coincide with the standard form of the Bessel equation, let us carry out the change of variable:

$$z = r\omega\sqrt{\frac{\rho h}{D}} .$$

Noting also that: $z = kr$ with $k = \omega\sqrt{\frac{\rho h}{D}}$, it follows:

$$\left(z^2 \frac{d^2}{dz^2} + z\frac{d}{dz} + z^2 - n^2\right)\left\{h_n(z)\right\} = 0 . \qquad [7.103]$$

Equation [7.103] admits two solutions: $J_n(z)$ and $Y_n(z)$: Bessel functions of the first and second type. Figure 7.9 illustrates the typical behavior of these functions for orders 0 and 1. The point to be emphasized is the oscillating character of these functions that can be approximated to the behavior of traveling waves appearing in the solution in Cartesian co-ordinates. Another characteristic aspect is the singularity of the Bessel function of second type at the origin; this non-physical characteristic leads to the suppression of this term in certain problems, as in section 7.7.2.

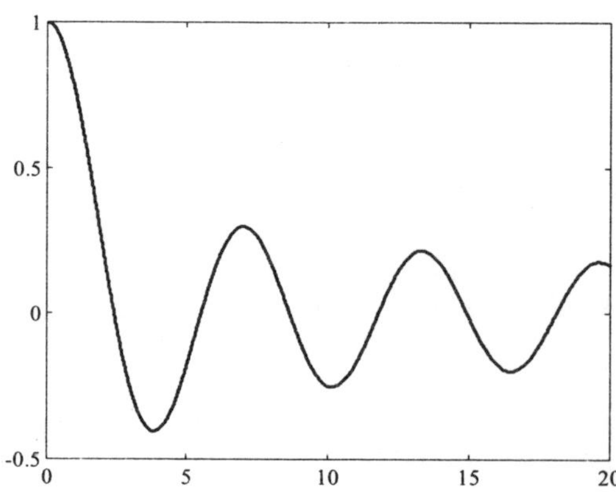

Figure 7.9a. *Bessel function of the first type* $J_0(z)$

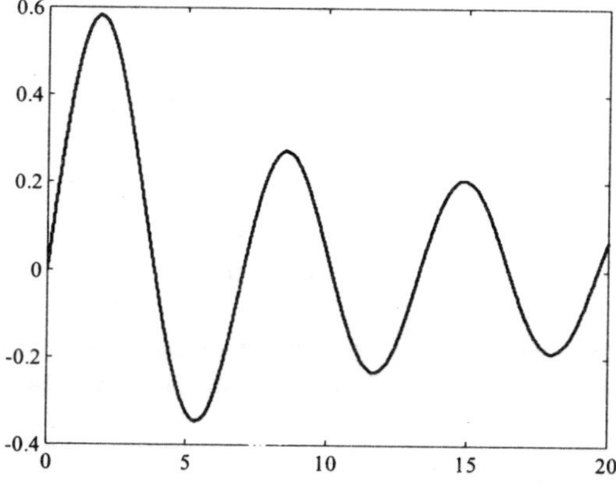

Figure 7.9b. *Bessel function of the first type* $J_1(z)$

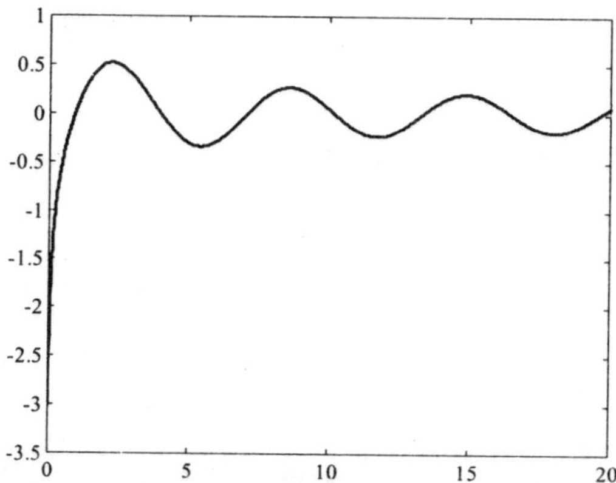

Figure 7.9c. *Bessel function of the second type* $Y_0(z)$

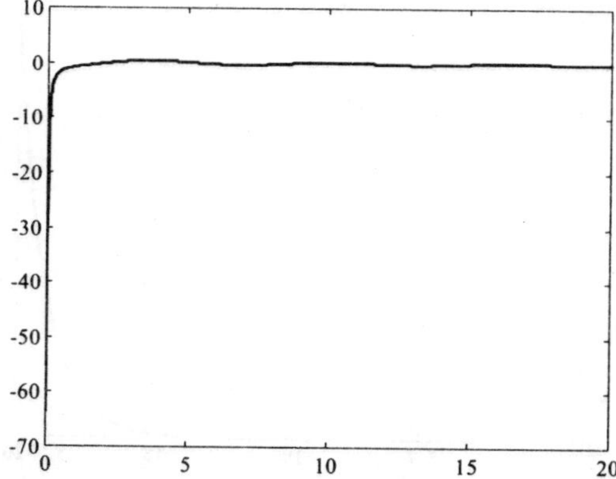

Figure 7.9d. *Bessel function of the second type* $Y_1(z)$

Proceeding in an identical manner for equation [7.98] we arrive at equation [7.103'], which is the modified Bessel equation:

$$\left(z^2 \frac{d^2}{dz^2} + z\frac{d}{dz} - (z^2 + n^2)\right)\left\{h_n(z)\right\} = 0. \qquad [7.103']$$

Equation [7.103'] admits the solutions $I_n(z)$ and $K_n(z)$ that are the modified Bessel functions of the first and second types respectively. Figure 7.10 illustrates the typical behavior of these functions at the order 0 and the order 1. These functions have a behavior that can be compared to that of vanishing waves of the solution in Cartesian co-ordinates. The $K_n(z)$ functions are singular at the origin and will have to be removed in the problems of full plates like the one studied in section 7.7.2.

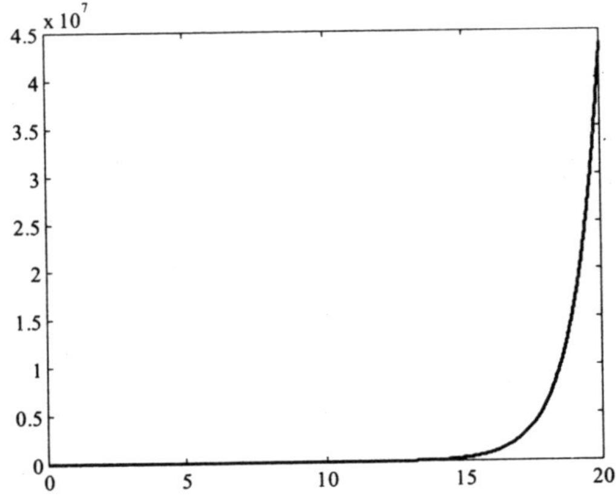

Figure 7.10a. *Modified Bessel function of the first type* $I_0(z)$

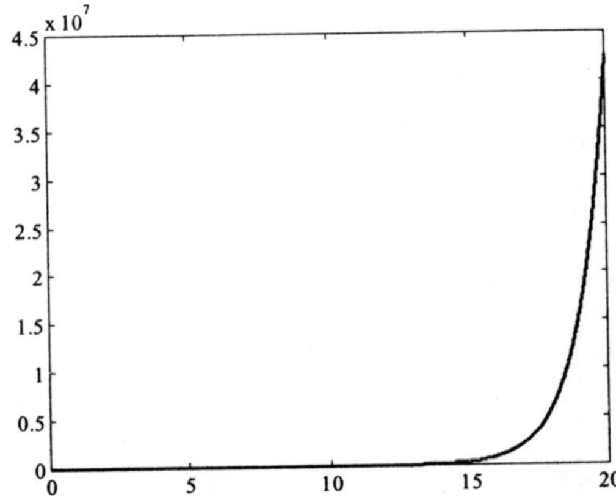

Figure 7.10b. *Modified Bessel function of the first type* $I_1(z)$

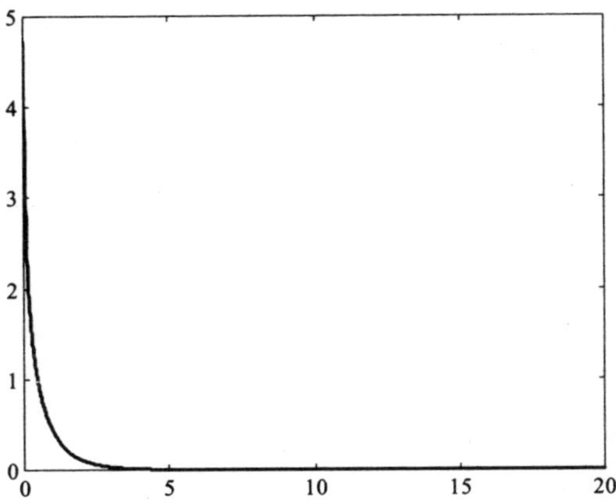

Figure 7.10c. *Modified Bessel function of the second type* $K_0(z)$

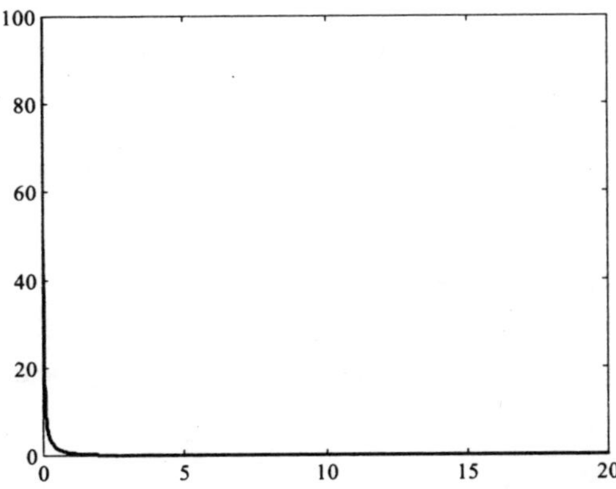

Figure 7.10d. *Modified Bessel function of the second type* $K_1(z)$

The general solution of $h_n(z)$ is obtained by cumulating the various solutions:

$$h_n(z) = A_n J_n(z) + B_n Y_n(z) + C_n I_n(z) + D_n K_n(z). \quad [7.104]$$

Naturally, it is advisable to fix the value of the constants using boundary conditions, but we will make this calculation in the following sections.

7.7.2. *Vibration modes of the full circular plate clamped at the edge*

The modal system stems from verifying the equation of motion and the boundary conditions. Using the solutions [7.104] obtained in section 7.7.1, we only have to verify the boundary conditions:

$$W(a,\theta,t)=0 \Rightarrow h_n(a)=0 \qquad [7.105]$$

and:

$$\frac{\partial W}{\partial r}(a,\theta,t)=0 \Rightarrow \frac{dh_n}{dr}(a)\,. \qquad [7.106]$$

Passing from writing the conditions over $W(r,\theta,t)$ to those over $h_n(r)$ in the preceding relations once again arises from the use of the properties of orthogonality of the functions $\cos n\theta$ and $\sin n\theta$.

The function $h_n(x)$ has four constants of integration and there are two boundary conditions; therefore, there is an apparent lack of information to calculate the constants. However, two of the functions of the solution [7.104] are not physical, since they lead to infinite displacements at the center of the plate and must thus be removed. We thus consider that:

$$h_n(r)=A_n J_n(kr)+C_n I_n(kr)\,. \qquad [7.107]$$

The application of the boundary conditions leads to the linear system:

$$\begin{pmatrix} J_n(\lambda) & I_n(\lambda) \\ \frac{dJ_n}{dz}(\lambda) & \frac{dI_n}{dz}(\lambda) \end{pmatrix}\begin{Bmatrix} A_n \\ C_n \end{Bmatrix}=\begin{Bmatrix} 0 \\ 0 \end{Bmatrix} \qquad [7.108]$$

$$\text{with: } \lambda = ka \quad \text{and} \quad k=\omega\sqrt{\frac{\rho h}{D}} \qquad [7.109]$$

To have a non-trivial solution, the linear system must have a zero determinant. We thus have:

$$J_n(\lambda)\frac{dI_n}{dz}(\lambda)-\frac{dJ_n}{dz}(\lambda)\; I_n(\lambda)=0\,. \qquad [7.110]$$

We can use the known relations for the derivation of the Bessel functions:

$$\lambda \frac{dJ_n}{d\lambda}(\lambda) = nJ_n(\lambda) - \lambda J_{n+1}(\lambda) \quad \text{and} \quad \lambda \frac{dI_n}{d\lambda}(\lambda) = nI_n(\lambda) + \lambda I_{n+1}(\lambda)$$

to rewrite the determinant:

$$J_n(\lambda)\ I_{n+1}(\lambda) + I_n(\lambda)\ J_{n+1}(\lambda) = 0. \qquad [7.111]$$

The solution of [7.111] is performed with the aid of a computer. The values have been tabulated in [LEI 93]; here we draw a short table thereof:

	$(\lambda_n^1)2$	$(\lambda_n^2)2$	$(\lambda_n^3)2$	$(\lambda_n^4)2$
n = 0	10.2158	39.771	89.104	158.183
n = 1	21.26	60.82	120.08	199.06
n = 2	34.88	84.58	153.81	242.71

Table 7.5. *First roots of equation [7.111] for various values of n. Normal angular frequencies are given by* $\omega_{nj} = \frac{1}{a^2}\sqrt{\frac{D}{\rho h}}(\lambda_n^j)^2$ *n for circumferential index, and j for radial index*

We obtain the solution for the radial shape thanks to equations [7.107] and [7.108] by normalizing the constant A_n to one:

$$h_n^j(r) = J_n(k_n^j r) + C_n^j\ I_n(k_n^j r) \qquad [7.112]$$

$$\text{with: } k_n^j = \frac{\lambda_n^j}{a} \quad \text{and} \quad C_n^j = -\frac{J_n(k_n^j\ a)}{I_n(k_n^j\ a)}. \qquad [7.113]$$

In short, the modal movements of clamped circular plates are provided by the following expressions:

– modes symmetrical in θ:

$$W_n^j(r,\theta,t) = (E_n^j \cos\omega_n^j t + F_n^j \sin\omega_n^j t)\ \cos n\theta\ (J_n\ (k_n^j r) + C_n^j I_n\ (k_n^j r)) \qquad [7.114]$$

for $n = 0, \ldots, \infty$;

– modes anti-symmetrical in θ:

$$W_n^j(r,\theta,t) = (E_n^j \cos\omega_n^j t + F_n^j \sin\omega_n^j t) \quad \sin n\theta \, (J_n(k_n^j r) + C_n^j I_n(k_n^j r))$$

[7.115]

for $n = 1, \ldots, \infty$.

The symmetrical and anti-symmetrical modes have the same normal angular frequencies given by expression [7.116]:

$$\omega_{nj} = \frac{1}{a^2}\sqrt{\frac{D}{\rho h}}\,(\lambda_n^j)2. \qquad [7.116]$$

The symmetrical and anti-symmetrical mode shapes are in fact identical; we obtain the anti-symmetrical ones by making the symmetrical ones turn by 90°. Mode 0 is particular, since it does not have a corresponding anti-symmetrical mode.

Mode shapes are provided by the two expressions:

$$f_{n,sym}^j(r,\theta) = \cos(n\theta)\left(J_n\left(k_n^j r\right) + C_n^j I_n\left(k_n^j r\right)\right) \qquad [7.117]$$

and:

$$f_{n,asym}^j(r,\theta) = \sin(n\theta)\left(J_n\left(k_n^j r\right) + C_n^j I_n\left(k_n^j r\right)\right). \qquad [7.118]$$

Each mode is thus characterized by two indices, j and n. The index n defines the number of nodal diameters, whilst the index j defines the number of nodal circles (equal to $j-1$). Table 7.6 gives the radius of nodal circles for some modes.

	$j=1$	$j=2$	$j=3$	$j=4$
$n=0$		0.379	0.255 0.583	0.191 0.439 0.688
$n=1$		0.489	0.350 0.640	0.272 0.497 0.721
$n=2$		0.559	0.414 0.679	0.330 0.540 0.746

Table 7.6. *Radii of nodal circles (R/a), for some modes of the clamped circular plate*

Figures 7.11 and 7.12 give a visualization of the mode shapes of the modes $(n = 1, j = 1)$ and $(n = 2, j = 2)$.

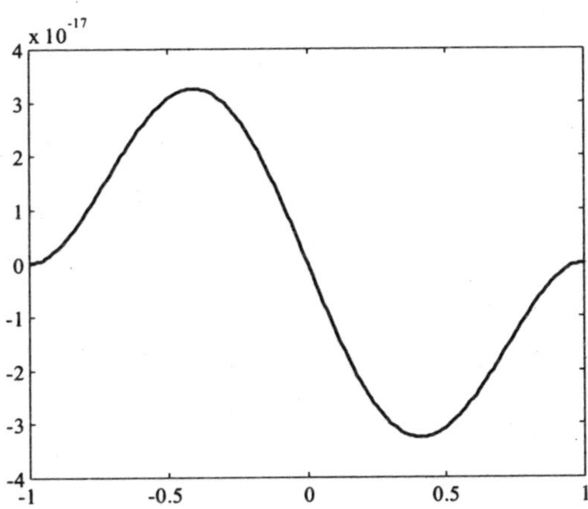

Figure 7.11. *Diametrical cross-section of the shape of the first mode n=1.* $f^{1}_{1,sym}$

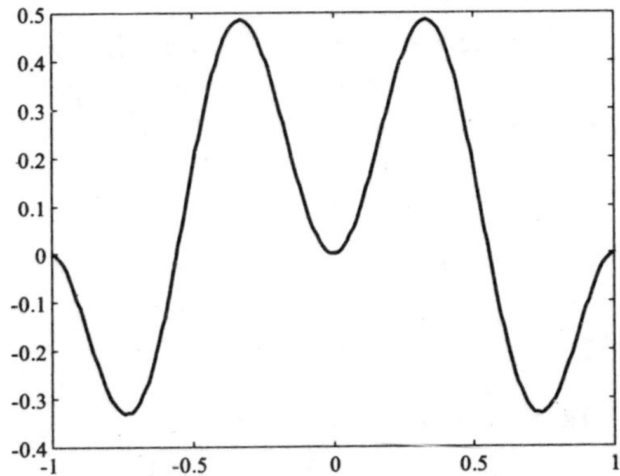

Figure 7.12. *Diametrical cross-section of the shape of the first mode n=2.* $f^{2}_{2,sym}$

Let us note, finally, that this technique of calculation can be used for all the types of boundary conditions: support, free edge, elastic support, guided edge, under the condition that the boundary condition remains the same in any point of the periphery of the plate. The difficulty for circular plates is due to the use of the Bessel functions. The use of tables of values used to be necessary but computers have completely eliminated this difficulty since the Bessel functions are now available in commercial softwares.

7.7.3. *Modal system of a ring-shaped plate*

We will cover this example very briefly in order to demonstrate the difference with the full circular plate. We will take the case where the boundary conditions are expressed in the simplest manner: clamped at the external circle, with a radius a, and at the internal circle, with a radius b, defining the limits of the plate.

The boundary conditions are thus:

$$W(a,\theta,t)=0 \;,\; W(b,\theta,t)=0 \;,\; \frac{\partial W}{\partial r}(a,\theta,t)=0 \text{ and } \frac{\partial W}{\partial r}(b,\theta,t)=0 . \qquad [7.119]$$

Again adopting the solutions [7.100], the boundary conditions [7.119] lead to:

$$h_n(a)=0 \;,\; h_n(b)=0 \;,\; \frac{dh_n}{dr}(a)=0 \text{ and } \frac{dh_n}{dr}(b)=0 . \qquad [7.120]$$

It now remains to replace $h_n(r)$ by its expression [7.104] which verifies the equation of motion to obtain the linear system [7.121]. It should be noted that for ring-shaped plates, the complete solution must be used as opposed to the case of full circular plates, where the singular functions were to be removed in [7.104] at the origin.

$$\begin{pmatrix} J_n(ka) & Y_n(ka) & I_n(ka) & K_n(ka) \\ J_n(kb) & Y_n(kb) & I_n(kb) & K_n(kb) \\ \frac{dJ_n}{dr}(ka) & \frac{dY_n}{dr}(ka) & \frac{dI_n}{dr}(ka) & \frac{dK_n}{dr}(ka) \\ \frac{dJ_n}{dr}(kb) & \frac{dY_n}{dr}(kb) & \frac{dI_n}{dr}(kb) & \frac{dK_n}{dr}(kb) \end{pmatrix} \begin{Bmatrix} A_n \\ B_n \\ C_n \\ D_n \end{Bmatrix} = \begin{Bmatrix} 0 \\ 0 \\ 0 \\ 0 \end{Bmatrix} . \qquad [7.121]$$

To obtain non-trivial solutions of equation [7.121], it is necessary that the determinant be nil. This condition provides the characteristic equation, which is satisfied for certain values of k, that is, for certain values of the angular frequency since k and ω are connected by the relation of dispersion:

$$k=\omega\sqrt{\frac{\rho h}{D}} . \qquad [7.122]$$

We provide some results obtained from published works ([LEI 93]), in Table 7.7.

Circumferential modal index	b/a = 0.1	b/a = 0.3	b/a = 0.5	b/a = 0.7
n = 0	27.3	45.2	89.2	248
n = 1	28.4	46.6	90.2	249
n = 2	36.7	51	93.3	251

Table 7.7. *Normal adimensional angular frequencies* $\omega a^2\sqrt{\frac{\rho}{D}}$ *for the first modes with circumferential indices 0, 1 and 2, according to the ratio of the internal and external radii of a ring-shaped plate*

7.8. Conclusion

In this chapter we have presented vibrations of rectangular and circular plates. Our discourse has been, however, mainly geared towards rectangular plates.

For rectangular plates, the method of separation of space variables provides analytical solutions in certain cases of boundary conditions, but only gives the trivial solution for the majority of cases. The problem of plates is thus of a superior order of difficulty to that of beams and even in *a priori* simple cases there is no analytical solution.

The classical techniques of approximate calculation are based on the Rayleigh-Ritz method. However, we did not exploit this path in this chapter because it will be the subject of a specific approach and instead we preferred a lesser known method of approximation, the edge effects method, which has the effect of better converging for the modes of a higher rank, as opposed to the methods of traditional discretization, which converge better for the first modes. This method can lead to incorrect results when two opposite edges are free; it would, therefore, be advisable to use it with care.

For circular plates a minor difficulty is due to the fact that the solutions are expressed by Bessel functions; however, modern data processing means make it possible to get rid of the tables of values and finally overcome the difficulty of calculation of these functions. The cases of circular and ring-shaped plates were demonstrated to highlight the parts of the equation of motion solution to be preserved in calculation.

From a general point of view, the vibrations of plates lead to the same phenomena as for beams: the existence of modes characterized by normal angular frequency and mode shapes. If the basic phenomena are identical, it is, however, advisable to note a quantitative difference: the density of modes in a given frequency band is much stronger for beams than for plates.

Chapter 8

Introduction to Damping: Example of the Wave Equation

8.1. Introduction

In the previous chapters we have described free vibratory movements of elastic solids. However, an important parameter was neglected: damping. The object of this chapter is to show its influence at the level of physical phenomena that it introduces, as well as at the level of mathematical difficulties that it raises with respect to the orthogonality of the modes.

The present discourse is based on the wave equation which describes the vibrations of beams in longitudinal or torsion movement, as well as of cords and sound pipes. The results naturally extend to more complicated systems, although it did not appear pertinent to us to present the calculations of complex cases considering how heavy they are.

Damping of a structure results from a loss of energy arising from several physical phenomena which are, generally, difficult to apprehend. Certain types of dissipation do not affect linearity, whereas others, such as solid friction, are strongly nonlinear. To draw the attention of the reader to the importance of damping for the free vibratory response of mechanical systems, we propose a small, easily realizable experiment, which clearly demonstrates the potential uses of dissipation.

We would need two stemmed crystal glasses (more ordinary glasses can be also used), a small spoon, a bottle of water and a champagne bottle. Fill three-quarters of the first glass with water, and the second one with champagne, then gently tap the two glasses with the spoon. We observe that the glass of water makes a ringing sound whereas the champagne glass makes a muted noise (Figure 8.1); this is the acoustic demonstration of very different vibratory states of the two glasses. Champagne dampens much more and stops the vibrations quickly, causing the muted noise. The explanation is obvious to anyone who has used a bicycle pump to inflate a tire. The compression of the volume of air heats up the pump and acts as thermal dissipation of the mechanical energy. It is the same phenomenon which explains the damping capacity of liquids with gas bubbles, the vibrations transmitted to the liquid acting on the bubbles as multiple small pumps. Let us note that if the reader does not have champagne, any carbonated beverage will be sufficient. A second characteristic fact must be stressed: heating effects related to damping are weak, and one would have to tap long and hard to heat the champagne. This tendency can be generalized to mechanical systems and explains the difficulty in measuring the loss factors precisely, considering their very low values.

This analysis of the phenomenon of two glasses is anecdotal; its purpose is to give a simple image of a much more complicated phenomenon. In reality, the presence of bubbles modifies the propagation velocity of waves, lowering it quite considerably. The spectrum of the response is modified by a downwards shift of the of resonance frequencies, which also participates in the modification of the acoustic output.

In this chapter, we will remain within the framework of linear damping: classical viscous damping, linear viscoelasticity and dissipation by absorbing limits. These three mechanisms will lead to the concept of complex normal angular frequency, characteristic of the vibrations of damped systems. We will then study the properties of orthogonality of the normal functions and will show how to introduce the initial conditions.

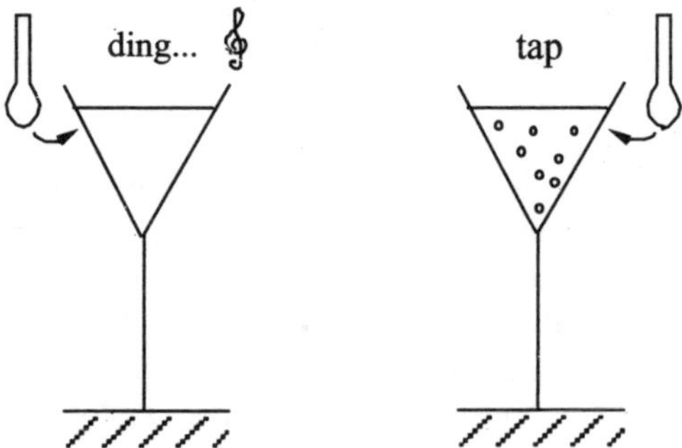

Figure 8.1. *The experiment with two glasses*

8.2. Wave equation with viscous damping

This model of damping is the simplest; it consists of introducing an additional term, proportional to speed, into the equation of motion. To be more specific, let us consider a beam in longitudinal vibration, whose equation of motion was provided in Chapter 3 (equation [3.21]) and let us introduce the term of viscous damping. The equation becomes:

$$\rho S \frac{\partial^2 U_1^0}{\partial t^2} + DS \frac{\partial U_1^0}{\partial t} - ES \frac{\partial^2 U_1^0}{\partial x^2} = 0 . \qquad [8.1]$$

Introducing the celerity of waves c and the variable U instead of U_1^0, to simplify the notation, the general equation, which interests us, takes the form:

$$\frac{\partial^2 U}{\partial t^2} + \delta \frac{\partial U}{\partial t} - c^2 \frac{\partial^2 U}{\partial x^2} = 0 . \qquad [8.2]$$

In this expression, $c = \sqrt{E/\rho}$ is the celerity of longitudinal waves and δ is the damping parameter homogeneous to the inverse of a time.

We solve equation [8.2] by separation of variables:

$$U(x,t) = f(x)\ g(t) . \qquad [8.3]$$

Introducing [8.3] into [8.2] and separating the variables, it follows:

$$\frac{d^2 g}{dt^2} + \delta \frac{dg}{dt} - K^2 g(t) = 0 , \qquad [8.4]$$

$$c^2 \frac{d^2 f}{dx^2} - K^2 f = 0 , \qquad [8.5]$$

where K^2 is a constant.

Let us solve the temporal equation [8.4] using the traditional method. After all the calculations it follows:

$$g(t) = A\,e^{-\delta/2\,t}\,e^{\left(\sqrt{\delta^2+4K^2}/2\right)t} + B\,e^{-\delta/2\,t}\,e^{\left(-\sqrt{\delta^2+4K^2}/2\right)t}. \quad [8.6]$$

A similar calculation leads to the solution of the space equation [8.5]:

$$f(x) = Ce^{-Kx/c} + De^{Kx/c}. \quad [8.7]$$

The space-time solution $U(x, t)$ is thus:

$$U(x,t) = \left(A\,e^{-\delta/2\,t}\,e^{\left(\sqrt{\delta^2+4K^2}/2\right)t} + B\,e^{-\delta/2\,t}\,e^{\left(-\sqrt{\delta^2+4K^2}/2\right)t}\right)\left(Ce^{-Kx/c} + De^{Kx/c}\right). \quad [8.8]$$

For the moment, the constant K which arises from the separation of variables is not fixed and may be complex. To determine it, we have to introduce boundary conditions for the beam. We choose non-dissipative boundary conditions, of the clamped type at two ends:

$$U(0,t) = 0 \quad \forall t, \quad [8.9a]$$

$$U(L,t) = 0 \quad \forall t. \quad [8.9b]$$

Applying these boundary conditions to the solution [8.8] yields:

$$C + D = 0, \quad [8.10]$$

$$Ce^{-KL/c} + De^{KL/c} = 0. \quad [8.11]$$

That is, if we are interested in a non-trivial solution:

$$e^{-KL/c} - e^{KL/c} = 0 \quad [8.12]$$

and:

$$C = -D \neq 0. \quad [8.13]$$

Let us consider a complex K of the form:

$$K = \alpha + j\omega . \qquad [8.14]$$

Equation [8.12] becomes:

$$-\cos\left(\frac{\omega}{c}L\right)\ \mathrm{sh}\left(\frac{\alpha}{c}L\right) - j\sin\left(\frac{\omega}{c}L\right)\ \mathrm{ch}\left(\frac{\alpha}{c}L\right) = 0 . \qquad [8.15]$$

Separating the real and imaginary parts:

$$\cos\left(\frac{\omega}{c}L\right)\ \mathrm{sh}\left(\frac{\alpha}{c}L\right) = 0 \quad \text{and} \quad \sin\left(\frac{\omega}{c}L\right)\ \mathrm{ch}\left(\frac{\alpha}{c}L\right) = 0 \cdot \qquad [8.16]$$

If α is different from 0, the system of equations [8.16] does not have a solution.

If $\alpha = 0$, the system admits an infinity of solutions ω such that:

$$\sin\left(\frac{\omega}{c}L\right) = 0 \Leftrightarrow \omega = \omega_n = \frac{n\pi}{L}c\ ,\ n = 1\ ,\ldots,\infty . \qquad [8.17]$$

In short, respecting non-dissipative boundary conditions [8.9] leads to purely imaginary values of K:

$$K_n = j\omega_n = jc\frac{n\pi}{L} . \qquad [8.18]$$

The constants C and D are opposed as indicated by [8.13] but are not defined in a unique fashion. Grouping all these results, the solution of space [8.7] is particularized for the mode shapes $f_n(x)$ defined for each mode n:

$$f_n(x) = C_n \sin\left(\frac{n\pi}{L}x\right) \qquad [8.19]$$

the constant C_n being arbitrary and non-nil.

The space-time solution $U(x, t)$ for these boundary conditions is obtained taking [8.18] into account in [8.8] and regrouping the possible solutions:

$$U(x,t) = \sum_{n=1}^{\infty} g_n(t)\ f_n(x)\,. \qquad [8.20]$$

$f_n(x)$ is the mode shape of mode n, given by [8.19].

$$g_n(t) = \left(A_n\ e^{\left(j\sqrt{-\delta^2+4\omega_n^2}/2\right)t} + B_n\ e^{\left(-j\sqrt{-\delta^2+4\omega_n^2}/2\right)t} \right) e^{-\delta/2\ t}\,.$$

The arbitrary value of the constant C_n is not important, since we can incorporate it in the constants A_n and B_n of the temporal solution. This situation makes it possible to simplify the expression of mode shape adopting $C_n = 1$ without loss of generality. This is the way in which we will proceed.

For weak damping, which is common in practice:

$$\delta < 2\omega_n \qquad [8.21]$$

the temporal solution is oscillating with the angular frequency Ω_n given in [8.22].

$$\Omega_n = \sqrt{\omega_n^2 - (\delta/2)^2}\,. \qquad [8.22]$$

It is the damped normal angular frequency as opposed to the non-damped normal angular frequency ω_n which characterizes the vibrations when $\delta = 0$ and equation [8.2] is reduced to the standard wave equation.

The temporal response of mode n, $g_n(t)$ can also be expressed in the form:

$$g_n(t) = A_n e^{j\lambda_n t} + B_n e^{-j\lambda_n^* t} \qquad [8.23]$$

with: $\lambda_n = \Omega_n + j\delta/2\,.$ [8.24]

The quantity λ_n is called the complex normal angular frequency of mode n: the real part represents the oscillating nature of the solution, while the imaginary part represents the dissipative character of the movement which weakens over time.

Note:

– if λ_n is a complex normal angular frequency, then its complex conjugate λ_n^* is also one;

– when $\delta = 0$, i.e. for a non-damped system, complex normal angular frequency becomes real. The concept of complex normal angular frequency is thus associated with that of damping;

– for the model of damping considered here, we observe that the imaginary part of λ_n given by [8.24] is independent of the mode. That will be different in the models of damping which we consider hereafter.

Let us examine the vibrations of the damped beam following these initial conditions:

$$\begin{cases} U(x,0) = d(x) \\ \dfrac{\partial U}{\partial t}(x,0) = 0. \end{cases} \qquad [8.25]$$

Let us use the general form of the solution [8.20] under the two initial conditions [8.25]. It follows:

$$U(x,0) = \sum_{n=1}^{\infty} (A_n + B_n) \sin\left(\frac{n\pi}{L} x\right) = d(x), \qquad [8.26]$$

$$\frac{\partial U}{\partial t}(x,0) = \sum_{n=1}^{\infty} j(A_n\lambda_n - \lambda_n^* B_n) \sin\left(\frac{n\pi}{L} x\right) = 0. \qquad [8.27]$$

To determine the constants A_n and B_n, it is enough to use the properties of orthogonality, which amounts to breaking up initial displacements and speeds into a Fourier series of sine and to identify then term by term. Equations [8.26] and [8.27] yield respectively:

$$A_n + B_n = \frac{2}{L}\int_0^L d(x) \sin\left(\frac{n\pi}{L} x\right) dx, \qquad [8.28]$$

$$A_n\lambda_n = B_n\lambda_n^*. \qquad [8.29]$$

Relation [8.29] leads to:

$$A_n = B_n \frac{\lambda_n^*}{\lambda_n}. \qquad [8.30]$$

From [8.28] we draw:

$$B_n = \frac{\lambda_n}{\Omega_n L} \int_0^L d(x) \sin\left(\frac{n\pi}{L} x\right) dx \qquad [8.31]$$

and:

$$A_n = \frac{\lambda_n^*}{\Omega_n L} \int_0^L d(x) \sin\left(\frac{n\pi}{L} x\right) dx . \qquad [8.32]$$

We note that if d(x) is real, $A_n = B_n^*$; vibratory displacement following the initial conditions [8.25] is, therefore:

$$U(x,t) = \sum_{n=1}^{\infty} e^{-\delta/2t} \int_0^L \sin\left(\frac{n\pi}{L} x\right) \frac{2d(x)}{L} dx \left(\cos(\Omega_n t) + \frac{\delta}{2\Omega_n} \sin(\Omega_n t)\right) \sin\left(\frac{n\pi}{L} x\right). \qquad [8.33]$$

The global response is obviously real since it represents true movement, even if at certain points in the calculation we introduce complex quantities (in particular, complex normal angular frequency).

Figure 8.2 illustrates the time history of mode n, which is a damped sinusoid representing a dissipation of energy during movement, which is stronger the larger δ is.

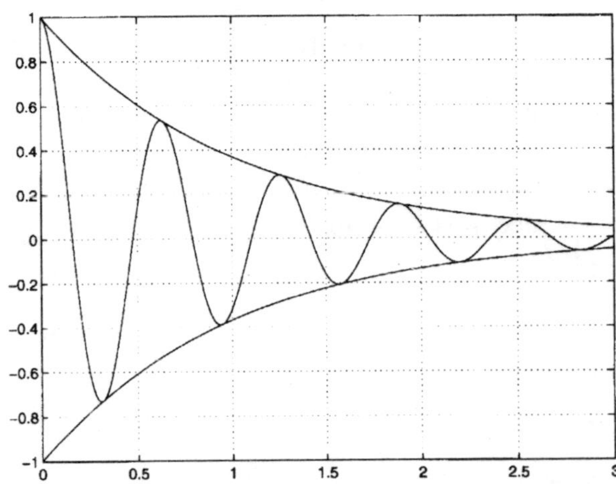

Figure 8.2. *Time history of mode n*

8.3. Damping by dissipative boundary conditions

8.3.1. *Presentation of the problem*

This type of dissipation in structures is very often encountered in practice, the losses at the boundaries often being predominant by comparison to other types of dissipation. The phenomena of losses at the boundaries result from the coupling with vibrating systems related to the medium considered. The exact description of these couplings is very complex; an approximate modeling, which we will examine, introducing the overall losses by a force of dissipation proportional to the vibratory speed of the boundaries is often preferred.

Thus, we consider a beam without damping, in longitudinal vibrations, embedded at 0 and having an absorbing boundary in L (see Figure 8.3).

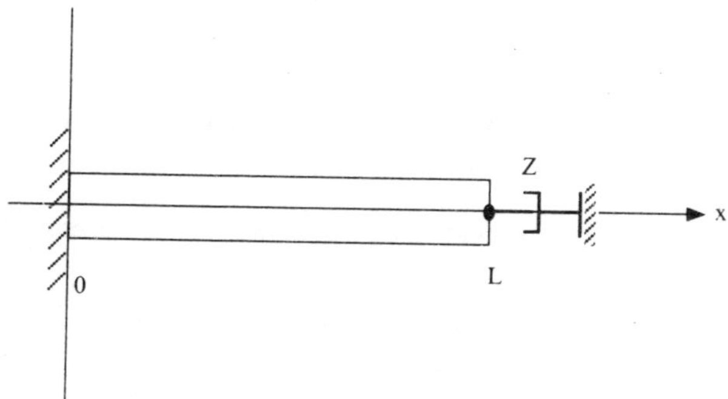

Figure 8.3. *Beam with a dissipative boundary*

The equations which govern the vibratory movement are:

$$\rho S \frac{\partial^2 U}{\partial t^2} - ES \frac{\partial^2 U}{\partial x^2} = 0\,, \qquad [8.34]$$

$$U(0,t) = 0\,, \qquad [8.35]$$

$$ES \frac{\partial U}{\partial x}(L,t) + Z \frac{\partial U}{\partial t}(L,t) = 0\,. \qquad [8.36]$$

The real constant Z translates the absorbing property of the boundary at L. It introduces neither elastic nor mass effects. To take these effects into account it would be necessary to replace [8.36] by:

$$ES\frac{\partial U}{\partial x}(L,t) + M\frac{\partial^2 U}{\partial t^2}(L,t) + Z\frac{\partial U}{\partial t}(L,t) + KU(L,t) = 0 \quad [8.37]$$

with M and K being respectively the mass and the stiffness of the boundary (see Figure 8.4).

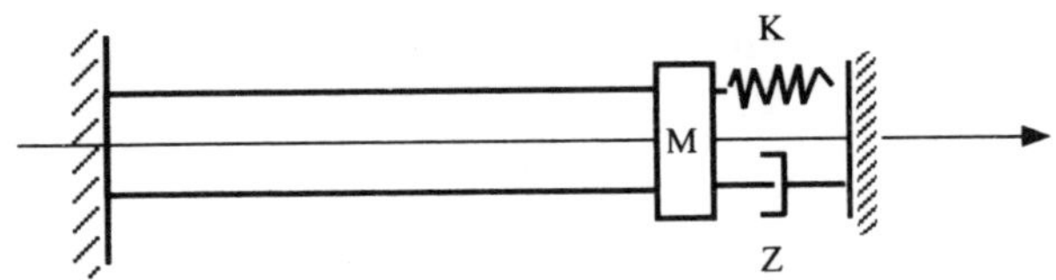

Figure 8.4. *Beam with a boundary condition, with stiffness K, mass M and damping Z.*

The solution would be very heavy, and thus we will only consider the purely dissipative boundary condition [8.36], which is sufficient to describe the effect of damping at the boundaries.

8.3.2. *Solution of the problem*

The general solution [8.34] obtained by separation of the variables is:

$$U(x,t) = f(x)\ g(t) \quad [8.38]$$

with:

$$f(x) = Ce^{-K\,x/c} + De^{K\,x/c} \quad [8.39]$$

and:

$$g(t) = Ae^{-Kt} + Be^{Kt}. \quad [8.40]$$

In these expressions, $c = \sqrt{E/\rho}$ is the speed of the longitudinal waves and K is a complex constant.

The boundary condition [8.36] implies:

$$C = -D, \quad [8.41]$$

$$A\left(\frac{ES}{c}\left(e^{-K\,L/c}+e^{K\,L/c}\right)+Z\left(e^{-K\,L/c}-e^{K\,L/c}\right)\right)C=0, \qquad [8.42]$$

$$B\left(\frac{ES}{c}\left(e^{-K\,L/c}+e^{K\,L/c}\right)-Z\left(e^{-K\,L/c}-e^{K\,L/c}\right)\right)C=0. \qquad [8.43]$$

First we will determine the solutions of [8.42], then those resulting from [8.43]; the total solution will be the sum of all the solutions.

Let us examine equation [8.42]. We have either A = 0, C = 0, or:

$$\frac{ES}{c}\left(e^{-K\,L/c}+e^{K\,L/c}\right)+Z\left(e^{-K\,L/c}-e^{K\,L/c}\right)=0.$$

The first two possibilities lead to the trivial solution. To discuss the third possibility, let us pose $K=\alpha+j\omega$. It follows:

$$\left(\frac{ES}{c}+Z\right)e^{-\alpha L/c}e^{-j\Omega L/c}+\left(\frac{ES}{c}-Z\right)e^{\alpha L/c}e^{j\Omega L/c}=0. \qquad [8.44]$$

By transforming the complex exponential and separating the real and the imaginary part, we obtain:

$$\cos(\Omega L/c)\left(\left(\frac{ES}{c}+Z\right)e^{-\alpha L/c}+\left(\frac{ES}{c}-Z\right)e^{\alpha L/c}\right)=0 \qquad [8.45]$$

and:

$$\sin(\Omega L/c)\left(-\left(\frac{ES}{c}+Z\right)e^{-\alpha L/c}+\left(\frac{ES}{c}-Z\right)e^{\alpha L/c}\right)=0. \qquad [8.46]$$

There are two possibilities to satisfy [8.45] and [8.46]:

a) First possibility:

$$\cos(\Omega L/c)=0 \qquad [8.47]$$

and:

$$\left(\left(\frac{ES}{c}+Z\right)e^{-\alpha L/c}=\left(\frac{ES}{c}-Z\right)e^{\alpha L/c}\right). \qquad [8.48]$$

Let us consider the case of weak damping, then $Z < ES/c$ and let us trace in Figure 8.5 the two curves $\left(\frac{ES}{c}+Z\right)e^{-\alpha L/c}$ and $\left(\frac{ES}{c}-Z\right)e^{\alpha L/c}$. The intersection shows that there is a value solving equation [8.48]: $\alpha = \delta/2 > 0$.

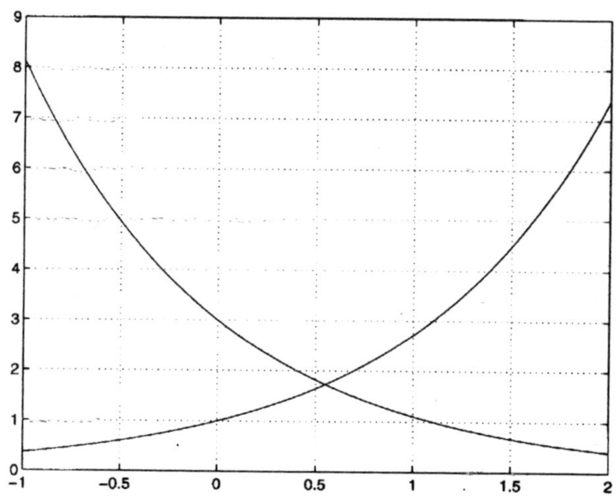

Figure 8.5. *Root of equation [8.48]*

Equation [8.47] has an infinite number of possible solutions, given by the relation:

$$\Omega\frac{L}{c}=\frac{2n-1}{2}\pi,$$

i.e. an infinite number of normal angular frequency:

$$\Omega_n = c\frac{2n-1}{2}\frac{\pi}{L}. \qquad [8.49]$$

In short, there exists a first set of non-trivial solutions. For $n = 1, \ldots, \infty$:

$$U(x,t) = A_n C_n e^{-j(\Omega_n - j\delta/2)\,t}\left(e^{-j(\Omega_n - j\delta/2)\,x/c} - e^{j(\Omega_n - j\delta/2)\,x/c}\right). \qquad [8.50]$$

b) Now let us consider the second possibility of solution of [8.45] and [8.46]:

$$\sin(\Omega L/c) = 0 \qquad [8.51]$$

and:

$$\left(\left(\frac{ES}{c} + Z\right)e^{-\alpha L/c} = -\left(\frac{ES}{c} - Z\right)e^{\alpha L/c}\right). \qquad [8.52]$$

Let us trace in Figure 8.6 the two curves $\left(\frac{ES}{c} + Z\right)e^{-\alpha L/c}$ and $-\left(\frac{ES}{c} - Z\right)e^{\alpha L/c}$. They do not have an intersection, and equation [8.52] cannot be satisfied. This second possibility does not give a solution.

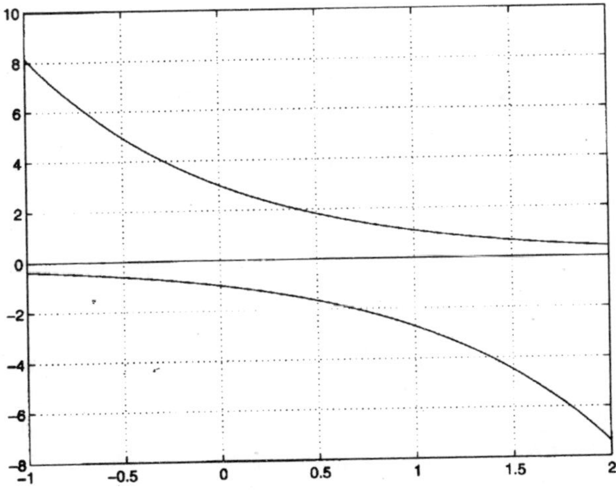

Figure 8.6. *Resolution of equation [8.51] (no solution)*

Let us examine the solutions of equation [8.43]. Using an approach similar to the discussion of equation [8.42], we obtain a range of solutions in the following form (it is enough to change Z into $(-Z)$ and α into $(-\alpha)$ in the previous discussion):

$$U(x, t) = B_n C_n \, e^{j(\Omega_n + j\delta/2)\, t} \left(e^{-j(\Omega_n + j\delta/2)\, x/c} - e^{j(\Omega_n + j\delta/2)\, x/c} \right) \qquad [8.53]$$

with Ω_n given by [8.49].

Combining all of the found solutions, we obtain the general solution (we have posed $A_n C_n = D_n$ and $B_n C_n = -E_n$):

$$U(x, t) = \sum_{n=1}^{\infty} D_n \, e^{-j(\Omega_n - j\delta/2)\, t} \left(e^{-j(\Omega_n - j\delta/2)\, x/c} - e^{j(\Omega_n - j\delta/2)\, x/c} \right) + E_n \, e^{j(\Omega_n + j\delta/2)\, t} \left(e^{j(\Omega_n + j\delta/2)\, x/c} - e^{-j(\Omega_n + j\delta/2)\, x/c} \right), \qquad [8.54]$$

i.e.:

$$U(x, t) = \sum_{n=1}^{\infty} D_n \, e^{-j\lambda_n^* t} f_n^*(x) + E_n \, e^{j\lambda_n t} f_n(x). \qquad [8.55]$$

Once again, as in the case covered in section 8.2, we find the complex normal angular frequency λ_n and λ_n^*, characteristic of the presence of damping .

$$\lambda_n = \Omega_n + j\frac{\delta}{2} \quad \text{and} \quad \lambda_n^* = \Omega_n - j\frac{\delta}{2}.$$

The imaginary parts of complex normal angular frequency are also independent of the mode here.

Also let us note that Ω_n, the real part of complex normal angular frequency, is independent of damping at the boundaries; this is a notable difference with the case of distributed damping covered in section 8.2. There is no damped normal angular frequency Ω_n here differing from the non-damped normal angular frequency ω_n .

However, there seems to be a considerable difference by comparison to the previous case, since the normal deformations $f_n(x)$ and $f_n^*(x)$, associated with the normal angular frequency λ_n and λ_n^*, are complex conjugate, whereas in the general solution [8.19] found in section 8.2, mode shapes were real and identical.

Figure 8.7 presents the real and imaginary parts of mode shape of the 2nd order, $f_2(x)$; the imaginary part is weak taking into account the considered case characteristic of a weak damping ($\delta = 0,02\ \Omega_2$). Figure 8.8 presents the real and imaginary parts of the same mode shape when damping is stronger. We note that the imaginary part adopts a much greater importance.

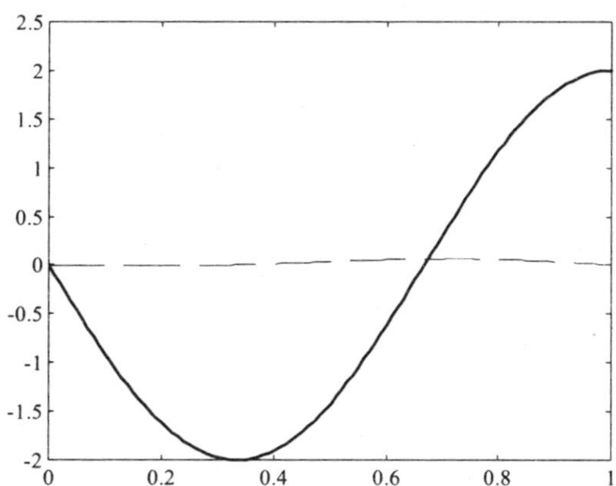

Figure 8.7. *Real ——— and imaginary - - - - . part of the mode shape of the second complex mode. 1 meter long beam, case of weak damping:* $\delta = 0.02\ \Omega_2$

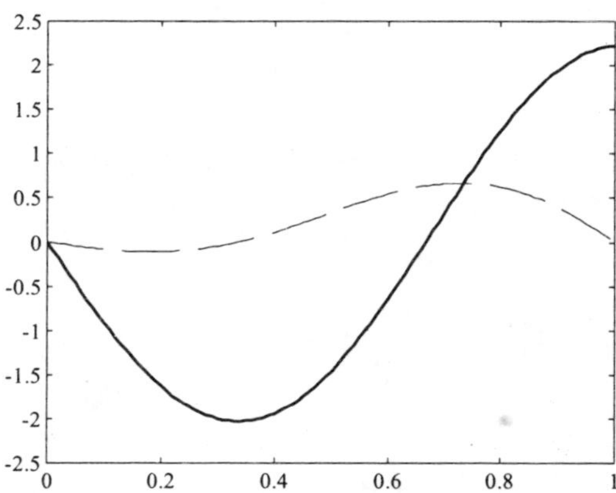

Figure 8.8. *Real ——— and imaginary - - - - . part of the mode shape of the second complex mode. 1 meter long beam, case of strong damping:* $\delta = 0.2\ \Omega_2$

What is the physical significance of complex mode shapes? To answer this question we point out the physical significance of the modes which we have already highlighted during the study of free vibrations of non-damped systems. A normal mode represents a state of displacement characterized by movements in the same or in opposed phase of all the points of the structure. When the modes are complex, the movement associated with a mode no longer occurs in the same or in opposed phase, but with a phase shift which could be characterized thanks to the real and imaginary parts of the mode shape. This phase is variable with the point of the beam considered; consequently, the nodes of vibration generally no longer exist, except if the real and imaginary parts of the mode shape are nil at the same points.

8.3.3. *Calculation of the vibratory response*

The calculation of the vibratory response to initial conditions poses the problem of orthogonality of the mode shape, which is here quite particular. We will come back to it in detail in section 8.5 while here we give its properties without demonstration:

$$(\lambda_p^* - \lambda_n) \int_0^L f_n(x)\, f_p^*(x)\, dx + j\frac{Zc^2}{ES} f_n(L)\, f_p^*(L) = 0, \qquad [8.56]$$

$$(\lambda_p - \lambda_n^*) \int_0^L f_n^*(x)\, f_p(x)\, dx - j\frac{Zc^2}{ES} f_n^*(L)\, f_p(L) = 0, \qquad [8.56']$$

$$(\lambda_p + \lambda_n) \int_0^L f_n(x)\, f_p(x)\, dx - j\frac{Zc^2}{ES} f_n(L)\, f_p(L) = 0 \ \text{ if } \ \lambda_n \neq \lambda_p, \qquad [8.57]$$

$$(\lambda_p^* + \lambda_n^*) \int_0^L f_n^*(x)\, f_p^*(x)\, dx + j\frac{Zc^2}{ES} f_n^*(L)\, f_p^*(L) = 0 \ \text{ if } \ \lambda_n \neq \lambda_p. \qquad [8.57']$$

Let us note that the properties [8.56] and [8.56'] are true even if $n = p$, whereas [8.57] and [8.57'] are not.

Let us suppose that the initial conditions given at $t = 0$ are:

$$U(x,0) = d(x), \qquad [8.58]$$

$$\frac{\partial U}{\partial t}(x,0) = v(x) . \qquad [8.59]$$

Let us use the decomposition [8.55] under the initial conditions [8.58] and [8.59]:

$$\sum_{n=1}^{\infty} D_n f_n^*(x) + E_n f_n(x) = d(x) , \qquad [8.60]$$

$$\sum_{n=1}^{\infty} - D_n \lambda_n^* f_n^*(x) + E_n \lambda_n f_n(x) = - j\, v(x) . \qquad [8.61]$$

To be able to use the properties of orthogonality, it is necessary to proceed in a rather special manner, as follows. Let us multiply [8.60] by $\lambda_p f_p(x)$ and [8.61] by $f_p(x)$:

$$\sum_{n=1}^{\infty} D_n \lambda_p f_p(x)\ f_n^*(x) + E_n \lambda_p f_p(x)\ f_n(x) = \lambda_p f_p(x)\ d(x) , \qquad [8.62]$$

$$\sum_{n=1}^{\infty} - D_n \lambda_n^* f_p(x)\ f_n^*(x) + E_n \lambda_n f_p(x)\ f_n(x) = - j f_p(x)\ v(x) . \qquad [8.63]$$

Let us add the two equations, then integrate the two members between 0 and L:

$$\begin{aligned} \sum_{n=1}^{\infty} D_n(\lambda_p - \lambda_n^*) \int_0^L f_p(x)\ f_n^*(x)\, dx + E_n(\lambda_n + \lambda_p) \int_0^L f_p(x)\ f_n(x)\, dx \\ = \lambda_p \int_0^L f_p(x)\ d(x)\, dx - j \int_0^L f_p(x)\ v(x)\, dx . \end{aligned} \qquad [8.64]$$

We recognize a part of the properties of orthogonality [8.56] and [8.57]; there are, however, several terms missing, which we will introduce noting that in x = L, the initial condition in displacement is verified:

$$U(L,0) = d(L) , \qquad [8.65]$$

i.e. using the decomposition [8.54]:

$$\sum D_n f_n^*(L) + E_n f_n(L) = d(L)\,. \tag{8.66}$$

Let us multiply [8.66] by $-j\dfrac{Zc^2}{ES} f_p(L)$ and sum up, member by member, with [8.64]:

$$\begin{aligned}
&\sum_{n=1}^{\infty} D_n \left((\lambda_p - \lambda_n^*) \int_0^L f_p(x)\, f_n^*(x)\, dx - j\frac{Zc^2}{ES} f_p(L)\, f_n^*(L) \right) \\
&\qquad + E_n \left((\lambda_n + \lambda_p) \int_0^L f_p(x)\, f_n(x)\, dx - j\frac{Zc^2}{ES} f_p(L)\, f_n(L) \right) \\
&= \lambda_p \int_0^L f_p(x)\, d(x)\, dx - j\int_0^L f_p(x)\, v(x)\, dx - j\frac{Zc^2}{ES} f_p(L)\, d(L).
\end{aligned} \tag{8.67}$$

Using the properties of orthogonality [8.56] and [8.57], it follows:

$$E_p = \frac{\displaystyle\int_0^L f_p(x) \left(\lambda_p d(x) - jv(x)\right) dx - j\frac{Zc^2}{ES} f_p(L)\, d(L)}{\displaystyle 2\lambda_p \int_0^L f_p^2(x)\, dx - j\frac{Zc^2}{ES} f_p^2(L)}\,. \tag{8.68}$$

To calculate D_p, it is necessary to proceed in a similar fashion, but multiplying [8.60] by $\lambda_p^* f_p^*(x)$, [8.61] by $f_p^*(x)$ and [8.66] by $j\dfrac{Zc^2}{ES} f_p^*(L)$. We then use the properties of orthogonality [8.56] and [8.57] to obtain:

$$D_p = \frac{\displaystyle\int_0^L f_p^*(x) \left(\lambda_p^*\, d(x) + jv(x)\right) dx + j\frac{Zc^2}{ES} f_p^*(L)\, d(L)}{\displaystyle 2\lambda_p^* \int_0^L f_p^{*2}(x)\, dx + j\frac{Zc^2}{ES} f_p^{*2}(L)}\,. \tag{8.69}$$

It now suffices to introduce the expressions of E_p and D_p in [8.54] to express the vibratory response of the beam with damping at the boundaries. We will not push this very technical calculation further, let us underline, however, that the final result must give a real displacement if we take real initial displacements and speeds, despite the appearance of complex calculation intermediaries (normal angular frequency and mode shapes).

Let us note in conclusion that this model of damping, localized at the boundaries, brings us to complex mode shape leads to much heavier calculations than distributed damping studied in section 8.2, which preserved real mode shapes independent of damping.

8.4. Viscoelastic beam

The hypothesis of linear viscoelasticity is that which provides the best approximation of internal dissipations in materials. It leads to a heavier formulation than those presented previously, because the stress-strain relation of material is defined by a product of convolution (see Chapter 1, equation [1.68]). In the case of a beam in longitudinal vibrations, the equations representative of the vibrations are the ones obtained in Chapter 3 (equations [3.21] and [3.12]), which we recall:

$$\rho S \frac{\partial^2 U}{\partial t^2} - \frac{\partial (S\sigma)}{\partial x} = 0 , \qquad [8.70]$$

$$\frac{\partial U}{\partial x} = \frac{\sigma}{E} . \qquad [8.71]$$

For a viscoelastic material, the stress-strain relation [8.71] is modified:

$$\frac{\partial U}{\partial x}(x,t) = \int_{-\infty}^{t} \Gamma(t-\tau) \frac{\partial \sigma}{\partial \tau}(x,\tau)\, d\tau . \qquad [8.72]$$

This law of viscoelasticity shows that longitudinal strain of the beam observed at the moment t depends on the state of stress at all prior moments. Thus, viscoelastic material has a memory effect.

We will seek the vibratory movements of the viscoelastic beam in the form:

$$U(x,t) = \overline{Y}(x)\, e^{j\lambda t} \qquad [8.73]$$

and:

$$\sigma(x, t) = \overline{\sigma}(x)\, e^{j\lambda t}. \qquad [8.74]$$

As in the previous examples, we will interpret λ as a complex normal angular frequency.

Introducing [8.73] and [8.74] into [8.70] and [8.72], we obtain:

$$-\lambda^2 \rho S \overline{Y}(x) - S \frac{d\overline{\sigma}}{dx}(x) = 0 \qquad [8.75]$$

and:

$$\frac{d\overline{Y}}{dx}(x)\, e^{j\lambda t} = j\lambda \int_{-\infty}^{t} \Gamma(t-\tau)\, e^{j\lambda\tau}\, d\tau\, \overline{\sigma}(x). \qquad [8.76]$$

To push the calculations further, it is necessary to make an assumption on the form of the function Γ. We choose:

$$\Gamma(u) = \begin{cases} \overline{\Gamma}(1 - e^{-\gamma u}) & \forall\ u > 0, \\ 0 \quad \forall\ u < 0. \end{cases} \qquad [8.77]$$

Note: the viscoelastic stress-strain relation introduces two parameters γ and $\overline{\Gamma}$. To illustrate the underlying physical properties, let us consider that a state of stress is applied abruptly at the moment T:

$$\sigma(x, \tau) = \tilde{\sigma}(x)\ H(\tau - T)$$

and thus:

$$\frac{\partial \sigma}{\partial \tau}(x, \tau) = \tilde{\sigma}(x)\ \delta(\tau - T).$$

In these expressions H is the step function and δ is the Dirac distribution.

The viscoelastic stress-strain relation yields the value of $\dfrac{\partial U}{\partial x}$:

$$\frac{\partial U}{\partial x}(x, t) = \int_{-\infty}^{t} \Gamma(t-\tau)\ \delta(\tau - T)\ d\tau\, \tilde{\sigma}(x), \qquad [8.78]$$

i.e.:

$$\frac{\partial U}{\partial x}(x, t) = \Gamma(t - T)\ \tilde{\sigma}(x) = \begin{cases} \left(1 - e^{-\gamma(t-T)}\right) \overline{\Gamma}\, \tilde{\sigma}(x) & \forall\, t > T, \\ 0 \quad \forall\, t < T. \end{cases} \qquad [8.79]$$

Figure 8.9 illustrates the phenomenon: non-abrupt strain corresponds to an abrupt application of stresses – it is the phenomenon of creep, which is more pronounced the larger γ is.

Let us introduce the function defined in [8.77] into equation [8.76]:

$$\frac{\partial \overline{Y}}{\partial x}(x)\, e^{j\lambda t} = j\lambda\overline{\Gamma} \int_{-\infty}^{t} \left(1 - e^{-\gamma(t-\tau)}\right) e^{j\lambda\tau}\, d\tau\, \overline{\sigma}(x), \qquad [8.80]$$

that is, after calculation:

$$\frac{\partial \overline{Y}}{\partial x}(x)\, e^{j\lambda t} = \overline{\Gamma}\frac{\gamma}{j\lambda + \gamma}\, \overline{\sigma}(x)\, e^{j\lambda\tau}. \qquad [8.81]$$

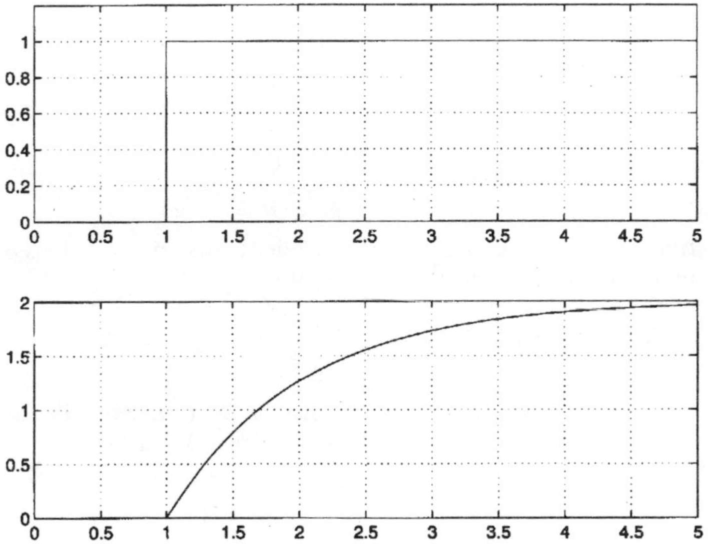

Figure 8.9. *Stress (top graph) and strain (bottom graph) of a viscoelastic beam in longitudinal vibrations versus time*

Note: in the integral [8.80], the limit at $\tau = -\infty$ which appears during integration by parts does not contribute insofar as λ has a negative imaginary part, as we will see later.

Using [8.75] and [8.81] we obtain the system of equations to be solved:

$$\begin{cases} \lambda^2 \rho S \overline{Y}(x) + S \dfrac{d\overline{\sigma}}{dx}(x) = 0 \\ \overline{\sigma}(x) = E^* \dfrac{d\overline{Y}}{dx}(x) \end{cases} \qquad [8.82]$$

where we introduced the complex Young modulus E^* defined by identification in [8.81]:

$$E^* = \frac{1 + j\lambda/\gamma}{\overline{\Gamma}}. \qquad [8.83]$$

The system [8.82] is reduced to the equation:

$$\lambda^2 \rho S \overline{Y}(x) + E^* S \frac{d^2\overline{Y}}{dx^2}(x) = 0\,. \qquad [8.84]$$

This differential equation is easily integrated:

$$\overline{Y}(x) = A e^{j\lambda\sqrt{\rho/E^*}\,x} + B e^{-j\lambda\sqrt{\rho/E^*}\,x}. \qquad [8.85]$$

To determine the values of λ, we will consider boundary conditions of the type which is clamped at both ends, that is:

$$\overline{Y}(0) = 0 \text{ and } \overline{Y}(L) = 0\,. \qquad [8.86]$$

The fact that [8.86] is verified by [8.85] leads to an infinite number of solutions of the form:

$$\sqrt{\frac{\rho}{E^*}}\,\lambda_n = \frac{n\pi}{L}\,,\ n = 1\,,\ldots,\infty\,, \qquad [8.87]$$

with:

$$\overline{Y}_n(x) = \sin\left(\frac{n\pi}{L}x\right). \qquad [8.88]$$

The relation [8.87] deserves thorough examination since E* depends on λ under the terms of [8.83]. It follows:

$$\lambda_n^2 \rho\overline{\Gamma} = \left(\frac{n\pi}{L}\right)^2 (1 + j\lambda_n/\gamma). \qquad [8.89]$$

Equation [8.89] is polynomial of the second degree:

$$\rho\overline{\Gamma}\lambda_n^2 - j\left(\frac{n\pi}{L}\right)^2 \frac{1}{\gamma}\lambda_n - \left(\frac{n\pi}{L}\right)^2 = 0. \qquad [8.90]$$

We deduce its two roots λ_n and λ_n^*:

$$\begin{cases} \lambda_n = j\left(\left(\frac{n\pi}{L}\right)^2 \frac{1}{2\rho\overline{\Gamma}\gamma}\right) + \sqrt{-\left(\frac{n\pi}{L}\right)^4 \left(\frac{1}{2\rho\overline{\Gamma}\gamma}\right)^2 + \left(\frac{n\pi}{L}\right)^2 \frac{1}{\rho\overline{\Gamma}}} \\ \lambda_n^* = -j\left(\left(\frac{n\pi}{L}\right)^2 \frac{1}{2\rho\overline{\Gamma}\gamma}\right) + \sqrt{-\left(\frac{n\pi}{L}\right)^4 \left(\frac{1}{2\rho\overline{\Gamma}\gamma}\right)^2 + \left(\frac{n\pi}{L}\right)^2 \frac{1}{\rho\overline{\Gamma}}}. \end{cases} \qquad [8.91]$$

Reintroducing the notations used previously, we will note that the non-damped normal angular frequency is obtained for values of γ tending towards infinity (which corresponds to abrupt strain when stress is applied abruptly: see Figure 8.9). In this case it follows:

$$\lambda_n = \sqrt{\frac{1}{\rho\overline{\Gamma}}}\frac{n\pi}{L}. \qquad [8.92]$$

This value corresponds to the non-damped normal angular frequency, which is generally given as $\omega_n = \lambda_n$.

The damped normal angular frequency in its turn is equal to:

$$\Omega_n = \omega_n\sqrt{1 - (\omega_n/2\gamma)^2}. \qquad [8.93]$$

Lastly, the complex normal angular frequency is expressed in the form:

$$\lambda_n = j\frac{\omega_n^2}{2\gamma} + \Omega_n$$

Note: expression [8.93] shows that for a fixed value of γ, there will always be a normal angular frequency, on the basis of which the radical will become imaginary. Under these conditions, the modal movements will no longer be vibratory but rather exponentially decreasing, since λ_n will be purely imaginary. This situation, characteristic of a super critical damping of the modes, will occur at higher frequencies the larger γ is, signifying that the material will be less dissipative.

In summary, on the basis of [8.73] and of the previous results [8.88] and [8.91], we obtain the general form of the vibratory movements of the viscoelastic beam as a combination of all the modal solutions:

$$U(x,t) = \sum_{n=1}^{\infty}\left(A_n\, e^{j\lambda_n t} + B_n\, e^{-j\lambda_n^* t}\right)\sin\left(\frac{n\pi}{L}x\right). \qquad [8.94]$$

It preserves the general form characteristic of damped systems with conjugated complex normal angular frequencies.

Normal stresses given in [8.88] are real: it is the consequence of a damping distributed uniformly over the entire beam (damping localized at the ends produced complex normal deformations).

A second form can be proposed for the vibratory displacement:

$$U(x,t) = \sum_{n=1}^{\infty} e^{-\left(\frac{\omega_n^2}{2\gamma}t\right)}\left(A'_n\cos(\Omega_n t) + B'_n\sin(\Omega_n t)\right)\sin\left(\frac{n\pi}{L}x\right). \qquad [8.95]$$

It is identical to the form [8.20] obtained for damping proportional to the speed, but this time the exponential decay of the amplitude over time depends on the mode. The complex Young modulus introduced in [8.83] can now be calculated taking into account the values of λ_n that solve the problem. After all the calculations it follows:

$$E_n^* = \frac{1}{\overline{\Gamma}}\left(1 - \frac{\omega_n^2}{2\gamma^2}\right) + j\frac{\omega_n}{\gamma}\sqrt{1 - \frac{\omega_n^2}{4\gamma^2}}. \qquad [8.96]$$

It should be noted that for our viscoelastic model, the complex Young modulus is variable with the mode. Consequently, we will index it.

In the limit case where γ tends towards infinity, we find again an elastic material; the complex Young modulus becomes purely real, independent of the mode and equal to:

$$E = \frac{1}{\overline{\overline{\Gamma}}}. \qquad [8.97]$$

The existence of an imaginary part for the Young modulus is thus characteristic of the phenomenon of viscoelastic damping. Very often the loss factor η is introduced, noting:

$$\overline{E}_n = E_n(1 + j\eta_n). \qquad [8.98]$$

In our case we obtain with these notations:

$$E_n = \frac{1}{\overline{\overline{\Gamma}}}\left(1 - \frac{\omega_n^2}{2\gamma^2}\right) \qquad [8.99]$$

and:

$$\eta_n = \frac{\Omega_n}{\gamma\left(1 - \frac{\omega_n^2}{2\gamma^2}\right)} = \frac{\omega_n}{\gamma}\,\frac{\sqrt{1 - \frac{\omega_n^2}{4\gamma^2}}}{\left(1 - \frac{\omega_n^2}{2\gamma^2}\right)}. \qquad [8.100]$$

In the case of slightly damped modes $\omega_n/\gamma << 1$, the Young modulus decreases as the normal angular frequency of the mode grows; the modal loss factor, close to ω_n/γ, thus increases with the normal angular frequency of the mode.

8.5. Properties of orthogonality of damped systems

As we saw, the general form of the vibratory response of a damped continuous medium is:

$$U(x,t) = \sum_{n=1}^{\infty} D_n\, e^{-j\lambda_n^* t} f_n^*(x) + E_n\, e^{j\lambda_n t} f_n(x). \qquad [8.101]$$

We can write the equations which must satisfy mode shapes $f_n(x)$ in a general form introducing each modal movement into the equation of motion and the boundary conditions of the problem. It follows:

$$\frac{d^2f_n}{dx^2}(x) = -a_n^2 f_n(x) \quad \forall x \in \left]0, L\right[, \qquad [8.102]$$

$$\alpha_n^0 \frac{df_n}{dx}(0) = f_n(0), \qquad [8.103]$$

$$\alpha_n^L \frac{df_n}{dx}(L) = f_n(L). \qquad [8.104]$$

In the problem covered in section 8.2, we can identify the constants:

$$a_n^2 = \left(\frac{n\pi}{L}\right)^2, \; \alpha_n^0 = 0, \; \alpha_n^L = 0. \qquad [8.105]$$

In the problem covered in section 8.3:

$$a_n^2 = \left(\frac{\lambda_n}{c}\right)^2, \; \alpha_n^0 = 0, \; \alpha_n^L = j\frac{ES}{Z}\frac{1}{\lambda_n}. \qquad [8.106]$$

In the problem covered in section 8.4:

$$a_n^2 = \left(\frac{n\pi}{L}\right)^2, \; \alpha_n^0 = 0, \; \alpha_n^L = 0. \qquad [8.107]$$

We thus write down the equations associated with [8.102] – [8.104] for the complex conjugate quantities:

$$\frac{d^2f_p^*}{dx^2}(x) = -a_p^{*2} f_p^*(x) \quad \forall x \in \left]0, L\right[, \qquad [8.108]$$

$$\alpha_p^{0*} \frac{df_p^*}{dx}(0) = f_p^*(0), \qquad [8.109]$$

$$\alpha_p^{L*} \frac{df_p^*}{dx}(L) = f_p^*(L). \qquad [8.110]$$

First property of orthogonality

Let us multiply equation [8.102] by the mode shape $f_p(x)$, then integrate the two members between 0 and L; it follows:

$$\int_0^L \frac{d^2 f_n}{dx^2}(x)\ f_p(x)\,dx = -a_n^2 \int_0^L f_n(x)\ f_p(x)\,dx.$$

Let us integrate the first member by parts and use equation [8.102] for the index p:

$$\int_0^L \frac{d^2 f_n}{dx^2}(x)\ f_p(x)\,dx = -a_p^2 \int_0^L f_n(x)\ f_p(x)\,dx + \left[\frac{df_n}{dx}(x)\ f_p(x)\right]_0^L - \left[\frac{df_p}{dx}(x)\ f_n(x)\right]_0^L.$$

Using the relations [8.103] and [8.104] and the relations symmetrical in n and p:

$$(a_n^2 - a_p^2)\int_0^L f_n(x)\ f_p(x)\,dx - \frac{df_p}{dx}(L)\frac{df_n}{dx}(L)(\alpha_n^L - \alpha_p^L) + \frac{df_p}{dx}(0)\frac{df_n}{dx}(0)(\alpha_n^0 - \alpha_p^0) = 0. \qquad [8.111]$$

Second property of orthogonality

Relation [8.111] is the first property of orthogonality of damped vibrating systems. There is a second one, which employs the complex conjugate mode shapes. Let us multiply equation [8.102] by $f_p^*(x)$, then integrate between 0 and L; it follows:

$$\int_0^L \frac{d^2 f_n}{dx^2}(x)\ f_p^*(x)\,dx = -a_n^2 \int_0^L f_n(x)\ f_p^*(x)\,dx.$$

Integrating anew the first member by parts, then using the relations [8.108] to [8.110], after all the calculations, we get:

$$(a_n^2 - a_p^{*2}) \int_0^L f_n(x)\ f_p^*(x)\, dx - \frac{df_p^*}{dx}(L) \frac{df_n}{dx}(L)(\alpha_n^L - \alpha_p^{L*}) + \frac{df_p^*}{dx}(0) \frac{df_n}{dx}(0)(\alpha_n^0 - \alpha_p^{0*}) = 0. \quad [8.112]$$

The relation [8.112] is the second property of orthogonality of damped systems.

Let us consider some particular cases:

a) If the constants α_n^0 and α_n^L are real and independent of the mode, the relations of orthogonality are reduced to:

$$(a_n^2 - a_p^2) \int_0^L f_n(x)\ f_p(x)\, dx = 0 \quad [8.113]$$

and:

$$(a_n^2 - a_p^{*2}) \int_0^L f_n(x)\ f_p^*(x)\, dx = 0. \quad [8.114]$$

Let us consider the case of two different modes n and p. It follows that $a_n \neq a_p$ and $a_n \neq a_p^*$, wherefrom we deduce:

$$\int_0^L f_n(x)\ f_p(x)\, dx = 0 \quad [8.115]$$

and:

$$\int_0^L f_n(x)\ f_p^*(x)\, dx = 0. \quad [8.116]$$

Note: it can so happen that for two different modes n and p we have an equality of a_n and a_p; it is a situation of degeneration where passing from [8.113] to [8.115] is no longer possible. We are then in a situation where several normal functions are associated with the same eigenvalue. However, thanks to Schmidt's orthogonalization process, it is always possible to make the mode shapes orthogonal among themselves.

Let us now consider the case where n = p; the relation [8.113] is automatically verified, and the relation [8.114] yields:

$$\text{Im}\left\{a_n^2\right\} \int_0^L \left|f_n(x)\right|^2 dx = 0,$$

i.e.:

$$\text{Im}\left\{a_n^2\right\} = 0.$$

The eigenvalues a_n^2 are purely real, the problems [8.102] – [8.104] and [8.108] – [8.110] coincide exactly and, consequently:

$$f_n^*(x) = f_n(x).$$

This amounts to stating that mode shapes are also purely real. The problems covered in sections 8.2 and 8.4 are observed in this particular situation ($\alpha_n^0 = \alpha_n^L = 0$).

b) In the case covered in section 8.3 we are dealing with the general situation. Replacing the various constants with the [8.106] values the two properties of orthogonality become:

$$\left(\frac{\lambda_n^2 - \lambda_p^2}{c^2}\right)\int_0^L f_n(x)\ f_p(x)\, dx + \frac{df_p}{dx}(L)\frac{df_n}{dx}(L)\ j\frac{ES}{Z}\left(\frac{\lambda_n - \lambda_p}{\lambda_n \lambda_p}\right) = 0, \quad [8.117]$$

$$\left(\frac{\lambda_n^2 - \lambda_p^{*2}}{c^2}\right)\int_0^L f_n(x)\ f_p^*(x)\, dx - \frac{df_p^*}{dx}(L)\frac{df_n}{dx}(L)\ j\frac{ES}{Z}\left(\frac{\lambda_n^* + \lambda_p}{\lambda_n \lambda_p^*}\right) = 0, \quad [8.118]$$

i.e. after some manipulation:

$$(\lambda_n + \lambda_p)\int_0^L f_n(x)\ f_p(x)\, dx - jf_p(L)\ f_n(L)\frac{Zc^2}{ES} = 0\ , \text{ if } \lambda_n \neq \lambda_p \quad [8.119]$$

and:

$$(\lambda_n - \lambda_p^*)\int_0^L f_n(x)\ f_p^*(x)\, dx - jf_p^*(L)\ f_n(L)\frac{Zc^2}{ES} = 0\ , \text{ if } \lambda_n \neq -\lambda_p^*. \quad [8.120]$$

These are the two properties and their conjugate expressions which we have used in section 8.3.

Let us note that we can define several other properties of orthogonality equivalent to [8.119] and [8.120]. It is enough, for example, to replace $f_n(x)$ by $-\frac{1}{a_n^2}\frac{d^2f_n}{dx^2}(x)$ under the terms of the equality [8.102] to obtain another form of the basic relations [8.111] and [8.112]. We leave it to the reader to look further into this aspect, taking into account the heaviness of the expressions. Following the problem, there exists an adapted form of the properties of orthogonality which it would be necessary to establish. Also, let us recall that these properties of orthogonality are key in the calculations of vibratory response because they offer the means of uncoupling the modes and, thus, of calculating the modal amplitudes separately.

8.6. Conclusion

The phenomenon of damping is related to very complex physical mechanisms which act inside the structures and at the level of their boundaries. They represent a conversion of mechanical energy into heat or a transfer of mechanical energy of the structure to its environment. Taking damping into account leads to the appearance of complex normal angular frequency. The imaginary part introduces the effect of exponential reduction of the vibratory amplitude over time, while the real part, as for the purely elastic systems, translates a sinusoidal movement over time. Mode shapes are in general also complex. They can, however, remain real, if the effect of damping is proportional to the effects of mass or stiffness. This situation occurs for distributed damping studied in sections 8.2 and 8.4, but not for damping localized at the boundaries considered in section 8.3.

Although complex quantities are introduced to characterize damping, the vibratory response resulting from real initial conditions is also real.

Calculations are considerably weighed down by taking damping into account; therefore, this study is undertaken only if the case of purely elastic systems is not sufficient to resolve the problem presented.

Chapter 9

Calculation of Forced Vibrations by Modal Expansion

9.1. Objective of the chapter

The problems of free vibrations which we have addressed in the previous chapter study vibratory movements following an initial disturbance of the state of equilibrium. Here we consider other vibratory movements caused by the application of a force variable in time. These problems are more complex since they superimpose the effect of the initial conditions and the application of force. They will be solved by the method of modal decomposition. This method is general and makes it possible to treat all types of forces: local or distributed, permanent or transitory.

The method is formulated on a reference example where the stages of calculation are well detailed. The amplitude associated with each vibration mode is the solution of the modal equation, i.e. the equation of a system with a degree of freedom, characterized by a generalized mass, a generalized stiffness, a generalized damping and a generalized force.

The difficulties of calculation, on the one hand, lie in the determination of these generalized quantities in the non-simple cases, in particular when the boundary conditions do not conform to the traditional cases (some examples will be provided). On the other hand, solving the modal equation is in itself a difficult problem when the force has a complicated temporal fluctuation. At this level it is necessary to break up the excitations into two distinct branches: deterministic forces, known at any moment, and random forces, known only in the sense of probabilities. Here we primarily consider the deterministic efforts, since the random vibrations require a specific treatment which will only be mentioned.

We outline the calculation in the two basic cases: harmonic excitation and impulse excitation. We then show how these two responses can be used to solve the case of any excitation in frequency domain thanks to the Fourier transform and in time domain by a convolution integral.

The response by modal decomposition is expressed by a series. The question of convergence of the series, which we study briefly, is then raised. We also provide a technique to accelerate the convergence of modal series.

The method of modal decomposition is general, which is its strength but also its weakness in the sense that its generality involves a heavy of calculation. There are also methods adapted to particular cases which offer faster processing. In Chapter 10 we will see the method of forced waves, which is very powerful in the problems of beams.

9.2. Stages of the calculation of response by modal decomposition

9.2.1. *Reference example*

The calculation that we are about to perform can be generalized, as shall be seen later. However, to avoid at least a heavy notation we will consider a rather simple reference example: a beam in bending supported at its two ends. This case lends itself particularly well to analytical calculation.

Stage 1. Presenting the problem

It is a matter of writing three groups of equations which completely define the problem of forced vibrations: the equation of motion, boundary conditions and initial conditions. Respecting all these equations is necessary to ensure uniqueness of the solution.

Equation of forced movement of the bending beam:

$$\rho S \frac{\partial^2 W}{\partial t^2} + EI \frac{\partial^4 W}{\partial x^4} = p(x,t) \quad \forall x \in]0, L[\,, \; t > 0 . \qquad [9.1]$$

It is the classic equation of bending beams when secondary effects are neglected (see Chapter 3, section 3.5). The second member represents the excitation: it is homogenous with force per unit of length and depends on space and time.

Boundary conditions (simple support at the ends):

$$W(0,t)=0 \qquad EI\frac{\partial^2 W}{\partial x^2}(0,t)=0,$$
$$W(L,t)=0 \qquad EI\frac{\partial^2 W}{\partial x^2}(L,t)=0. \qquad [9.2]$$

Initial conditions:

$$W(x,0)=d_0(x),$$
$$\frac{\partial W}{\partial t}(x,0)=v_0(x). \qquad [9.3]$$

The quantities $d_0(x)$ and $v_0(x)$ are respectively initial displacement and initial speed at any point x of the beam.

Stage 2. Calculation of the vibration modes, orthogonality of mode shapes

The calculation of the forced vibratory response is based on the preliminary knowledge of the vibration modes, that is, of the solution of the problem of free vibrations.

The normal vibration modes are solutions in the form of (see Chapter 6):

$$W_n(x,t)=(A_n\cos\omega_n t+B_n\sin\omega_n t)\ f_n(x). \qquad [9.4]$$

where ω_n is the normal angular frequency and $f_n(x)$ is the mode shape of the order n. $W_n(x,t)$ is the modal displacement.

Introducing this solution into the equation of free movement (equation [9.1] with the second member being nil) and for the boundary conditions [9.2], we note that:

$$-\omega_n^2\rho S f_n+EI\frac{d^4 f_n}{d x^4}=0 \quad x\in\,]0,L[, \qquad [9.5]$$

$$\begin{cases} f_n(0) = 0 & \dfrac{d^2 f_n}{dx^2}(0) = 0 \\ f_n(L) = 0 & \dfrac{d^2 f_n}{dx^2}(L) = 0. \end{cases} \qquad [9.6]$$

Equations [9.5] and [9.6] are at the origin of the properties of orthogonality of mode shape.

Let us examine the symmetry of the operators of mass and stiffness of equation [9.5]. These properties, which we want to establish, are defined by:

$$\int_0^L EI \frac{d^4 f_n}{dx^4} f_p \, dx = \int_0^L EI \frac{d^4 f_p}{dx^4} f_n \, dx, \qquad [9.7]$$

$$\int_0^L \rho S f_n f_p \, dx = \int_0^L \rho S f_p f_n \, dx. \qquad [9.8]$$

Equation [9.8] is obviously verified, the demonstration of [9.7] is carried out by integration by parts of the second member of [9.7] and taking into account the boundary conditions [9.6]. Indeed:

$$\int_0^L EI \frac{d^4 f_p}{dx^4} f_n \, dx = -\int_0^L EI \frac{d^3 f_p}{dx^3} \frac{d f_n}{dx} dx + \left[EI \frac{d^3 f_p}{dx^3} f_n \right]_0^L.$$

The terms at the boundaries are nil, since mode shapes verify [9.6]. Repeating integration by parts, it follows:

$$\int_0^L EI \frac{d^4 f_p}{dx^4} f_n \, dx = \int_0^L EI \frac{d^2 f_p}{dx^2} \frac{d^2 f_n}{dx^2} dx - \left[EI \frac{d^2 f_p}{dx^2} \frac{d f_n}{dx} \right]_0^L.$$

Here the terms at the boundaries are still nil, taking into account [9.6]. Applying the same procedure once again, it follows:

$$\int_0^L EI \frac{d^4 f_p}{dx^4} f_n \, dx = -\int_0^L EI \frac{d^3 f_n}{dx^3} \frac{d f_p}{dx} dx + \left[EI \frac{d^2 f_n}{dx^2} \frac{d f_p}{dx} \right]_0^L.$$

The terms at the boundaries are still nil. Finally, a last integration by parts produces:

$$\int_0^L EI \frac{d^4 f_p}{dx^4} f_n \, dx = \int_0^L EI \frac{d^4 f_n}{dx^4} f_p \, dx - \left[EI \frac{d^3 f_n}{dx^3} f_p \right]_0^L .$$

It thus suffices to note that the terms at the boundaries are nil in order to declare the verifying of [9.7].

These properties of symmetry are at the basis of the orthogonality of mode shapes. Let us note that they are not only related to the operators of mass and stiffness but also to the boundary conditions. Finally, let us underline that the terms at the boundaries would disappear for all other cases of standard boundary conditions (clamped, free end, etc.) and that consequently the properties of symmetry are identical for all these boundary conditions.

To get to the properties of orthogonality, let us proceed as follows:

Let us multiply equation [9.5] by f_p and integrate between 0 and L:

$$- \omega_n^2 \int_0^L \rho S f_n f_p \, dx + \int_0^L EI \frac{d^4 f_n}{dx^4} f_p \, dx = 0 . \qquad [9.9]$$

A symmetrical formula is obtained by inverting the indices n and p in equation [9.9]:

$$- \omega_p^2 \int_0^L \rho S f_p f_n \, dx + \int_0^L EI \frac{d^4 f_p}{dx^4} f_n \, dx = 0 . \qquad [9.10]$$

Using the properties of symmetry [9.7] and [9.8] in [9.9] we obtain:

$$- \omega_n^2 \int_0^L \rho S f_n f_p \, dx + \int_0^L EI \frac{d^4 f_p}{dx^4} f_n \, dx = 0 ,$$

i.e. by subtracting member by member with [9.10]:

$$(\omega_n^2 - \omega_p^2) \int_0^L \rho\, S\, f_n\, f_p\, dx = 0\,.$$

If $\omega_n = \omega_p$ the equation is automatically verified.

If $\omega_n \neq \omega_p$ the equality to zero implies:

$$\int_0^L \rho\, S\, f_n\, f_p\, dx = 0\,. \qquad [9.11]$$

This is the property of orthogonality with respect to the operator of mass. With [9.9] we immediately deduce from it a second property of orthogonality with respect to the operator of stiffness:

$$\text{if} \quad \omega_n \neq \omega_p \Rightarrow \int_0^L EI \frac{d^4 f_n}{dx^4} f_p\, dx = 0. \qquad [9.12]$$

At this point it is interesting to introduce two quantities which will play an important part in the calculation of the modal response.

We denote the following integral as generalized mass M_n:

$$M_n = \int_0^L \rho\, S\, f_n^2\, dx\,. \qquad [9.13]$$

We also denote the following integral as generalized stiffness K_n:

$$K_n = \int_0^L EI \frac{d^4 f_n}{dx^4} f_n\, dx\,. \qquad [9.14]$$

With [9.5], [9.13] and [9.14] we note that generalized mass and generalized stiffness enjoy the remarkable property:

$$\sqrt{K_n / M_n} = \omega_n\,. \qquad [9.15]$$

Stage 3. Modal decomposition of the response, modal equations

We can demonstrate that the set of normal functions $f_n(x)$ is a basis of the functional space where the solution of the problem defined by equations [9.1], [9.2] and [9.3] lies. We admit this result without demonstration; the reader may refer to the work of M. Roseau [ROS 84] for its mathematical aspects. Let us note that this property has already been exploited for the free response of beams in Chapters 5, 6 and 7, since the response is expressed as a series of normal functions. This is the idea is used in modal decomposition; we seek the solution of the problem of forced vibrations in the form:

$$W(x,t) = \sum_{n=1}^{+\infty} a_n(t)\ f_n(x)\,. \qquad [9.16]$$

In expression [9.16] the amplitudes $a_n(t)$ are unknown and thus need to be calculated in order to solve the problem. These amplitudes must adapt so that the modal expansion [9.16] verifies the three groups of equations [9.1], [9.2] and [9.3].

Let us note, first of all, that the modal expansion verify the boundary conditions [9.2] by construction, since each normal function verifies them separately (equation [9.6]). Thus, we have, for example:

$$W(0,t) = \sum_{n=1}^{+\infty} a_n(t)\ f_n(0) = 0 \quad (\text{since } f_n(0) = 0 \quad \text{according to [9.6]})\,.$$

The same applies to the other three boundary conditions.

Let us now examine the equation of forced movement [9.1] and substitute the form [9.16] in [9.1]:

$$\rho S \sum_{n=1}^{+\infty} \ddot{a}_n(t)\ f_n(x) + EI \sum_{n=1}^{+\infty} a_n(t) \frac{d^4 f_n}{dx^4}(x) = p(x,t)\,. \qquad [9.17]$$

This form is sterile since this equation has an infinite number of unknowns a_n. The key to the solution consists of uncoupling the modes by using the orthogonality of normal functions. For that it suffices to multiply equation [9.17] by a normal function $f_p(x)$ and to integrate it between 0 and L; it follows:

$$\int_0^L \rho S \sum_{n=1}^{+\infty} \ddot{a}_n(t)\ f_n(x)\ f_p(x)\,dx + \int_0^L EI \sum_{n=1}^{+\infty} a_n(t) \frac{d^4 f_n}{dx^4} f_p(x)\,dx = \int_0^L p(x,t)\, f_p(x)\,dx\,.$$

Let us invert the summations and the integrals:

$$\sum_{n=1}^{+\infty} \ddot{a}_n(t) \int_0^L \rho S f_n f_p \, dx + \sum_{n=1}^{+\infty} a_n(t) \int_0^L EI \frac{d^4 f_n}{dx^4} f_p(x) \, dx = \int_0^L p(x,t) \; f_p(x) \, dx . \qquad [9.18]$$

We recognize the properties of orthogonality with respect to the operators of mass and stiffness in [9.18], which implies that all the terms of the sums are nil except for the single index n = p. It follows:

$$\ddot{a}_n(t) \; M_p + a_p(t) \; K_p = F_p(t) . \qquad [9.19]$$

M_p and K_p are respectively the generalized mass and the generalized stiffness of the mode p, defined by equations [9.13] and [9.14]. $F_p(t)$ is the generalized force of the mode p given by [9.20]:

$$F_p(t) = \int_0^L p(x,t) \, f_p(x) \; dx . \qquad [9.20]$$

Equation [9.18] is the modal equation associated to the mode p; its solution will provide the unknown for the $a_p(t)$ problem. The remarkable point is that this equation is that of a system vibrating with one degree of freedom (see Figure 9.1), which, taking into account [9.15], has the same normal angular frequency as the vibration mode studied.

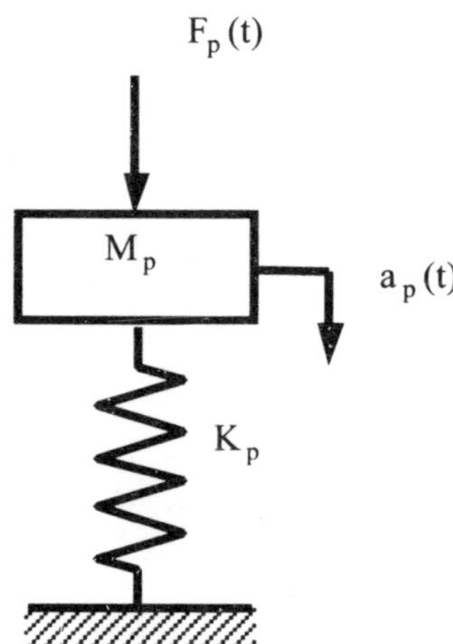

Figure 9.1. *System with one degree of freedom, representative of the modal equation*

To ensure the uniqueness of the solution of [9.19], it is necessary to provide this equation with initial conditions; these will of course stem from equations [9.3]. Let us use modal decomposition in these equations:

$$\sum_{n=1}^{+\infty} a_n(0)\ f_n(x) = d_0(x)\,, \tag{9.21}$$

$$\sum_{n=1}^{+\infty} \dot{a}_n(0)\ f_n(x) = v_0(x)\,. \tag{9.22}$$

To exploit these equations, it is also necessary to uncouple the modes using the properties of orthogonality: let us use the orthogonality with respect to the operator of mass, multiplying [9.21] and [9.22] by $\rho S\, f_p(x)$ and integrating it between 0 and L. (we could also use orthogonality with respect to stiffness multiplying [9.21] and [9.22] by $EI\dfrac{d^4 f_p}{dx^4}$).

After using orthogonality we obtain:

$$a_p(0) = \int_0^L \rho S\, f_p(x)\, d_0(x)\ dx/M_p\,, \tag{9.23}$$

$$\dot{a}_p(0) = \int_0^L \rho S\ f_p(x)\, v_0(x)\ dx/M_p\,. \tag{9.24}$$

The solution of the modal equation [9.19] equipped with the initial conditions [9.23] and [9.24] provides the unknown modal amplitudes. It is enough then to introduce them into the initial expression [9.16] to find the solution of the problem. We will provide some examples later on.

9.2.2. *Overview*

The method which was been highlighted with a reference example can be easily generalized. Let us consider an equation of forced motion of the type:

$$J\left(\frac{\partial^2 W}{\partial t^2}\right) + L(W) = p(x,t) \quad \forall\, x \in \left]0, L\right[\ ,\ t > 0 \tag{9.25}$$

where J is the operator of mass and L is the operator of stiffness. J and L are two differential space operators.

Let us suppose the existence of normal modes verifying the analog of equation [9.5]:

$$-\omega_n^2 J(f_n) + L(f_n) = 0. \qquad [9.26]$$

Finally, let us suppose that the two operators are symmetrical. (The boundary conditions which we did not write down must be defined so that symmetry is verified.) We show then, as previously, the properties of orthogonality:

$$\int_0^L f_p J(f_n)\, dx = \begin{pmatrix} 0 & \text{if } n \neq p \\ M_n & \text{if } n = p, \end{pmatrix} \qquad [9.27]$$

$$\int_0^L f_p L(f_n)\, dx = \begin{pmatrix} 0 & \text{if } n \neq p \\ K_n & \text{if } n = p. \end{pmatrix} \qquad [9.28]$$

Under these conditions, the modal amplitudes associated to the decomposition of the solution in the form [9.4] are obtained by the solution of modal equations of the [9.19] type, where the generalized masses and stiffness are provided by [9.27] and [9.28].

The method of modal decomposition is very generally applicable to problems of vibration of beams, but also of plates, shells, etc. The procedure is exactly identical to the one employed in the reference example; the difficulty that can appear in certain cases is the description of the properties of orthogonality when they are complicated by non-standard boundary conditions.

Let us take two examples to consolidate our ideas:

a) The non-homogenous bending beam. The equation of forced movement is as follows:

$$\rho S \frac{\partial^2 W}{\partial t^2} + \frac{\partial^2}{\partial x^2}\left(EI \frac{\partial^2 W}{\partial x^2} \right) = p(x, t). \qquad [9.29]$$

The generalized mass and the generalized stiffness of mode n are:

$$\begin{cases} M_n = \int_0^L \rho S \, f_n^2 \, dx \\ K_n = \int_0^L f_n \frac{d^2}{dx^2}\left(EI \frac{d^2 f_n}{dx^2} \right) dx. \end{cases} \quad [9.30]$$

Standard boundary conditions (support, embedding, free or guided) are assumed.

b) Longitudinal vibrations of a beam with nonstandard boundary conditions.

We consider the vibrations of a non-homogenous beam defined in Figure 9.2. It is clamped at 0 and attached to a mass M at the L end. It is the mass added at the end L which makes the problem non-standard.

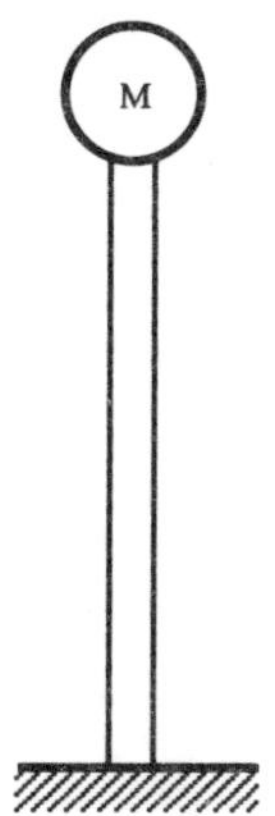

Figure 9.2. *Beam studied in case b)*

The equation of motion and the boundary conditions are as follows:

$$\rho S \frac{\partial^2 U}{\partial t^2} - \frac{\partial}{\partial x}\left(ES \frac{\partial U}{\partial x} \right) = p(x,t) \quad x \in \left] 0, L \right[, \quad [9.31]$$

$$\begin{cases} U(0,t)=0 \\ -ES\dfrac{\partial U}{\partial x}(L,t)=M\dfrac{\partial^2 U}{\partial t^2}(L,t). \end{cases} \quad [9.32]$$

Mode shapes verify the following equations:

$$-\omega_n^2 \rho S\ f_n - \frac{d}{dx}\left(ES\frac{df_n}{dx}\right)=0, \quad [9.33]$$

$$\begin{cases} f_n(0)=0 \\ ES\dfrac{df_n}{dx}(L)=+\omega_n^2 f_n(L)\,M. \end{cases} \quad [9.34]$$

We leave it to the reader to demonstrate the two properties of orthogonality of our problem:

$$\int_0^L \rho S\ f_n\ f_p\ dx + M\, f_n(L)\, f_p(L)=0 \quad \text{if } n \neq p, \quad [9.35]$$

$$\int_0^L \frac{d}{dx}\left(ES\frac{df_n}{dx}\right) f_p\ dx + \omega_n^2\ M\, f_n(L)\, f_p(L)=0 \quad \text{if } n \neq p, \quad [9.36]$$

and to deduce from it that the modal equation is of the [9.19] type with generalized quantities equal to:

$$M_n = \int_0^L \rho S\ f_n^2\, dx + M\ f_n^2(L), \quad [9.37]$$

$$K_n = \int_0^L \frac{d}{dx}\left(ES\frac{df_n}{dx}\right) f_n\ dx + \omega_n^2\ M\ f_n^2(L), \quad [9.38]$$

$$F_n(t) = \int_0^L f_n\ p(x,t)\, dx. \quad [9.39]$$

Note: for the calculation of generalized masses and stiffness, it is necessary to use equation [9.33] and the boundary condition in L [9.34].

9.2.3. *Taking damping into account*

The problem that we have treated previously does not take an important parameter into account: damping. It is an oversimplification since under certain conditions of excitation damping is the parameter which dominates the forced vibratory response. It is necessary, therefore, to take it into account in the calculations. Two methods present themselves: consider an equation of motion or dissipative boundary conditions (in the case of free vibrations, we have examined various possibilities in Chapter 8), or pragmatically introduce generalized modal damping in the modal equation. Although less elegant, it is very often this second option which is chosen in practice, because it meets two requirements of an engineer: simplicity and correspondence with the damping measurement technique.

In this chapter we will consider only the introduction of generalized damping λ_n into the modal equation:

$$M_n \ddot{a}_n(t) + \lambda_n \dot{a}_n(t) + K_n a_n(t) = F_n(t). \qquad [9.40]$$

It is a viscous type damping modeled in equation [9.40]. Other models can be considered; the reader will be able to find many descriptions thereof in other works concerning systems with one degree of freedom, which represent the modal equation in short form.

A reduced form of the modal equation [9.40] is often introduced:

$$\ddot{a}_n(t) + 2\varepsilon_n \omega_n \dot{a}_n(t) + \omega_n^2 a_n(t) = \frac{F_n(t)}{M_n}, \qquad [9.41]$$

$$\omega_n = \sqrt{\frac{K_n}{M_n}} \text{ : normal angular frequency,} \qquad [9.42]$$

$$\varepsilon_n = \lambda_n / 2 M_n \omega_n \text{: rate of damping}. \qquad [9.43]$$

The advantage of this reduced form is that it characterizes damping by the parameter ε_n, which in general is slowly variable with the mode. We admit that for a standard mechanical system $\varepsilon_n \approx 0.01$, for a very slightly damped mechanical system $\varepsilon_n \approx 0.001$, for a heavily damped mechanical system $\varepsilon_n \approx 0.1$ (such a strong value requires the use of a sandwich with a highly damped core).

9.3. Examples of calculation of generalized mass and stiffness

9.3.1. *Homogenous, isotropic beam in pure bending*

Let us consider a homogenous, isotropic beam in pure bending simply supported at its ends. We have demonstrated in Chapter 6 that the vibration modes were given by:

$$\omega_n = \sqrt{\frac{EI}{\rho S}\frac{n^2\pi^2}{L^2}}, \qquad [9.44]$$

$$f_n(x) = D_n \sin\left(\frac{n\pi}{L}x\right). \qquad [9.45]$$

The constant D_n is arbitrary and can be fixed at one without losing the generality of the expression; however, let us preserve it here in calculations.

Generalized mass has the expression:

$$M_n = \int_0^L \rho S\, D_n^2 \sin^2\left(\frac{n\pi}{L}x\right)dx, \qquad [9.46]$$

i.e. after calculations:

$$M_n = \rho S \frac{L}{2} D_n^2. \qquad [9.47]$$

Generalized stiffness has the expression:

$$K_n = \int_0^L D_n^2\, EI \frac{d^4 \sin\frac{n\pi}{L}x}{dx^4} \sin\frac{n\pi}{L}x\, dx, \qquad [9.48]$$

that is, after calculations:

$$K_n = EI\frac{n^4\pi^4}{L^4} D_n^2 \frac{L}{2}. \qquad [9.49]$$

Finally, generalized force is given by:

$$F_n(t) = \int_0^L D_n \sin \frac{n\pi}{L} x \ p(x,t)\,dx \,.$$

If the type of force causing the excitation is not specified, we cannot proceed further with the calculation of the generalized force.

A first observation must be made: generalized values do not have intrinsic physical significance, since they depend on the normalization of mode shapes by the arbitrary factor D_n.

For $D_n = \sqrt{2}$ the generalized mass is equal to the mass of the beam, but for $D_n = 1{,}000\sqrt{2}$ the generalized mass is a million times more than the real mass. Therefore, physical significance should not be attached to generalized quantities, except, however, for the K_n / M_n ratio, which is independent of normalization and equal to ω_n^2.

If one takes $D_n = 1$ for all the modes, as is common in practice, we note that generalized mass is constant with the mode: it is not a general property but a characteristic of this particular case. Generalized stiffness increases with the index of the mode meaning that the dynamic stiffness of the beam increases when the wavelength of the mode shapes decreases.

9.3.2. *Isotropic homogenous beam in pure bending with a rotational inertia effect*

It is a slightly more complicated case since the operator of mass has an additional term by comparison to the previous case:

$$J = \rho S - \rho I \frac{d^2}{dx^2}, \qquad [9.50]$$

$$L = EI \frac{d^4}{dx^4}. \qquad [9.51]$$

We determined the vibration modes in the case of boundary conditions of the simple support type in Chapter 6. Mode shapes keep the simple form:

$$f_n(x) = \sin \frac{n\pi}{L} x \,. \qquad [9.52]$$

From that we deduce that generalized mass and generalized stiffness given by [9.13] and [9.14] are equal to:

$$M_n = \int_0^L \left[\rho S \sin^2 \frac{n\pi}{L} x - \rho I \frac{d^2}{dx^2}\left(\sin \frac{n\pi}{L} x \right) \right] \sin \frac{n\pi}{L} x \, dx , \qquad [9.53]$$

$$K_n = \int_0^L EI \frac{d^4 \sin \dfrac{n\pi}{L} x}{dx^4} \sin \frac{n\pi}{L} x \, dx , \qquad [9.54]$$

that is, after all calculations:

$$M_n = \left(\rho S + \rho I \frac{n^2 \pi^2}{L^2} \right) \frac{L}{2} , \qquad [9.55]$$

$$K_n = EI \frac{n^4 \pi^4}{L^4} \frac{L}{2} . \qquad [9.56]$$

Generalized mass increases with the mode owing to the effect of rotational inertia, which becomes dominating for high ranking modes. Generalized stiffness remains identical to the case in 9.3.1 (equation [9.49] when $D_n = 1$).

9.4. Solution of the modal equation

We will consider the two basic cases: harmonic excitation and impulse excitation, then we will show how these two cases make it possible to treat the general case.

9.4.1. *Solution of the modal equation for a harmonic excitation*

To consolidate our ideas let us consider the example of reference from the beginning of the chapter, i.e. the vibrations of bending of a simply supported beam. The generalization of the results which we are going to highlight is rather obvious.

Let us, moreover, consider that the excitation is harmonic of angular frequency ω:

$$p(x, t) = \overline{p}(x) \, e^{j\omega t} . \qquad [9.57]$$

The modal equation with damping [9.40] is particularized on the basis of [9.20] for our excitation to a generalized force of the type:

$$F_n(t) = \int_0^L \overline{p}(x)\ f_n(x)\ dx\ e^{j\omega t}, \tag{9.58}$$

that is, $F_n(t) = \overline{F}_n\ e^{j\omega t}$. [9.59]

Taking into account the time-space separation [9.57], the temporal form of the generalized force is identical to that of the excitation.

The modal equation is thus:

$$\ddot{a}_n(t) + 2\,\varepsilon_n\,\omega_n\,\dot{a}_n(t) + \omega_n^2\,a_n(t) = \frac{\overline{F}_n}{M_n}\,e^{j\omega t}. \tag{9.60}$$

It is a second-order differential equation with constant coefficients, which is integrated in very classical fashion. The solution is the sum of the general solution of the homogenous equation and of a particular solution of the equation with a second member.

The general solution of the homogenous equation is given by:

$$a_n(t) = (A_n \sin \Omega_n t + B_n \cos \Omega_n t)\, e^{-\omega_n \varepsilon_n t} \tag{9.61}$$

where $\Omega_n = \omega_n \sqrt{1-\varepsilon_n^2}$. [9.62]

Ω_n is the damped normal angular frequency. It coincides with the normal angular frequency ω_n if damping is nil; in the contrary case it is slightly weaker. In the standard cases $(\varepsilon_n = 0.01)$, the shift is negligible.

The particular solution is of the type:

$$a_n(t) = \overline{a}_n\ e^{j\omega t} \tag{9.63}$$

with $\overline{a}_n = \dfrac{\overline{F}_n}{M_n} \dfrac{1}{\omega_n^2 - \omega^2 + 2j\ \varepsilon_n\ \omega_n\ \omega}$. [9.64]

The amplitude of the forced response is the product of the amplitude of the excitation $(\overline{F}_n / M_n)$ by a term called frequency response $H_n(\omega)$ which represents the amplitude of the vibrating system under unitary excitation. We have:

$$H_n(\omega) = 1/(\omega_n^2 - \omega^2 + 2j\,\varepsilon_n\,\omega_n\omega)\,. \qquad [9.65]$$

Joining [9.61] and [9.63] we obtain the desired solution. It is not defined in a unique manner since the constants A_n and B_n are not fixed. To eliminate the uncertainty, it is necessary to use the initial conditions [9.21] and [9.22]. Let us adopt initial conditions of rest to reduce calculations, i.e. $d_0(x) = 0$ and $v_0(x) = 0$ it follows with [9.21] and [9.22]:

$$a_n(0) = 0 \quad \text{and} \quad \dot{a}_n(0) = 0\,.$$

The first condition leads to:

$$B_n + \overline{a}_n = 0\,.$$

The second is more complicated:

$$-\omega_n\,\varepsilon_n\,B_n + \Omega_n\,A_n + j\,\omega\,\overline{a}_n = 0\,,$$

from which we draw:

$$B_n = -\,\overline{a}_n \quad \text{and} \quad A_n = -\,\overline{a}_n\,\frac{j\omega + \omega_n\,\varepsilon_n}{\Omega_n}\,.$$

The solution of the problem is thus:

$$a_n(t) = \overline{a}_n\left[\left(-\frac{j\omega + \omega_n\,\varepsilon_n}{\Omega_n}\sin\Omega_n\,t - \cos\Omega_n\,t\right)e^{-\omega_n\,\varepsilon_n t} + e^{j\omega t}\right] \qquad [9.66]$$

where $\overline{a}_n$ is given by [9.64].

This expression of modal amplitude shows several basic phenomena that are crucial for the comprehension of vibratory phenomena forced by a harmonic excitation. The modal response breaks up into two parts:

– the transitory state, which stems from the general solution of the equation without a second member, and which occurs with the damped normal angular frequency Ω_n that is completely independent of the excitation and characteristic of the structure;

– the forced response, which comes from the particular solution of the equation with a second member, which occurs with the angular frequency of the force.

There is, therefore, a basic difference between these two parts of the solution since they are carried out at different frequencies – that is the first remarkable phenomenon. The second is linked to the exponential decay of the transitory state with time, which is all the stronger the stronger the damping is and the higher the normal angular frequency is. The transitory state, in fact, only has real importance at the very beginning of the phenomenon, at the moments immediately after the initial moment. Figure 9.3 gives an illustration of the phenomenon.

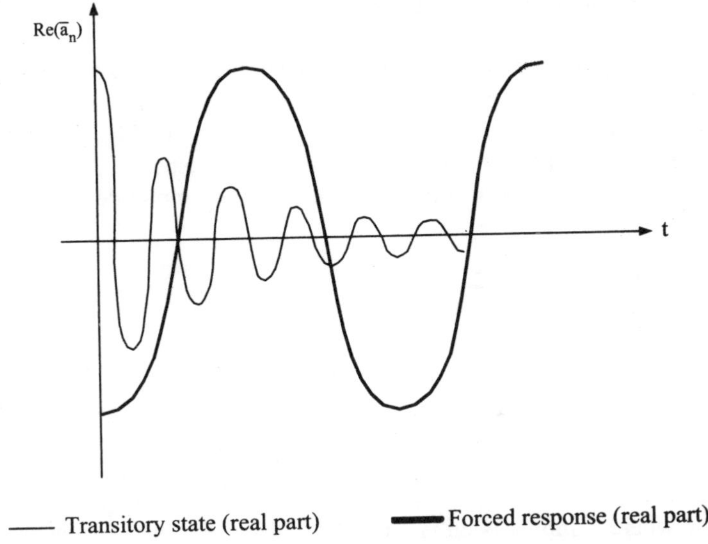

Figure 9.3. *Transitory and forced responses of the vibratory response*

Insofar as we are generally interested in the established response, and that in any event the initial conditions are often badly known in practice, calculations are generally limited to the forced response alone. We thus approach the response given in [9.66]:

$$a_n(t) \approx \bar{a}_n \, e^{j\omega t}$$

The calculation is simplified, since the initial conditions no longer intervene. It is, in fact, a very good approximation of the modal response as long as the transitory signal is weakened. For a given damping, this intervenes all the quicker the larger the normal angular frequency of the mode is. The most slowly weakening mode is, thus, the first. As an example, if the first normal frequency is 100 Hz and the rate of damping is 10^{-2}, as is common in mechanics, the exponential decrease after 1 second is given by $e^{-2\pi}$. This represents a reduction in the vibratory amplitude by a factor of 1,86 10^{-3} after 1 second.

Let us note that joining the two parts of the solutions together can have a constructive effect leading to maximum vibratory amplitude at the start of the phenomenon. In certain borderline cases, this effect leads to the rupture of the structure at the beginning of the excitation.

The third notable phenomenon is the resonance. It characterizes the amplification of vibratory amplitude at a particular frequency of excitation. To reveal the resonance, it suffices to study the factor $\bar{a}_n$ given by [9.64], which appears in calculation as the complex amplitude of the forced response. It is preferable to introduce the module and the phase associated with this amplitude:

$$\left|\bar{a}_n\right| = \frac{\left|F_n\right|}{M_n} \frac{1}{\sqrt{(\omega_n^2 - \omega^2 + 4\,\varepsilon_n^2\,\omega_n^2\,\omega^2)}}, \tag{9.67}$$

$$\phi_n = \text{Arctg}\left(\frac{2\,\varepsilon_n\,\omega_n\,\omega}{\omega^2 - \omega_n^2}\right). \tag{9.68}$$

The case where damping is nil is interesting because it demonstrates the extreme limit of the phenomenon of resonance. We note in Figure 9.4 that amplitude tends towards infinity when ω tends towards ω_n; it is the phenomenon of resonance characterized by an amplification of vibratory movement when the angular frequency of excitation is close to the normal angular frequency of the mode. We note that the phase ϕ_n is then equal to $\pi/2$. For excitation angular frequencies that are much weaker than the pulsation of resonance $\omega << \omega_n$, we note with [9.67] and [9.68] that:

$$\left|\bar{a}_n\right| \approx \frac{\left|F_n\right|}{K_n} \quad \text{and} \quad \phi_n \approx 0.$$

The vibratory behavior of the generalized vibrating system is dominated by stiffness, as is the static behavior.

For excitation angular frequency much higher than the angular frequency of resonance $\omega >> \omega_n$, we note that:

$$\left|\bar{a}_n\right| \approx \frac{F_n}{\omega^2 M_n} \quad \text{and} \quad \phi_n \approx \pi \cdot$$

It is a vibratory behavior dominated by the generalized mass of the vibrating system. The case of non-zero damping is not very different: only the behavior with resonance is notably modified, the maximum amplitude being reached when the pulsation of excitation ω is equal to the angular frequency resonance $\overline{\Omega}_n$:

$$\overline{\Omega}_n = \omega_n \sqrt{1 - 2\,\varepsilon_n^2}\,. \qquad [9.69]$$

The frequency of resonance thus depends on the damping of the system which provokes it. Let us note that the angular frequency of damped resonance is different from the damped normal angular frequency given by [9.62]: $\Omega_n \neq \overline{\Omega}_n$. This variation of the frequency of the maximum amplitude is, however, very weak for the current case $(\varepsilon_n = 0.01)$. The passing of the phase to zero, on the other hand, always occurs for $\omega = \omega_n$, whatever the value of damping. This is why we sometimes speak of $\overline{\Omega}_n$ as the angular frequency of amplitude resonance and of ω_n as of the phase resonance. The amplitude resonance $\overline{\Omega}_n$ is given by:

$$\left|\bar{a}_n\right| = \frac{\left|F_n\right|}{M_n} \frac{1}{2\,\omega_n^2\,\varepsilon_n \sqrt{1 - \varepsilon_n^2}}\,. \qquad [9.70]$$

The forced vibratory behavior of a vibration mode is dominated by mass or stiffness as soon as ω moves away from $\overline{\Omega}_n$; in this case it is impossible to measure the damping since its effect is masked by those of mass or stiffness. On the other hand, during resonance, damping dominates the phenomenon as shown by [9.70], and can thus be measured. This is why the introduction of damping mode by mode as we did in [9.40] is coherent with the reality of measurements of damping. The frequency zone where damping dominates is rather weak. In practice, we can show that the 3dB bandwidth, Δ_n, associated with the peak of resonance, for weak damping is equal to:

$$\Delta_n = \omega_n\, 2\,\varepsilon_n\,. \qquad [9.71]$$

For modal rate of damping of $\varepsilon_n = 0.01$ we have a bandwidth of $\Delta_n = 2.10^{-2}\,\omega_n$. A decrease of 3dB is observed when we deviate by 1% from the frequency of resonance.

Equation [9.71] provides a means of measuring damping; it is the technique of the bandwidth.

Let us note, finally, that for very strong damping, unrealistic for normal mechanical systems, the phenomenon of resonance disappears. Indeed, equation [9.70] shows that the frequency of the maximum of amplitude is nil if ε_n is equal to $1/\sqrt{2}$ and the maximum no longer exists if $\varepsilon_n > 1/\sqrt{2}$.

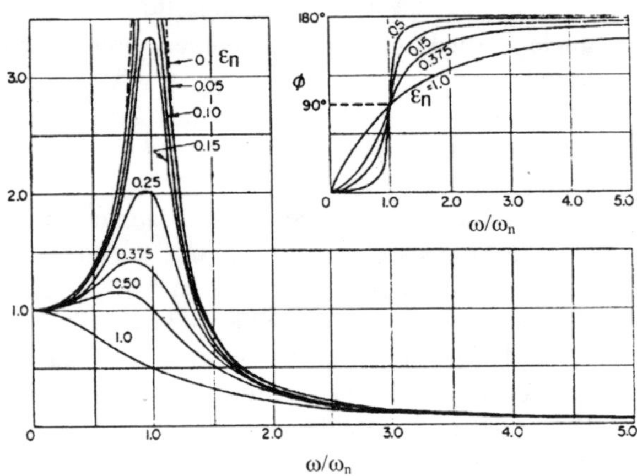

Figure 9.4. *Normalized amplitude and phase of the response in frequency of the generalized vibrating system*

9.4.2. *Solution of the modal equation for an impulse excitation*

To consolidate the ideas, let us look again at the example of reference of the bending beam supported at its ends. The force of excitation per unit of length has the form:

$$p(x,t) = \tilde{p}(x)\ \delta(t)\,. \qquad [9.72]$$

The quantity $\tilde{p}(x)$ is the space distribution of the external effort and $\delta(t)$ is the Dirac distribution, representing an ideal impulse excitation.

The generalized force associated with the impulse excitation [9.72] is also an impulse:

$$F_n(t) = \tilde{F}_n \; \delta(t) \qquad [9.73]$$

with $\tilde{F}_n = \int_0^L f_n(x) \; \tilde{p}(x) \, dx$.

The modal equation takes the form:

$$\ddot{a}_n(t) + 2\,\varepsilon_n\,\omega_n\,\dot{a}_n(t) + \omega_n^2\,a_n(t) = \frac{\tilde{F}_n}{M_n}\,\delta(t)\,. \qquad [9.74]$$

The solution of this equation is simple to obtain. We give the result directly:

$$a_n(t) = \begin{cases} 0 \quad \text{if } t < 0, \\ \dfrac{\tilde{F}_n}{M_n \Omega_n}\; e^{-\omega_n \varepsilon_n t} \sin \Omega_n t \quad \text{if } t \geq 0\,. \end{cases} \qquad [9.75]$$

This solution coincides with the transitory state following an initial condition of speed:

$$\dot{a}_n(0) = \tilde{F}_n \,/\, M_n\,.$$

The impulse excitation thus generates a free vibratory state, characterized by the damped normal angular frequency Ω_n and an exponential decrease that is stronger the stronger $\varepsilon_n\,\omega_n$ is. We can introduce the modal impulse response $h_n(t)$ as the particular case of [9.75] where the amplitude of the modal impulse force of equation [9.74] equals unity:

$$h_n(t) = \begin{cases} 0 \quad \text{if } t < 0, \\ \dfrac{1}{\Omega_n}\; e^{-\omega_n \varepsilon_n t} \sin \Omega_n t \quad \text{if } t \geq 0 \end{cases} \qquad [9.76]$$

This elementary solution will be used as a basis for calculation of the response to any excitation in section 9.4.4.

9.4.3. *Unspecified excitation, solution in frequency domain*

Let us now consider an unspecified excitation p(x, t) that is still sufficiently regular so that the Fourier transform exists:

$$P(x, \omega) = \int_{-\infty}^{+\infty} e^{-j\omega t}\, p(x, t)\, dt\,. \qquad [9.77]$$

We can naturally obtain by inverse transformation:

$$p(x, t) = \frac{1}{2\pi} \int_{-\infty}^{+\infty} P(x, \omega)\, e^{+j\omega t}\, d\omega\,. \qquad [9.78]$$

The expression [9.78] shows that any excitation can be broken up into a harmonic sum of excitation whose amplitude with the angular frequency ω is equal to $P(x, \omega) / 2\pi$. From a physical point of view, we conceive that the response will be the sum of the responses to the various harmonic excitations, since the system is linear. To give shape to this idea, it is sufficient to use the Fourier transformation of the modal equation [9.41]:

$$-\omega^2 A_n(\omega) + 2\, j\omega\, \varepsilon_n\, \omega_n\, A_n(\omega) + \omega_n^2\, A_n(\omega) = F_n(\omega) / M_n \qquad [9.79]$$

where $A_n(\omega) = \int_{-\infty}^{+\infty} a_n(t)\, e^{-j\omega t}\, dt$ [9.80]

and:

$$F_n(\omega) = \int_{-\infty}^{+\infty} \int_0^L f_n(x)\; p(x, t)\, dx\, e^{-j\omega t}\, dt = \int_0^L f_n(x)\; P(x, \omega)\, dx\,. \qquad [9.81]$$

We can draw the value of $A_n(\omega)$ from equation [9.79]:

$$A_n(\omega) = \frac{F_n(\omega)}{M_n}\, \frac{1}{\omega_n^2 - \omega^2 + 2\, j\omega\, \varepsilon_n\, \omega_n}\,. \qquad [9.82]$$

In this expression we recognize the frequency response of the mode n, $H_n(\omega)$, characteristic of the harmonic response with the angular frequency ω which we have introduced in section 9.4.1, equation [9.65].

$$H_n(\omega) = \frac{1}{\omega_n^2 - \omega^2 + 2\, j\omega\, \varepsilon_n\, \omega_n},$$

$$A_n(\omega) = \frac{F_n(\omega)}{M_n} H_n(\omega). \qquad [9.83]$$

The modal vibratory amplitude is obtained by taking the inverse Fourier transformation:

$$a_n(t) = \int_{-\infty}^{+\infty} \frac{F_n(\omega)}{2\pi M_n} H_n(\omega)\, e^{+j\omega t}\, d\omega \qquad [9.84]$$

The expression [9.84] demonstrates that which has been suggested physically by the linearity of the system: the response $a_n(t)$ is equal to the sum of the responses to harmonic interferences with an amplitude of $F_n(\omega) / 2\pi M_n$.

The first way to solve the modal equation in the case of any excitation consists in using the Fourier transform, thus, working in frequency domain. The vibrating system is characterized by its frequency response $H_n(\omega)$, the excitation is characterized by the Fourier transform of the generalized force.

There is a second possibility to calculate the vibratory response working in time domain. We will set out its form in the next section.

9.4.4. *Unspecified excitation, solution in time domain*

Let us consider an unspecified excitation p (x, t). The associated generalized force is $F_n(t)$. We have to find the modal response $a_n(t)$ verifying:

$$\ddot{a}_n(t) + 2\, \varepsilon_n\, \omega_n\, \dot{a}_n(t) + \omega_n^2\, a_n(t) = \frac{F_n(t)}{M_n}. \qquad [9.85]$$

Let us suppose, moreover, that the generalized system has the following initial conditions at the time t_0:

$$a_n(t_0) = d_0, \qquad [9.86]$$

$$\dot{a}_n(t_0) = v_0. \qquad [9.87]$$

To solve this problem we will use the impulse response h(t) defined by equation [9.76]. Let us pose the integral I_n :

$$I_n = \int_{-\infty}^{+\infty} \left(\ddot{a}_n(t) + 2\,\varepsilon_n\,\omega_n\,\dot{a}_n(t) + \omega_n^2\,a_n(t)\right) h_n(\tau - t)\;dt\,.$$

We suppose, moreover, that before the initial moment t_0, the system is at rest and that, therefore:

$$a_n(t) = 0 \quad \text{if} \quad t < t_0\,.$$

The integral I_n can thus be written:

$$I_n = \int_{t_0}^{+\infty} \left(\ddot{a}_n(t) + 2\,\varepsilon_n\,\omega_n\,\dot{a}_n(t) + \omega_n^2\,a_n(t)\right) h_n(\tau - t)\;dt\,. \qquad [9.88]$$

Taking into account equation [9.85], we also have:

$$I_n = \int_{t_0}^{+\infty} h_n(\tau - t)\;\frac{F_n(t)}{M_n}\,dt\,. \qquad [9.89]$$

Let us carry out integration by parts of [9.88]; after all the calculations it follows:

$$I_n = \int_{t_0}^{+\infty} a_n(t)\left(\ddot{h}_n(\tau - t) + 2\,\varepsilon_n\,\omega_n\,\dot{h}_n(\tau - t) + \omega_n^2\,h_n(\tau - t)\right) dt$$
$$+ \left[\left(\dot{a}_n(t) + 2\,\varepsilon_n\omega_n\,a_n(t)\right) h_n(\tau - t)\right]_{t_0}^{\infty} - \left[a_n(t)\,\dot{h}_n(\tau - t)\right]_{t_0}^{\infty}. \qquad [9.90]$$

The expression [9.90] is simplified by taking into account the properties of the impulse response, which verifies the equation of motion:

$$\ddot{h}_n(\tau - t) + 2\,\varepsilon_n\,\omega_n\,\dot{h}_n(\tau - t) + \omega_n^2\,h_n(\tau - t) = \delta\,(\tau - t)\,.$$

The integral of [9.90] is thus equal to:

$$\int_{t_0}^{+\infty} a_n(t)\;\delta\,(\tau - t)\,dt = a_n(\tau)\,. \qquad [9.91]$$

Noting, furthermore, that the law of causality implies that the impulse response is nil before the force is applied, we have:

$$h_n(\tau - t) = 0 \quad \text{if} \quad \tau < t.$$

The terms at the boundaries of [9.90] become:

$$-\left(\dot{a}_n(t_0) + 2\,\varepsilon_n\,\omega_n\,a_n(t_0)\right) h_n(\tau - t_0) + a_n(t_0)\;\dot{h}_n(\tau - t_0)\,. \qquad [9.92]$$

Finally, gathering [9.89], [9.91] and [9.92], we obtain:

$$a_n(\tau) = \int_{t_0}^{\tau} h_n(\tau - t)\frac{F_n(t)}{M_n}\,dt - a_n(t_0)\;\dot{h}_n(\tau - t_0) + \left(\dot{a}_n(t_0) + 2\,\varepsilon_n\,\omega_n\,a_n(t_0)\right) h_n(\tau - t_0)\,. \qquad [9.93]$$

This expression gives the modal amplitude at any moment τ according to the initial conditions and the force applied between the initial moment and the moment of observation. The impulse response is given by [9.76], that is:

$$h_n(\tau - t) = \begin{cases} 0 \quad \text{if} \quad \tau < t, \\ \dfrac{1}{\Omega_n} e^{-\omega_n \varepsilon_n (\tau - t)} \sin\Omega_n(\tau - t) \quad \text{if} \quad t \le \tau. \end{cases} \qquad [9.94]$$

In many practical cases, the vibrating system is at rest when the force is applied to it; it follows that:

$$a_n(t_0) = 0 \quad \text{and} \quad \dot{a}(t_0) = 0\,.$$

The expression [9.93] is simplified into:

$$a_n(\tau) = \int_{t_0}^{\tau} h_n(\tau - t)\frac{F_n(t)}{M_n}\,dt\,. \qquad [9.95]$$

The [9.95] form of the response has a simple physical explanation. Let us note, first of all, that $h_n(\tau - t)$ is the impulse response at the moment τ when the generalized vibrating system is excited at the moment t. The integral [9.95] indicates that the force $F_n(t)/M_n$ can be broken up into a succession of impulses producing impulse responses whose superposition gives the total vibration.

Expression [9.95] is very simple but requires going backwards until the moment of rest to calculate the action of all the forces applied. For numerical calculations that render the integral [9.95] discrete, the number of calculation steps can be very large. We may then find it beneficial to carefully use [9.93] taking $t_0 \doteq \tau - \Delta$ as the initial moment where Δ is the temporal step of calculation.

9.5. Example response calculation

9.5.1. *Response of a bending beam excited by a harmonic force*

a) Point excitation

First of all, let us present the problem. We consider a homogenous beam in pure bending with support boundary conditions. The equation of motion and the boundary conditions are thus:

$$\rho S \frac{\partial^2 W}{\partial t^2} + EI \frac{\partial^4 W}{\partial x^4} = P\,\delta(x - x_0)\, e^{j\omega t} ,$$

$$W(0, t) = 0 \ , \ W(L, t) = 0 \ ,$$

$$\frac{\partial^2 W}{\partial x^2}(0, t) = 0 \ , \ \frac{\partial^2 W}{\partial x^2}(L, t) = 0 .$$

The excitation force has an amplitude P, an angular frequency ω and is localized in x_0.

The vibration modes of the beam were calculated in the chapter on free vibrations of beams in bending. It is given by:

$$f_n(x) = \sin \frac{n\pi}{L} x \ , \ \omega_n = \sqrt{\frac{EI}{\rho S}} \left(\frac{n\pi}{L} \right)^2 .$$

The vibratory response is, thus, sought in the form of a modal series:

$$W(x, t) = \sum_{n=1}^{+\infty} a_n(t) \sin \frac{n\pi}{L} x ,$$

modal amplitudes $a_n(t)$ being solutions of modal equations:

$$\ddot{a}_n(t) + 2\,\varepsilon_n\,\omega_n\,\dot{a}_n(t) + \omega_n^2\,a_n(t) = \frac{F_n(t)}{M_n}.$$

The generalized mass and force are given respectively by:

$$M_n = \int_0^L \rho S\,\sin^2\left(\frac{n\pi}{L}x\right)\,dx = \rho S\frac{L}{2}$$

and:

$$F_n(t) = \int_0^L \sin\left(\frac{n\pi}{L}x\right)\,P\,\delta\left(x - x_0\right)\,dx\;e^{j\omega t} = P\sin\left(\frac{n\pi}{L}x_0\right)\,e^{j\omega t}.$$

All intermediate calculations were provided in section 9.2.1 since this case constituted our example of reference. However, we have not given a particular space form to the excitation force distribution.

The solution of the modal equation has been provided in section 9.4.1. When the transitory effects are neglected, we have:

$$a_n(t) = \frac{P\sin\frac{n\pi}{L}x_0}{\rho S\,L/2}\;\frac{1}{\omega_n^2 - \omega^2 + 2\,j\,\varepsilon_n\,\omega_n\,\omega}\;e^{j\omega t}.$$

The forced response is thus provided by:

$$W(x,t) = \sum_{n=1}^{+\infty}\frac{P}{\rho S\,L/2}\;\frac{\sin\frac{n\pi}{L}x_0\;\sin\frac{n\pi}{L}x}{\omega_n^2 - \omega^2 + 2j\,\varepsilon_n\,\omega_n\,\omega}\;e^{j\omega t}. \qquad [9.96]$$

Expression [9.96] clearly shows the influence of a point excitation force. Let us take, for example, an effort applied in $x_0 = L/2$. We observe that the responses of modes 2, 4, 6… are nil. This result is explained by the fact that the mid-point of the beam is a vibration node for the even modes. A point transverse force applied to a vibration node does not get modal response. Conversely, the response of modes 1, 3, 5, … etc. is maximum since $\left|\sin(\frac{n\pi}{L}x_0)\right|$ is equal to one when $x_0 = L/2$. This result is linked to the fact that the mid-point of the beam is an antinode of vibration for the odd modes. This constitutes the second remarkable tendency for point excitation: if the force is applied to an antinode it maximizes the modal response.

A way of reducing modal response is thus to localize the force at one of its vibration nodes.

Note: the described tendency is very often verified: excitation at a node does not produce modal response. It is, however, not general; an excitation of transverse beam vibrations by a localized torque produces exactly the opposite effect and it is the excitation at an antinode that cancels modal response.

b) Multi-point excitation

Let us consider the excitation force applied at two points:

$$p(x,t) = \left(P\,\delta(x - x_0) + P'\,\delta(x - x'_0)\right) e^{j\omega t}.$$

The forced response of the beam is obtained by applying the principle of superposition, a simple consequence of the linearity of the problem. Using the previous calculation (equation [9.96]), it follows:

$$W(x,t) = \sum_{n=1}^{+\infty} \frac{1}{\rho S\, L/2} \frac{P \sin\frac{n\pi}{L} x_0 + P' \sin\frac{n\pi}{L} x'_0}{\omega_n^2 - \omega^2 + 2j\,\varepsilon_n\,\omega_n\,\omega} \sin\frac{n\pi}{L} x\; e^{j\omega t}. \qquad [9.97]$$

This result shows that the response of a mode can be canceled by adding a secondary force (P'); for that it suffices to satisfy the reduction:

$$P \sin\frac{n\pi}{L} x_0 + P' \sin\frac{n\pi}{L} x'_0 = 0.$$

We thus have:

$$P' = \frac{-P \sin\frac{n\pi}{L} x_0}{\sin\frac{n\pi}{L} x'_0}.$$

Canceling the modal response is always possible as long as the point of application x'_0 is not at a vibration node.

The result [9.97] can be generalized to K excitation points in an obvious manner by summing up the separately calculated responses.

c) Uniformly distributed excitation

Let us take a constant distributed force:

$$p(x, t) = B\,e^{j\omega t}.$$

In the previous calculations, only the generalized force is modified and becomes:

$$F_n(t) = \int_0^L B \sin\frac{n\pi}{L}x\,dx\;e^{j\omega t} = B\frac{1-\cos n\pi}{n\pi/L}\,e^{j\omega t}.$$

Let us note immediately that generalized force is nil for all the even modes, which implies a zero response for these modes. Physically this tendency is related to the symmetry of loading which cannot excite asymmetrical modes. We may also interpret this result as cancelation of work resulting from the symmetrical force applied to asymmetric modal deformations. This concept of work of the force applied to modal deformations gives a physical image of the generalized force; we may then better understand the reduction of the generalized force of the odd modes as the order of the mode increases (see Figure 9.5).

To sum up, the vibratory response of the beam is equal to:

$$W(x, t) = \sum_{n=1,3,5\ldots} B\frac{2}{n\pi/L}\,\frac{1}{\rho S\,L/2}\,\frac{\sin\frac{n\pi}{L}x}{\omega_n^2-\omega^2+2j\,\varepsilon_n\,\omega_n\,\omega}\,e^{j\omega t}. \qquad [9.98]$$

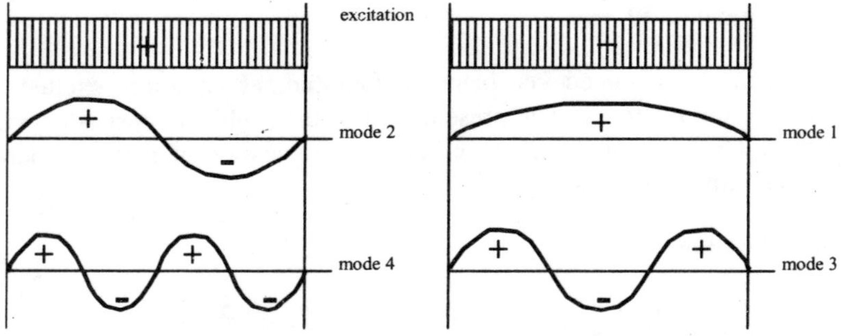

Figure 9.5. *Work of a uniform excitation applied to various modes of a supported-supported beam. Modes 2 and 4: total compensation of positive and negative work. Mode 3: partial compensation. Mode 1: no compensation*

d) Sinusoidally distributed excitation

Let us consider an excitation of the type:

$$p(x, t) = \sin\frac{k\pi}{L}x\ e^{j\omega t}.$$

The calculation of the generalized force leads to the result:

$$F_n(t) = \begin{cases} 0 \text{ if } n \neq k, \\ (L/2)e^{j\omega t} \text{ if } n = k. \end{cases}$$

It follows that the series is reduced to a single non-zero term corresponding to the index n = k:

$$W(x, t) = \frac{1}{\rho S}\frac{\sin\frac{k\pi}{L}x}{\omega_k^2 - \omega^2 + 2j\,\varepsilon_k\,\omega_k\,\omega}\,e^{j\omega t}. \qquad [9.99]$$

This notable property is linked to the fact that loading coincides with the normal deformation of the order k. The orthogonality of normal functions leads to the cancelation of all generalized forces except for a single one for the index $n = k$.

From a physical point of view, we may interpret this cancelation by the phenomenon of compensation of generalized forces highlighted in section c). Here the compensation is complete for all modes n which are different from k.

9.5.2. *Response of a beam in longitudinal vibration excited by an impulse force (time domain calculation)*

Let us consider a clamped-free beam in longitudinal vibration, excited by a shock applied to the point x_0. The presentation of the problem consists of writing down the three types of equations given below: the equation of motion, boundary conditions and initial conditions.

Equation of motion:

$$\rho S\frac{\partial^2 U}{\partial t^2} - ES\frac{\partial^2 U}{\partial x^2} = \delta(x - x_0)\ f(t).$$

Boundary conditions:

$$U(x,0) = 0 \ , \ ES\frac{\partial U}{\partial x}(L,t) = 0 \, .$$

Initial conditions:

$$U(x,0) = 0 \ , \ \frac{\partial U}{\partial t}(x,0) = 0 \, .$$

The function f(t) appearing in the equation of motion is homogenous to a force. It has constant amplitude, equal to one, during the period of force application T:

$$f(t) = \begin{cases} 0 & \text{if } t \le 0 \, , \\ 1 & \text{if } 0 < t < T \, , \\ 0 & \text{if } t \ge T \, . \end{cases}$$

Let us apply the method of modal decomposition to solve this problem. The modal system of the clamped-free beam has been calculated in Chapter 6:

$$\omega_n = \sqrt{\frac{E}{e}}\,\frac{2n-1}{2}\,\frac{\pi}{L} \qquad f_n(x) = \sin\left(\frac{2n-1}{2}\,\frac{\pi}{L}\,x\right).$$

We seek the response in the form of a modal series:

$$U(x,t) = \sum_{n=1}^{+\infty} a_n(t)\ \sin\frac{2n-1}{2}\,\frac{\pi}{L}\,x \, .$$

Amplitudes are provided by resolving the modal equation:

$$\ddot{a}_n(t) + 2\,\varepsilon_n\,\omega_n\,\dot{a}_n(t) + \omega_n^2\,a_n(t) = \frac{F_n(t)}{M_n} \, .$$

In this equation, the generalized mass and force are given respectively by the two equations below:

$$M_n = \int_0^L \rho S \sin^2\left(\frac{2n-1}{2}\frac{\pi}{L}x\right)dx = \rho S\frac{L}{2},$$

$$F_n(t) = f(t)\int_0^L \delta(x-x_0)\,\sin\left(\frac{2n-1}{2}\frac{\pi}{L}x\right)\,dx = f(t)\sin\left(\frac{2n-1}{2}\frac{\pi}{L}x_0\right).$$

The modal equation can be resolved using two approaches, time and frequency domains, described in section 9.4. The one that is more appropriate here is the time domain approach, which we provide in detail.

The solution is given by [9.95], which leads to two different expressions, during and after the moment T.

During the shock $(\tau < T)$:

$$a_n(\tau) = \frac{2}{\rho SL}\sin\left(\frac{2n-1}{2}\frac{\pi}{L}x_0\right)\int_0^\tau \frac{1}{\Omega_n}e^{-\omega_n\varepsilon_n(\tau-t)}\sin\Omega_n(\tau-t)\,dt. \qquad [9.100]$$

For the moments following the application of shock $(\tau > T)$, the formula changes a little, since the force is nil after the moment T:

$$a_n(\tau) = \frac{2}{\rho SL}\sin\left(\frac{2n-1}{2}\frac{\pi}{L}x_0\right)\int_0^T \frac{1}{\Omega_n}e^{-\omega_n\varepsilon_n(\tau-t)}\sin\Omega_n(\tau-t)\,dt. \qquad [9.101]$$

This integral is not difficult to calculate; however, the expression obtained is very cumbersome, so in order to simplify matters, we will consider the case of a non-damped beam $(\varepsilon_n = 0)$. The integral [9.101] then leads to the result:

$$a_n(\tau) = \frac{2}{\rho SL}\frac{\cos\left(\omega_n(\tau-T)\right)-\cos\omega_n\tau}{\omega_n^2}\sin\left(\frac{2n-1}{2}\frac{\pi}{L}x_0\right). \qquad [9.102]$$

This expression is valid for $\tau > T$. For $\tau \leq T$, i.e. during the shock, we obtain from [9.100]:

$$a_n(\tau) = \frac{2}{\rho SL} \frac{1 - \cos \omega_n \tau}{\omega_n^2} \sin\left(\frac{2n-1}{2} \frac{\pi}{L} x_0\right). \quad [9.103]$$

The vibratory amplitude of the beam is obtained by using these expressions of $a_n(t)$ in the modal decomposition. For example, during the application of the shock, the vibratory amplitude is given by:

$$U(x,t) = \sum_{n=1}^{+\infty} \frac{2}{\rho SL} \frac{1 - \cos(\omega_n t)}{\omega_n^2} \sin\left(\frac{2n-1}{2} \frac{\pi}{L} x_0\right) \sin\left(\frac{2n-1}{2} \frac{\pi}{L} x\right). \quad [9.104]$$

9.5.3. *Response of a beam in longitudinal vibrations subjected to an impulse force (frequency domain calculation)*

We consider the general case of the solution in frequency domain. Modal vibratory amplitude is provided by equation [9.84] which we recall:

$$a_n(t) = \frac{1}{2\pi M_n} \int_{-\infty}^{+\infty} \frac{F_n(\omega)}{\omega_n^2 - \omega^2 + 2j\omega\, \varepsilon_n \omega_n} e^{j\omega t}\, d\omega. \quad [9.105]$$

The calculation of the inverse Fourier transform which yields $a_n(t)$ can be carried out numerically. In the calculation of the integral, it would then be necessary to take into account the angular frequencies close to the normal angular frequency ω_n for which the denominator is very small and, consequently, the integrand takes its maximum value, which is characteristic of the resonance phenomenon. These numerical calculations are, however, very long and we can sometimes avoid them by an integration in the complex plane.

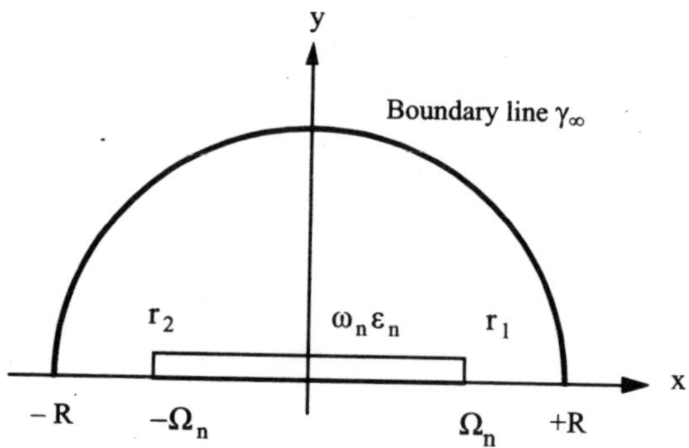

Figure 9.6. *Path of integration (z=x+jy)*

Let us consider the path of integration γ of the form defined in Figure 9.6 and calculate the integral:

$$\int_\gamma f(z)\,dz = \int_{-R}^{+R} f(x)\,dx + \int_0^\pi f\left(R\,e^{j\theta}\right) R\,d\theta .$$

Limits of integrals when R tends towards infinity show that:

$$\int_{-\infty}^{+\infty} f(x)\,dx = \int_{\gamma_\infty} f(z)\,dz - \lim_{R\to\infty} \int_0^\pi f\left(R\,e^{j\theta}\right) R\,d\theta .$$

If we write:

$$f(z) = \frac{1}{2\pi M_n} \frac{F_n(z)\,e^{jzt}}{\omega_n^2 - z^2 + 2jz\,\varepsilon_n\,\omega_n},$$

we then have:

$$a_n(t) = \int_{\gamma_\infty} f(z)\,dz - \lim_{R\to\infty} \int_0^\pi f\left(R\,e^{j\theta}\right) R\,d\theta . \qquad [9.106]$$

Applying the residue theorem makes it possible to calculate the first integral of the second member:

$\int_{\gamma_\infty} f(z)\,dz = 2j\,\pi \sum$ residues located in the complex half-plane with an imaginary positive part.

Let us suppose that the function $F_n(z)$ does not have a pole in the field of integration. The function:

$$\frac{1}{2\pi M_n} \frac{F_n(z)\, e^{jzt}}{\omega_n^2 - z^2 + 2jz\, \varepsilon_n\, \omega_n}$$

then has two of them defined by the zeros of the denominators r_1 and r_2 :

$$r_1 = \Omega_n + j\,\omega_n\, \varepsilon_n \ , \ r_2 = -\Omega_n + j\,\omega_n\, \varepsilon_n \,.$$

These two poles are located in the vicinity of the damped normal angular frequency Ω_n and its opposite $(-\Omega_n)$.

Damping creates a positive imaginary part with two poles, which are thus both in the field of integration.

These are two simple poles; the residues are calculated using the two expressions:

$$\text{Residue in } r_1 = \lim_{z \to r_1} (z - r_1)\ f(z) = -\frac{1}{2\pi M_n} \frac{F_n(r_1)\, e^{j\, r_1 t}}{r_1 - r_2},$$

$$\text{Residue in } r_2 = \lim_{z \to r_2} (z - r_2)\ f(z) = -\frac{1}{2\pi M_n} \frac{F_n(r_2)\ e^{j\, r_2 t}}{r_2 - r_1}.$$

We thus draw from it the following expression of $a_n(t)$:

$$a_n(t) = -\frac{j}{M_n} \frac{F_n(r_1)\ e^{j r_1 t} - F_n(r_2)\ e^{j r_2 t}}{2\Omega_n} - \lim_{R \to \infty} \int_0^{\pi} f(R\, e^{j\theta})\ R d\theta\,. \qquad [9.107]$$

The simple practical cases are those where the integral of the second member is nil; we thus have:

$$a_n(t) = -\frac{j}{2\Omega_n\, M_n}\ e^{-\omega_n \varepsilon_n t} \left[F_n(r_1)\ e^{j\Omega_n t} - F_n(r_2)\ e^{-j\Omega_n t} \right]. \qquad [9.108]$$

It is a solution of the free modal vibration type, which represents the solution at the moments when the force is no longer applied and, thus, when the structure vibrates freely.

We can now apply the result [9.108] to find the solution by temporal calculation of the problem highlighted to section 9.5.2.

The Fourier transform of the generalized force is given by the integral:

$$F_n(\omega) = \int_0^T e^{-j\omega T} \sin\left(\frac{2n-1}{2}\frac{\pi}{L}x\right) dt .$$

That is, after calculation:

$$F_n(\omega) = \frac{1-e^{-j\omega t}}{j\omega} \sin\left(\frac{2n-1}{2}\frac{\pi}{L}x_0\right).$$

It is enough to introduce this expression into [9.108] to obtain the vibratory response. In order to compare the results with temporal calculations, we will consider the borderline case of zero damping. Under these conditions $r_1 = -r_2 = \omega_n$ and equation [9.108] gives:

$$a_n(t) = \frac{1}{\omega_n^2 M_n}\left(-\cos\omega_n t + \cos\omega_n(t-T)\right) \sin\left(\frac{2n-1}{2}\frac{\pi}{L}x_0\right).$$

This expression coincides exactly with [9.102] which gave the vibratory amplitude for the moments $t > T$.

On the other hand, the vibratory amplitude during shock [9.103] is not provided by this result. This is due to the integral which appears in the second member of [9.107], which is effectively zero in the case of calculations leading to [9.108], but is no longer nil if $\tau \le T$.

The expression of the integral appearing in the second member of [9.107] is:

$$\lim_{R\to\infty} \int_0^\pi \frac{1}{2\pi M_n} \frac{e^{jR\,e^{j\theta}\,t} - e^{-jR\,e^{j\theta}(T-t)}}{jR\,e^{j\theta}\left(\omega_n^2 - R^2 e^{2j\theta} + 2j\,\varepsilon_n\,\omega_n\,R\,e^{j\theta}\right)} R\,d\theta .$$

The behavior of the integral when R tends to infinity is related to the exponentials of the numerator; after transformation, the numerator can be written:

$$-e^{+jR\cos\theta(t-T)}\,e^{-R\sin\theta(t-T)}+e^{jR\cos\theta\, t}\,e^{-R\sin\theta\, t}.$$

Between 0 and π sin θ is always positive and the integrand tends towards 0 when R tends towards infinity, if $t > T$. The integral is thus null if $t > T$; it is a consequence of Jordan's lemma, since in this case we can affirm that the boundary of $|zf(z)|$ is nil when R tends towards infinity. If $\tau \le T$, the preceding property is no longer verified, which explains why our calculation no longer leads to the result, since it presumes that the integral is nil. A complete calculation taking the value of the integral into account is possible; we will not perform this here so as to avoid weighing down the text.

Generally the application of [9.108] would require the nullity of the integral to be verified over the half-circle with an infinite radius. The form of [9.108] is characteristic of a free vibratory response and thus cannot represent the vibratory state when the force applied is not nil. This expression will thus be interesting for impulse excitations and will provide the answer after the moment of the shock.

9.6. Convergence of modal series

The method of modal decomposition expresses the vibratory response in the form of a series, which leads to the problem of convergence. We may, of course, find it beneficial to accumulate the least number of terms possible in order to accelerate calculations and certain techniques are sometimes used to improve convergence.

9.6.1. *Convergence of modal series in the case of harmonic excitations*

Let us consider the case covered in section 9.5.1, point a). We considered a bending beam excited at the point x_0 by a harmonic force with the angular frequency ω. The vibratory response provided by equation [9.96] was:

$$W(x,t)=\left(\sum_{n=1}^{\infty}\frac{P}{\rho SL/2}\,\frac{\sin\frac{n\pi}{L}x_0}{\omega_n^2-\omega^2+2j\,\varepsilon_n\,\omega_n\,\omega}\,\sin\frac{n\pi}{L}x\right)e^{j\omega t}.$$

This expression is characteristic of the calculation of response by decomposition in modal series. To accelerate its convergence it is, therefore, necessary to study the series of the following type:

$$\sum_{n=1}^{\infty} a_n \sin \frac{n\pi}{L} x .$$

These are series of the Fourier type whose convergence is well-known; if the amplitude of the term of the order n is in the form of $a_n = O(1/n^r)$ when $n \to \infty$, then if $r \geq 1$, the series converges, and if $r < 1$, the series diverges.

We may, moreover, show that the derivative $\dfrac{\partial^{r-1} W}{\partial x^{r-1}}$ is discontinuous. Thus, the function broken up into series has the following regularity: if r = 2, W(x,t) has a discontinuity of slope, if r = 1, it has a discontinuity, and if r = 0, it is a Dirac distribution (the series diverges).

In the case considered here, the normal angular frequency is given by $\omega_n = \sqrt{EI/\rho S}\, n^2 \pi^2 / L^2$, the generic term of the series being consequently proportional to $1/n^4$ when $n \to \infty$. The series is thus convergent and the response continues according to x. The variation of a_n in $1/n^4$ means, in fact, that it is the third derivative of the function which is discontinuous. On the physical level, this is quite coherent with our knowledge of bending internal efforts, since a localized force introduces a discontinuity of the shearing force $EI\, \partial^3 W / \partial x^3$.

For the uniformly distributed excitation studied in section 9.5.1, point c), the response presents a generic term with an even faster decrease $a_n = O(1/n^5)$. On the one hand, the series will converge quicker and, on the other hand, the shearing force will be continuous this time.

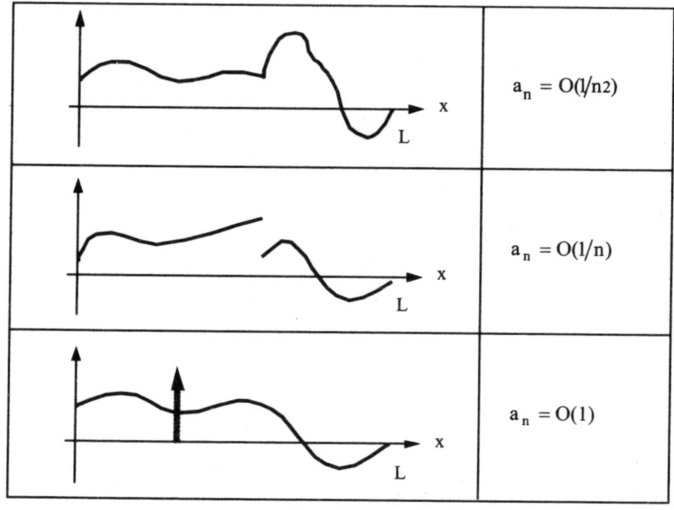

Figure 9.7. *Evolution of the generic term of the modal series according to the regularity of the response*

To sum up, the convergence of the modal series is linked to the regularity of the force applied, i.e. to the second member of the equation of motion. If the load is continuous, all the quantities with a physical significance and expressed by derivation of the displacement will be continuous. In the case of a bending beam considered in section 9.5.1, point c), the bending moment and the shearing force can be calculated using term by term derivation of the modal series and will converge.

If the load is irregular, in particular, if it is a Dirac distribution, the modal series giving the displacement will converge and the calculation of its successive derivatives using term by term derivation of the series is legitimate as long as the generic term decreases at least in $1/n^2$. When the decrease is in $1/n$, the represented function is discontinuous, and its term by term derivation within the framework of the theory of distributions leads to a divergent series. In this case, we may either work with the distributions, although the functions constructed on the basis of divergent series do not have local sense, or if we wish to preserve the local sense, we may no longer carry out term by term derivation but do it in the sense of the decomposition into a Fourier series of discontinuous functions (the reader may refer to any good mathematical work on the Fourier series). An example of this type of behavior is the calculation of the shearing force when the beam is excited by localized torque. Using term by term derivation of the series representing displacement, we obtain a divergent series for the shearing force.

9.6.2. *Acceleration of the convergence of modal series of forced harmonic responses*

Once again let us take the calculation performed in section 9.5.1, point a), in the case of the excitation of a beam bending under a harmonic force with the angular frequency ω localized in x_0. The response is provided by [9.96]. We observe that the modes can be assembled into three groups:

– those responding in mass; we then have: $\omega_n << \omega$. Their responses in the first approximation are given by:

$$a_n = \frac{P}{\rho SL/2} \frac{\sin \frac{n\pi}{L} x_0}{-\omega^2}; \quad [9.109]$$

– resonant modes verifying $\omega_p \approx \omega$. Their responses are given by the general form:

$$a_p = \frac{P}{\rho SL/2} \frac{\sin \frac{p\pi}{L} x_0}{\omega_p^2 - \omega^2 + 2j\,\varepsilon_p\,\omega_p\,\omega}; \quad [9.110]$$

– modes responding in stiffness; we then have: $\omega_r >> \omega$. Their responses in the first approximation are of the form:

$$a_r = \frac{P}{\rho SL/2} \frac{\sin \frac{r\pi}{L} x_0}{\omega_r^2}. \quad [9.111]$$

This last category of modes is the most numerous (there is an infinite number of them) and determines the speed of convergence.

In fact, the amplitude [9.111] is characteristic of static modal response; it is enough to make $\omega = 0$ in the general form [9.110] to obtain it. It is this property which is at the root of accelerated convergence.

Let us consider the static problem resulting from the previous case: it is the same problem, but the angular frequency of excitation is nil. The solution of this problem noted $W_S(x)$ is given by the expression [9.96] particularized to $\omega = 0$, that is, [9.112]:

$$W_S(x) = \sum_{n=1}^{\infty} \frac{P}{\rho SL/2} \frac{\sin \frac{n\pi}{L} x_0}{\omega_r^2} \sin \frac{n\pi}{L} x . \qquad [9.112]$$

The displacement $W_S(x)$ verifies equation [9.113] as well as the conditions of support at the ends:

$$EI \frac{\partial^4 W_S}{dx^4} = P \delta (x - x_0) . \qquad [9.113]$$

The acceleration of convergence is achieved by making the static solution $W_S(x)$ take into account the modes responding in stiffness (see the articles of M.A. Akgün [AKG 93] and D. Williams [WIL 45] for more information).

We write [9.114] where $W_D(x)$ is the dynamic contribution to determine:

$$W(x,t) = \left(W_D(x) + W_S(x)\right) e^{j\omega t} . \qquad [9.114]$$

Let us use the [9.114] decomposition of the solution in the equation of motion (the dependence in $e^{j\omega t}$ is omitted to simplify matters):

$$-\omega^2 \rho S \left(W_D(x) + W_S(x)\right) + EI \left(\frac{d^4 W_D}{dx^4} + \frac{d^4 W_S}{dx^4} \right) = P \delta (x - x_0) . \qquad [9.115]$$

Taking into account [9.113] we obtain:

$$-\omega^2 \rho S \, W_D(x) + EI \frac{d^4 W_D}{dx^4} = \omega^2 \rho S \, W_S(x) . \qquad [9.116]$$

We return to a standard equation to calculate $W_D(x)$ but the excitation is no longer the one actually applied but is linked to the static solution. Taking into account the regularity of $W_D(x)$, the calculation leads to a rapid convergence.

Applying modal decomposition to equation [9.116] we find:

$$W_D(x) = \sum_{n=1}^{+\infty} \frac{1}{M_n} \frac{F_{Sn}}{\omega_n^2 - \omega^2 + 2j\,\varepsilon_n\,\omega_n\,\omega} \sin\frac{n\pi}{L}x\,, \qquad [9.117]$$

with:

$$F_{Sn} = \omega^2 \int_0^L \rho S\, W_S(x)\, \sin\frac{n\pi}{L}x\; dx\,. \qquad [9.118]$$

Generally the vibratory solution obtained by traditional modal decomposition is given by:

$$W_D = \sum_{n=1}^{+\infty} \frac{1}{M_n} \frac{F_n}{\omega_n^2 - \omega^2 + 2j\,\varepsilon_n\,\omega_n\,\omega} \sin\frac{n\pi}{L}x$$

with: $F_n = \int_0^L p(x)\, \sin\frac{n\pi}{L}x\; dx\,,$

where p(x) is the distribution of the excitation force.

Noting that the static solution verifies:

$$EI\frac{d^4\,W_S}{dx^4} = p(x)\,,$$

we can demonstrate that:

$$F_{Sn} = F_n\frac{\omega^2}{\omega_n^4}\,. \qquad [9.119]$$

Thus, we note the acceleration of convergence of the calculation of $W_D(x)$, since for large n indices, modal amplitudes decrease in $1/\omega_n^4$ and not in $1/\omega_n^2$ as in the classical solution. The static solution can be obtained in analytical form in many cases and at any rate can be calculated much more easily than the vibratory solution.

To illustrate this method, let us take the case from section 9.5.1, point c), that is, a uniform distributed excitation. The static solution is obtained easily since it is the solution of the problem:

$$EI\frac{\partial^4 W_S}{\partial x^4} = 1\,.$$

The solution of this equation verifying the boundary conditions of support is given by:

$$W_S(x) = \frac{1}{24\,EI}\left(x^4 - 2L\,x^3 + L^3\,x\right). \qquad [9.120]$$

The dynamic part is now calculated with [9.117]. Using [9.119] in [9.118], we obtain:

$$F_{Sn} = \omega^2\frac{\rho S}{EI}\frac{1-(-1)^n}{\left(\frac{n\pi}{L}\right)^5} = \frac{\omega^2}{\omega_n^2}\frac{1-(-1)^n}{\frac{n\pi}{L}}\,.$$

The dynamic response is thus equal to:

$$W_D(x) = \sum_{n=1}^{+\infty}\frac{1}{M_n}\frac{\omega^2}{\omega_n^2}\frac{1-(-1)^n}{\frac{n\pi}{L}}\frac{1}{\omega_n^2-\omega^2+2j\,\varepsilon_n\,\omega_n\,\omega}\sin\frac{n\pi}{L}x\,. \qquad [9.121]$$

The vibratory response is the joining of $W_S(x)$ and $W_D(x)$ given respectively by [9.120] and [9.121]. The improvement of convergence is spectacular; for the initial calculation in section 9.5.1, point c), the amplitude of mode n was proportional to $1/n^5$; for the present case, the amplitude of mode n of the dynamic solution is proportional to $1/n^9$ (which is what was indicated by [9.119]).

9.7. Conclusion

This chapter has presented the calculation of the forced vibratory response by modal decomposition. It is a general method which introduces modal amplitudes as unknowns. They are determined through the resolution of the modal equation, which is that of a system with one degree of freedom where the generalized characteristics of the mode appear: generalized mass, stiffness and force. Generalized damping is also introduced at the level of this equation with a viscous model.

The description of the modal equation results from the application of the properties of orthogonality, which are at the foundation of the decoupling of modes. In the simple cases, the properties of orthogonality are fairly easy to determine; in the case of complicated boundary conditions, they may be difficult to pinpoint.

The resolution of the modal equation may be carried out in time domain or frequency domain. In time domain, the modal impulse response is the basic tool of calculation, since the modal response is obtained by the product of convolution of the impulse response and the force. In frequency domain, the modal harmonic response is used multiplied by the Fourier transform of the force: we obtain the modal response according to the frequency and then the inverse Fourier transform yields the time history.

The end of the chapter presents the problem of the convergence of modal series and especially of its acceleration by using the static response of the vibrating system.

The disadvantage of the modal method is that it expresses the response in the form of a series which presents convergence problems and leads to heavy calculations. It should, however, be noted that we can also see an advantage in this approach, because the answer is split into elementary movements (modes) and can thus be easily understood, offering a course of action to reduce vibrations.

We can consider another approach, not based on modal decomposition, which in certain cases makes it possible to obtain the response in analytical form: this will be discussed in the next chapter.

Chapter 10

Calculation of Forced Vibrations by Forced Wave Decomposition

10.1. Introduction

In Chapter 9 we provided a method of calculation of the vibratory response of structures subjected to dynamic stresses by modal decomposition. This method is general, since it is applicable to any structure and any type of excitation. This generality, however, costs us, since the response is expressed in the form of a series which presents problems of calculation related to the convergence of modal series.

The method that we are going to develop is more restrictive, since it is primarily applicable to mono-dimensional structures, excited at a point by a harmonic force. Its biggest advantage is that it offers analytical solutions. The fundamental element of the method is the concept of the forced wave, which is the solution of the homogenous equation of motion. As we will see, the technique of calculation rests on a sub-structuring of the vibrating system, the solutions will thus be defined by parts.

The discourse will be based on some examples: torsion and bending of beams, the extension of which to more complex cases is quite straightforward. At the end of the chapter we will present the generalization that can be made for distributed and non- homogenous excitations, which removes the initial restriction of the method on harmonic localized excitations.

In short, the method of response calculation by decomposition in forced waves is applicable to mono-dimensional structures homogenous by parts. It is based on a sub-structuring, each section being delimited by two singular points (discontinuity of structure or point of excitation); the solution is thus provided by parts.

The method extends to plates via semi-modal decomposition of the vibratory response. At the end of the chapter we present this approach, which, however, remains of limited use for rectangular or circular plates and for particular boundary conditions.

10.2. Introduction to the method on the example of a beam in torsion

10.2.1. *Example: homogenous beam in torsion*

To give form to the method we will consider the simplest case, in order to avoid the technical difficulties and, thus, to better outline the fundamental idea of calculation.

The wave equation, representative of the vibrations of torsion of beams (but also of the longitudinal vibrations of beams, as well as vibrations of the cords and pipes), will provide the example we are looking for.

The decomposition of the vibratory solutions into forced waves appears naturally when a harmonic excitation applied to a point is considered. To formulate the method, let us consider a homogenous beam with a constant cross-section excited at the point x_0 by harmonic torque. In the example, the beam is clamped at both ends (see Figure 10.1).

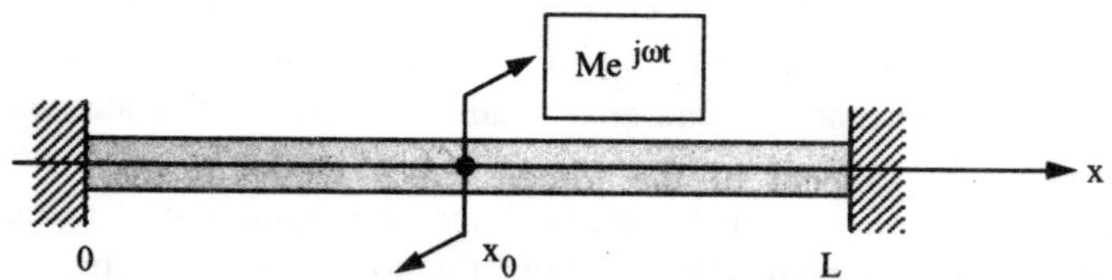

Figure 10.1. *Beam in torsion excited by harmonic torque* $Me^{j\omega t}$ *localized in* x_0

Presenting the problem so as to reveal the decomposition into a forced wave has its particularities. In fact, it is necessary to sub-structure it so as to have to solve only homogenous equations of motion. In our case we will divide the beam into two sections: SS1 and SS2.

The subsystem SS1 is the section of beam defined by the open segment $]0, x_0[$. In this part, the unknown representing the angle of torsion will be noted $\alpha_1(x, t)$.

The subsystem SS2 is the section of beam defined by the open segment $]x_0, L[$. In this part the unknown representing the angle of torsion will be noted $\alpha_2(x,t)$. The equations that must be satisfied by these two unknowns are:

$$\text{SS1}\quad \begin{aligned} & x \in \left]0, x_0\right[\\ & \rho I_0 \frac{\partial^2}{\partial t^2}\alpha_1(x,t) - GI_0 \frac{\partial^2}{\partial x^2}\alpha_1(x,t) = 0, \end{aligned} \qquad [10.1]$$

$$\text{SS2}\quad \begin{aligned} & x \in \left]x_0, L\right[\\ & \rho I_0 \frac{\partial^2}{\partial t^2}\alpha_2(x,t) - GI_0 \frac{\partial^2}{\partial x^2}\alpha_2(x,t) = 0. \end{aligned} \qquad [10.2]$$

The boundary conditions of clamped type are:

$$\alpha_1(0,t) = 0, \qquad [10.3]$$

$$\alpha_2(L,t) = 0. \qquad [10.4]$$

The conditions of connection to the interface $(x = x_0)$ are:

– continuity of displacements:

$$\alpha_1(x_0, t) = \alpha_2(x_0, t); \qquad [10.5]$$

– discontinuity of moments of torsion due to the localized torque applied:

$$GI_0 \frac{\partial \alpha_1}{\partial x}(x_0, t) - GI_0 \frac{\partial \alpha_2}{\partial x}(x_0, t) = Me^{j\omega t}. \qquad [10.6]$$

The vibratory movement solution is the superposition of the forced vibration and the free vibration. Decomposition into forced waves applies only to the forced vibration and is, therefore, only representative of the response once the transitory state weakens.

We will thus only seek the forced response, which has the form:

$$\alpha_1(x,t) = \overline{\alpha}_1(x)\, e^{j\omega t} \qquad [10.7]$$

$$\alpha_2(x, t) = \overline{\alpha}_2(x)\, e^{j\omega t} . \qquad [10.8]$$

The quantities $\overline{\alpha}_1(x)$ and $\overline{\alpha}_2(x)$ represent the complex amplitudes of harmonic vibratory movements.

Using the expressions [10.7] and [10.8], equations [10.1] – [10.6] become:

$$\rho I_0\, \omega^2\, \overline{\alpha}_1(x) + GI_0 \frac{d^2\, \overline{\alpha}_1}{dx^2} = 0 \ , \ x \in \,]0, x_0[\, , \qquad [10.9]$$

$$\rho I_0\, \omega^2\, \overline{\alpha}_2(x) + GI_0 \frac{d^2\, \overline{\alpha}_1}{dx^2} = 0 \ , \ x \in \,]x_0 , L[\, , \qquad [10.10]$$

$$\overline{\alpha}_1(0) = 0 \, , \qquad [10.11]$$

$$\overline{\alpha}_2(L) = 0 \, , \qquad [10.12]$$

$$\overline{\alpha}_1(x_0) = \overline{\alpha}_2(x_0) \, , \qquad [10.13]$$

$$GI_0 \frac{d\overline{\alpha}_2}{dx}(x_0) - GI_0 \frac{d\overline{\alpha}_1}{dx}(x_0) = M \, . \qquad [10.14]$$

Let us note that it is not necessary to write down the initial conditions, since those are used only for the calculation of the transitory state, which is not taken into account here.

10.2.2. *Forced waves*

The solutions of equations [10.9] and [10.10] can be easily calculated, since we are dealing with standard differential equations. We obtain:

$$\overline{\alpha}_1(x) = A_1\, e^{jkx} + B_1\, e^{-jkx} \qquad [10.15]$$

with: $k = \omega / c_T$ and $c_T = \sqrt{G/\rho}$; [10.16] similarly:

$$\overline{\alpha}_2(x) = A_2\, e^{jkx} + B_2\, e^{-jkx}\,. \qquad [10.17]$$

These solutions can be interpreted in terms of traveling waves propagating in both directions: of increasing and decreasing x. They are forced waves in the sense that the angular frequency ω and, therefore, the wave number k provided in [10.16] are determined by the force applied.

The equivalent form is often used:

$$\overline{\alpha}_1(x) = a_1 \sin kx + b_1 \operatorname{cox} kx \qquad [10.18]$$

$$\overline{\alpha}_2(x) = a_2 \sin kx + b_2 \operatorname{cox} kx\,. \qquad [10.19]$$

10.2.3. *Calculation of the forced response*

It suffices to make the solutions defined in each section respect the boundary and connection conditions. Verifying [10.11] and [10.12] yields:

$$\overline{\alpha}_1(x) = a_1 \sin kx\,, \qquad [10.20]$$

$$\overline{\alpha}_2(x) = a_2 \sin k(x - L)\,. \qquad [10.21]$$

Respecting [10.13] and [10.14] leads to the linear system [10.22]:

$$\begin{bmatrix} \sin kx_0 & -\sin\left(k(x_0 - L)\right) \\ \cos kx_0 & -\cos\left(k(x_0 - L)\right) \end{bmatrix} \begin{bmatrix} a_1 \\ a_2 \end{bmatrix} = \begin{bmatrix} 0 \\ M/k\, GI_0 \end{bmatrix}. \qquad [10.22]$$

The solutions are:

$$a_1 = -\frac{M}{k\, GI_0}\,\frac{\sin k(x_0 - L)}{\det}\,, \qquad [10.23]$$

$$a_2 = -\frac{M}{k\, GI_0}\,\frac{\sin kx_0}{\det}\,. \qquad [10.24]$$

where det, the determinant of the system, is equal to:

$$\det = \sin kL\,. \tag{10.25}$$

The forced response is obtained immediately replacing a_1 and a_2 in [10.20] and [10.21]. We obtain an expression by substructure:

$$\overline{\alpha}_1(x) = -\frac{M}{k\,GI_0}\,\frac{\sin k(x_0 - L)}{\sin kL}\,\sin kx\ ,\ x \in \,]0, x_0[\,, \tag{10.26}$$

$$\overline{\alpha}_2(x) = -\frac{M}{k\,GI_0}\,\frac{\sin kx_0}{\sin kL}\,\sin k(x-L)\ ,\ x \in \,]x_0, L[\,. \tag{10.27}$$

As opposed to the method of modal decomposition, the response is obtained in an analytical form, which does away with the problem of series calculation. However, the same phenomena are present, in particular, the concept of resonance, which appears here when the determinant of the linear system [10.22] is nil. That is, when $k = \frac{n\pi}{L}$ and, thus, when the angular frequency of excitation takes the values $\omega_n = \sqrt{\frac{G}{\rho}}\frac{n\pi}{L}$.

These values correspond to the normal angular frequency of the clamped-clamped beam. Thanks to the expressions of the response [10.26] and [10.27], we observe that for these frequencies, vibratory amplitude is infinite. This tendency is coherent with the fact that the beam is non-damped.

The analytical expression of the forced response, furthermore, highlights the phenomena that are difficult to identify by modal decomposition. In fact, the effects of anti-resonance block the response of a section of the beam.

Let us examine the response of the SS1 section. It is nil in any point when:

$$\sin k(x_0 - L) = 0\,,$$

i.e. with the angular frequency of excitation:

$$\omega_p = \sqrt{\frac{G}{\rho}}\,\frac{p\pi}{L - x_0}\,.$$

These angular frequencies correspond to resonances of the SS2 section clamped at both ends. Everything occurs as if the SS2 section, which has the ability to vibrate with great amplitude at these frequencies, absorbed the whole of the excitation, thus blocking the other section.

Of course, the situation changes with the angular frequency of resonance of the SS1 section:

$$\omega_q = \sqrt{\frac{G}{\rho}}\,\frac{q\pi}{L}.$$

10.2.4. *Heterogenous beam*

Decomposition into forced waves can be used in the case of beams with abruptly variable heterogenity. Let us take the case of Figure 10.2 to illustrate the method.

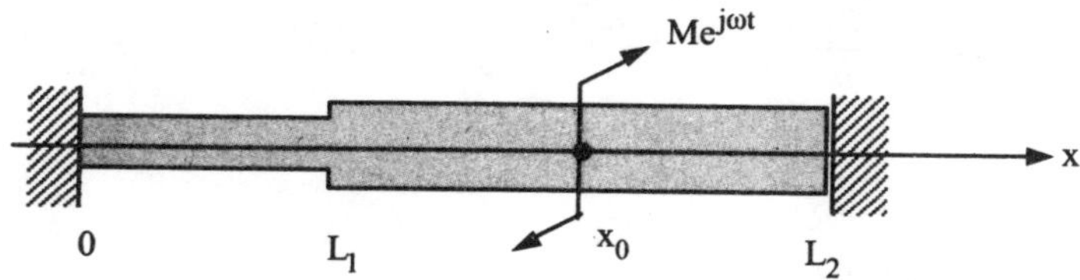

Figure 10.2. *Vibrations of torsion of a beam with abruptly variable inertia*

Sub-structuring must reveal three sections where three unknown functions are defined:

$$\overline{\alpha}_1(x) \quad \text{when} \quad x \in]0, L_1[,$$

$$\overline{\alpha}_2(x) \quad \text{when} \quad x \in]L_1, x_0[,$$

$$\overline{\alpha}_3(x) \quad \text{when} \quad x \in]x_0, L_2[.$$

The three functions verify the following equations of motion:

$$\rho_1\, I_1\, \omega^2\, \overline{\alpha}_1 + G_1\, I_1 \frac{d^2\, \overline{\alpha}_1}{dx^2} = 0, \qquad [10.28]$$

$$\rho_2 I_2 \omega^2 \overline{\alpha}_2 + G_2 I_2 \frac{d^2 \overline{\alpha}_2}{dx^2} = 0, \qquad [10.29]$$

$$\rho_2 I_2 \omega^2 \overline{\alpha}_3 + G_2 I_2 \frac{d^2 \overline{\alpha}_3}{dx^2} = 0. \qquad [10.30]$$

These three equations have classical solutions of the type [10.18]:

$$\overline{\alpha}_1(x) = a_1 \sin k_1 x + b_1 \cos k_1 x, \qquad [10.31]$$

$$\overline{\alpha}_2(x) = a_2 \sin k_2 x + b_2 \cos k_2 x, \qquad [10.32]$$

$$\overline{\alpha}_3(x) = \alpha_3 \sin k_2 x + b_3 \cos k_2 x, \qquad [10.33]$$

with:

$$k_1 = \omega\sqrt{\frac{G_1}{\rho_1}}$$
$$k_2 = \omega\sqrt{\frac{G_2}{\rho_2}}. \qquad [10.34]$$

Now it is sufficient to write down the connection and boundary conditions which link these solutions:

$$\begin{cases} \overline{\alpha}_1(0) = 0, \\ \overline{\alpha}_1(L_1) = \overline{\alpha}_2(L_1), \\ G_1 I_1 \dfrac{d\overline{\alpha}_1}{dx}(L_1) = G_2 I_2 \dfrac{d\overline{\alpha}_2}{dx}(L_1), \\ \overline{\alpha}_2(x_0) = \overline{\alpha}_3(x_0), \\ G_2 I_2 \dfrac{d\overline{\alpha}_3}{dx}(x_0) - G_2 I_2 \dfrac{d\overline{\alpha}_2}{dx}(x_0) = M, \\ \overline{\alpha}_3(L_2) = 0. \end{cases} \qquad [10.35]$$

The use of the solutions of the equation of motion under the [10.35] conditions yields a non- homogenous linear system which when resolved provides the unknown amplitudes $(a_1, b_1, ..., b_3)$ and, thus, the vibratory response in each section with [10.31] – [10.33]. We leave it to pursue the calculations later on.

Decomposition into forced waves requires sub-structuring aimed at isolating sections of the homogenous beam with constant inertia whose solution for the equation of motion is known. The sections would thus be delimited by points of singularity (discontinuity of structure and applied force); here two singularities give three sections. Generally, N singularities will give $N+1$ sections.

10.2.5. *Excitation by imposed displacement*

In certain problems we know the amplitude of vibrations at the point of excitation while the applied force is unknown. The method of decomposition in forced waves can be used in this case in a completely simple manner. Let us take once again the case of section 10.2.1. Replacing the excitation by torque by an imposed angle of torsion:

$$\begin{cases} \alpha_1(x_0, t) = \bar{\gamma}\, e^{j\omega t} \\ \alpha_2(x_0, t) = \bar{\gamma}\, e^{j\omega t}. \end{cases} \qquad [10.36]$$

The solution is identical to the case of section 10.2.1. Equations [10.9] – [10.12] remain unchanged. The conditions of connection are different here; it is necessary to replace [10.13] and [10.14] by the two conditions:

$$\bar{\alpha}_1(x_0) = \bar{\gamma}, \qquad [10.37]$$

$$\bar{\alpha}_2(x_0) = \bar{\gamma}. \qquad [10.38]$$

Vibratory movement is calculated in a similar way and after all the calculations we obtain:

$$\bar{\alpha}_1(x) = \bar{\gamma}\frac{\sin kx}{\sin kx_0}, \qquad [10.39]$$

$$\overline{\alpha}_2(x) = \overline{\gamma}\frac{\sin k(x - L)}{\sin k(x_0 - L)}. \qquad [10.40]$$

We can deduce the torque applied to produce this vibratory movement thanks to the relation [10.14]:

$$M = GI\left(\frac{d\overline{\alpha}_1}{dx}(x_0) - \frac{d\overline{\alpha}_2}{dx}(x_0)\right), \qquad [10.41]$$

that is:

$$M = GIk\overline{\gamma}\left(\frac{\sin kL}{\sin kx_0 \, \sin k(x_0 - L)}\right). \qquad [10.42]$$

The expression [10.42] makes it possible to highlight two remarkable properties of torque that have to be applied to a beam to produce a given angle $\overline{\gamma}$:

1) the torque that has to be applied tends towards 0 at the beam's resonance angular frequencies. At these frequencies $\omega_n = \sqrt{G/\rho}\,\frac{n\pi}{L}$, the numerator is nil;

2) the torque tends towards infinity at anti-resonance angular frequencies when the denominator is nil. The angular frequencies of anti-resonance are given by the two following expressions: $\omega_p = \sqrt{\frac{G}{\rho}}\frac{p\pi}{L}$ and $\omega_q = \sqrt{\frac{G}{\rho}}\frac{q\pi}{L - x_0}$.

Figure 10.3 illustrates the responses of the same structure excited either by an imposed angular displacement constant with the frequency (equation [10.39]), or by a localized torque constant with the frequency (equation [10.26]). The curves are very different, in particular, beam resonances no longer appear as maxima for the excitation in displacement; in fact, it is at the subsystem anti-resonance angular frequency that the amplitude is infinite. We see here all the difficulty of interpretation of frequency response when the excitation is barely known.

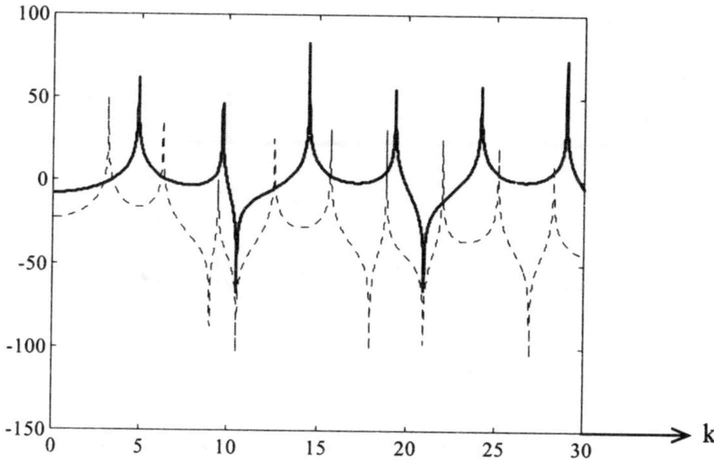

Figure 10.3. *Level of vibratory displacement according to the wave number* k *of a 1 m long beam, observed at the 0.3 m point for an excitation of torsion in 0.65 m by:*
- - - - - *Constant imposed angular displacement* ($\overline{\gamma} = 1$)
——— *Constant imposed excitation torque* ($M/(GI_0) = 1$)

10.3. Resolution of the problems of bending

10.3.1. *Example of an excitation by force*

In principle, the problem of bending is not different from that of torsion; at the technical level, the difficulty is greater since the solution of the equation of motion contains vanishing waves and because it is possible to excite bending beams in two distinct ways: by force and by torque.

We will initially consider a clamped-free beam excited by a harmonic transverse force applied at x_0 (see Figure 10.4). We seek the solution of the forced vibration in the form:

$$W(x, t) = \begin{cases} \overline{W}_1(x)\, e^{j\omega t} \quad , \; x \in \left]0, x_0\right[\\ \overline{W}_2(x)\, e^{j\omega t} \quad , \; x \in \left]x_0, L\right[. \end{cases} \qquad [10.43]$$

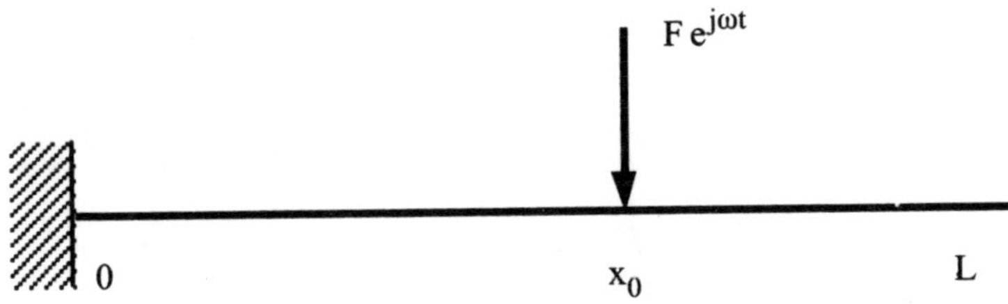

Figure 10.4. *Bending vibrations of a clamped-free beam, excited by a harmonic force* $F\,e^{j\omega t}$

It is necessary, of course, to sub-structure the beam into two sections. Vibratory amplitudes $W_1(x)$ and $W_2(x)$ verify the equations of motions:

$$-\omega^2 \rho S\,\overline{W}_1(x) + EI\frac{\partial^4 \overline{W}_1}{\partial x^4} = 0 \;,\; x \in \left]0, x_0\right[, \qquad [10.44]$$

$$-\omega^2 \rho S\,\overline{W}_2(x) + EI\frac{\partial^4 \overline{W}_2}{\partial x^4} = 0 \;,\; x \in \left]x_0, L\right[. \qquad [10.45]$$

These differential equations are easily integrated:

$$\overline{W}_1(x) = A_1 \sin kx + B_1 \cos kx + C_1\,\mathrm{sh}(kx) + D_1\,\mathrm{ch}(kx)\,, \qquad [10.46]$$

$$\overline{W}_2(x) = A_2 \sin kx + B_2 \cos kx + C_2\,\mathrm{sh}(kx) + D_2\,\mathrm{ch}(kx)\,, \qquad [10.47]$$

with: $$k = \sqrt{\omega} \Big/ \sqrt[4]{\frac{EI}{\rho S}}\,. \qquad [10.48]$$

The solutions reveal traveling waves through the sine and cosine, and also vanishing waves through the hyperbolic sine and cosine.

Writing down the boundary and connection conditions will allow, as previously, calculating the constants $A_1, B_1, \ldots, C_2, D_2$ appearing in the solutions [10.46] and [10.47].

Boundary conditions:

$$\begin{cases} \overline{W}_1(0) = 0\,, \\ \dfrac{d\overline{W}_1}{dx}(0) = 0\,, \\ EI\dfrac{d^2\,\overline{W}_2}{dx^2}(L) = 0\,, \\ EI\dfrac{d^3\,\overline{W}_2}{dx^3}(L) = 0\,. \end{cases} \qquad [10.49]$$

Conditions of connection:

$$\overline{W}_1(x_0) = \overline{W}_2(x_0) \qquad \text{continuity of displacements,}$$

$$\frac{d\overline{W}_1}{dx}(x_0) = \frac{d\overline{W}_2}{dx}(x_0) \qquad \text{continuity of slopes,}$$

$$EI\frac{d^2\overline{W}_1}{dx^2}(x_0) = EI\frac{d^2\overline{W}_2}{dx^2}(x_0) \qquad \text{continuity of bending moments,} \qquad [10.50]$$

$$EI\frac{d^3\overline{W}_2}{dx^3}(x_0) - EI\frac{d^3\overline{W}_1}{dx^3}(x_0) = F \quad \text{discontinuity of the shearing forces.}$$

Introducing solutions [10.46] and [10.47] under the 8 boundary and connection conditions leads to the calculation of the 8 constants. We will not proceed further with this extremely heavy calculation, which requires computerized processing. To give an example, we take the case of a supported-free beam where the point of excitation is at the $x_0 = L$ end of the beam. In this case only one section is necessary and the boundary conditions are:

$$\overline{W}_1(0) = 0\,, \qquad [10.51]$$

$$EI\frac{d^2\,\overline{W}_1}{dx^2}(0) = 0\,, \qquad [10.52]$$

$$EI\frac{d^2\overline{W}_1}{dx^2}(L)=0\,, \qquad [10.53]$$

$$EI\frac{d^3\overline{W}_1}{dx^3}(L)=F\,. \qquad [10.54]$$

The introduction of [10.46] into [10.51] and [10.52] shows that $B_1 = D_1 = 0$.

The conditions [10.53] and [10.54] give:

$$\begin{bmatrix} -\sin kL & \text{sh}\, kL \\ -\cos kL & \text{ch}\, kL \end{bmatrix}\begin{bmatrix} A_1 \\ C_1 \end{bmatrix}=\begin{bmatrix} 0 \\ F/(EI\,k^3) \end{bmatrix}, \qquad [10.55]$$

that is:

$$\begin{cases} A_1=\dfrac{-\text{sh}(kL)}{\text{sh}(kL)\ \cos(kL)-\sin(kL)\ \text{ch}(kL)}F/(EI\,k^3) \\ C_1=\dfrac{-\sin(kL)}{\text{sh}(kL)\ \cos(kL)-\sin(kL)\ \text{ch}(kL)}F/(EI\,k^3) \end{cases}. \qquad [10.56]$$

The vibratory response is thus:

$$\overline{W}_1(x)=\frac{-F/(EI\,k^3)}{\text{sh}\, kL\ \cos kL-\sin kL\ \text{ch}\, kL}\left[\text{sh}\, kL\ \sin kx+\sin kL\ \text{sh}\, kx\right]. \qquad [10.57]$$

The denominator of [10.57], when equal to zero, corresponds to the characteristic equation of a supported-free beam; vibratory amplitude will thus tend towards infinity at angular frequencies of resonance of the supported-free beam, which satisfy the characteristic equation.

10.3.2. *Excitation by torque*

Let us once again take the example of Figure 10.4 replacing the excitation force by a harmonic couple $Me^{j\omega t}$. The solution of the problem is obtained in a manner similar to the case of section 10.3.1. Vibratory movements in the two sections have the general form [10.46] and [10.47], while boundary conditions are those described in [10.49]. Conditions of connection are slightly modified, the two last equations of [10.50] becoming:

$$\begin{cases} EI\dfrac{d^2\overline{W}_2}{dx^2}(x_0) - EI\dfrac{d^2\overline{W}_1}{dx^2}(x_0) = M, \\ \\ EI\dfrac{d^3\overline{W}_2}{dx^3}(x_0) = EI\dfrac{d^3\overline{W}_1}{dx^3}(x_0). \end{cases} \qquad [10.58]$$

To identify the constants appearing in the solutions of the equations of motion [10.46] and [10.47], we use the 8 boundary and connection conditions.

10.4. Damped media (case of the longitudinal vibrations of beams)

10.4.1. *Example*

We will consider the longitudinal vibrations of a clamped-free damped beam excited by a harmonic force localized in x_0.

Damping is introduced via a complex Young modulus, since we are dealing with harmonic movement:

$$E^* = E(1 + j\eta),$$

where η is the loss factor of material.

We will sub-structure the beam into two sections where longitudinal displacements are respectively:

$\overline{U}_1(x)$ when $x \in \left]0, x_0\right[$

and $\overline{U}_2(x)$ when $x \in \left]x_0, L\right[$.

The equations to verify are the following:

$$\rho S\,\omega^2\,\overline{U}_1 + E^*\,S\frac{d^2\overline{U}_1}{dx^2} = 0\ ,\ x \in \left]0, x_0\right[, \qquad [10.59]$$

$$\rho S\,\omega^2\,\overline{U}_2 + E^*\,S\frac{d^2\,\overline{U}_2}{dx^2} = 0\ ,\ x \in \left]x_0\,,L\right[, \tag{10.60}$$

$$\overline{U}_1(0) = 0\,, \tag{10.61}$$

$$E^*\,S\frac{d\overline{U}_2}{dx}(L) = 0\,, \tag{10.62}$$

$$\overline{U}_1(x_0) = \overline{U}_2(x_0)\,, \tag{10.63}$$

$$E^*\,S\frac{d\overline{U}_2}{dx}(x_0) - E^*\,S\frac{d\overline{U}_1}{dx}(x_0) = F\,. \tag{10.64}$$

The solutions of the equations of motion [10.59] and [10.60] are obtained in a traditional fashion:

$$\overline{U}_1(x) = A_1\,e^{jk^*x} + B_1\,e^{-jk^*x}\,, \tag{10.65}$$

$$\overline{U}_2(x) = A_2\,e^{jk^*x} + B_2\,e^{-jk^*x}\,, \tag{10.66}$$

with: $$k^* = \omega \Big/ \left(\sqrt{\frac{E}{P}}\,\sqrt{1+j\eta}\right). \tag{10.67}$$

The wave number being complex, the propagation is carried out with a weakening of amplitude. For weak damping, which generally is the case in practice, there is the following approximation:

$$k^* \approx \omega \Big/ \left(\sqrt{\frac{E}{P}}\right)\left(1 - j\frac{\eta}{2}\right) = k - j\gamma\,. \tag{10.68}$$

The calculation of vibratory response is straightforward; it suffices to calculate the integration constants appearing in [10.65] and [10.66] so that they verify the conditions [10.61] – [10.64]. However, in order to avoid weighing down the discussion, we will take a simpler case supposing that $x_0 = L$, i.e. the excitation is at the end of the beam. Only one section is necessary; the solution [10.65] must verify the boundary conditions:

$$\overline{U}_1(0) = 0 \qquad \text{and} \quad E^* S \frac{dU_1}{dx}(L) = F, \qquad [10.69]$$

that is:

$$\begin{cases} A_1 + B_1 = 0 \Rightarrow A_1 = -B_1 \\ \text{and} \\ jk^* \left(e^{jk^*L} + e^{-jk^*L}\right) A_1 = F / (E^*S). \end{cases} \qquad [10.70]$$

Thus, we obtain:

$$\overline{U}_1(x) = \frac{F/(E^*S)}{jk^*\left(e^{jk^*L} + e^{-jk^*L}\right)} \left(e^{jk^*L} - e^{-jk^*L}\right). \qquad [10.71]$$

Taking damping into account, the denominator of equation [10.71] cannot be zero; we are therefore witnessing damped resonances.

Taking the damping account into account is done without difficulty, thanks to the introduction of complex elasticity moduli. The forced waves then present a decrease during their propagation and the calculation of the vibratory response no longer reveals any infinite amplitudes at resonance angular frequencies.

10.5. Generalization: distributed excitations and non-harmonic excitations

10.5.1. *Distributed excitations*

The decomposition in forced waves was established for localized excitations. We can use these solutions as "solvers" for more complex cases. Let us take, for example, the case of the torsion of beams from section 10.2.1, but suppose that the beam is excited by a distributed harmonic moment: $m(x)\, e^{j\omega t}$.

The equation to be satisfied is thus:

$$\rho I_0\, \omega^2\, \alpha(x) + G I_0 \frac{d^2\alpha}{dx^2} = m(x)\ ,\ x \in \left]0, L\right[, \tag{10.72}$$

$$\alpha(0) = 0\ , \tag{10.73}$$

$$\alpha(L) = 0\ . \tag{10.74}$$

Let us consider the solution obtained in [10.26] and [10.27], for a localized excitation by a torque placed in x_0. Let us suppose, moreover, that the torque has a unit amplitude $(M = 1)$:

$$\overline{\alpha}_1(x) = -\frac{1}{k\, G I_0} \frac{\sin k(x_0 - L)}{\sin kL} \sin kx\ ,\ x \in \left]0, x_0\right[, \tag{10.75}$$

$$\overline{\alpha}_2(x) = -\frac{1}{k\, G I_0} \frac{\sin k(x_0)}{\sin kL} \sin k(x - L)\ ,\ x \in \left]x_0, L\right[. \tag{10.76}$$

This elementary solution is related to the Green function of the problem and corresponds to the solution of the problem of torsion of a beam excited by a Dirac distribution placed in x_0. To show that the solution described by [10.75] and [10.76] makes it possible to solve the problem defined by equations [10.72], [10.73] and [10.74], the step is a little long. Let us carry out the integral:

$$I = \int_0^{x_0} \left(\rho I_0\, \omega^2\, \alpha(x) + G I_0 \frac{d^2\alpha}{dx^2} \right) \overline{\alpha}_1(x)\ dx + \int_{x_0}^{L} \left(\rho I_0\, \omega^2\, \alpha(x) + G I_0 \frac{d^2\alpha}{dx^2} \right) \overline{\alpha}_2(x)\ dx\ . \tag{10.77}$$

Taking into account [10.72], we have on the one hand:

$$I = \int_0^{x_0} \overline{\alpha}_1(x)\ m(x)\ dx + \int_{x_0}^{L} \overline{\alpha}_2(x)\ m(x)\ dx\ . \tag{10.78}$$

In addition, let us integrate by parts the expression [10.77]; we obtain:

$$
\begin{aligned}
I = & \int_0^{x_0} \left(\rho I_0\, \omega^2\, \overline{\alpha}_1(x) + GI_0 \frac{d^2 \overline{\alpha}_1}{dx^2} \right) \alpha(x)\ dx \\
& + \int_{x_0}^{L} \left(\rho I_0\, \omega^2\, \overline{\alpha}_2(x) + GI_0 \frac{d^2 \overline{\alpha}_2}{dx^2} \right) \alpha(x)\ dx \\
& + \left[GI_0 \frac{d\alpha}{dx}\ \overline{\alpha}_1(x) \right]_0^{x_0} - \left[GI_0 \frac{d\overline{\alpha}_1}{dx}\ \alpha(x) \right]_0^{x_0} \\
& + \left[GI_0 \frac{d\alpha}{dx}\ \overline{\alpha}_2(x) \right]_{x_0}^{L} - \left[GI_0 \frac{d\overline{\alpha}_2}{dx}\ \alpha(x) \right]_{x_0}^{L} .
\end{aligned}
\qquad [10.79]
$$

Taking into account [10.9] and [10.10], the integrals are nil; taking into account [10.11] – [10.14], [10.72] and [10.73], the non-nil terms at the boundaries are summarized by:

$$
I = \alpha(x_0) \left[-\,GI_0 \frac{d\overline{\alpha}_1}{dx}(x_0) + GI_0 \frac{d\overline{\alpha}_2}{dx}(x_0) \right]. \qquad [10.80]
$$

When the excitation moment is unitary $(M = 1)$, equation [10.14] indicates with [10.80] that the term between brackets is equal to unity, and thus:

$$
I = \alpha(x_0) .
$$

After grouping with [10.78], it follows:

$$
\alpha(x_0) = \int_0^{x_0} \overline{\alpha}_1(x)\, m(x)\, dx + \int_{x_0}^{L} \overline{\alpha}_2(x)\, m(x)\, dx ,
$$

that is:

$$
\begin{aligned}
\alpha(x_0) = & -\int_0^{x_0} \frac{m(x)}{kGI_0}\ \frac{\sin k(x_0 - L)}{\sin kL}\ \sin(kx)\ dx \\
& - \int_{x_0}^{L} \frac{m(x)}{kGI_0}\ \frac{\sin kx_0}{\sin kL}\ \sin\big(k(x - L)\big)\ dx .
\end{aligned}
\qquad [10.81]
$$

The response to a localized unitary excitation allows the calculation of the response to a distributed excitation via the integral equation [10.81].

To illustrate this expression, let us first consider an excitation torque of the Dirac distribution type $m(x) = \delta(x - x')$ and suppose that $x' < x_0$. The application of [10.81] leads to:

$$\alpha(x_0) = -\int_0^{x_0} \frac{\delta(x - x')}{kGI_0} \frac{\sin k(x_0 - L)}{\sin kL} \sin kx \; dx = -\frac{1}{kGI_0} \frac{\sin k(x_0 - L)}{\sin kL} \sin kx' .$$

We find again the solution [10.75] with notations reversed between x and x'.

If $x' > x_0$, we obtain:

$$\alpha(x_0) = -\int_{x_0}^{L} \frac{\delta(x - x')}{kGI_0} \frac{\sin kx_0}{\sin kL} \sin kx \; dx = -\frac{1}{kGI_0} \frac{\sin kx_0}{\sin kL} \sin k(x' - L) .$$

We also find again the relationship [10.76], still with reversed notations. Thus the Green function defined by the expressions [10.75] and [10.76] appears as the response at the point x to a Dirac distribution placed in x_0.

Let us now consider an excitation consisting of a set of torque as shown in Figure 10.5. This excitation corresponds to an approximation of the distributed moment $m(x)$:

$$m(x) \approx \sum_{i=1}^{N} M(x_i) \; \delta(x - x_i) , \qquad [10.82]$$

with: $M(x_i) = m(x_i) \; \Delta_i$ where Δ_i is the length of application of the torque $m(x_i)$.

The response at the point x_0 to these N excitation torques is by linearity the sum of the responses to each excitation torque. The expression of this response has two forms depending on whether the point of observation is to the right [10.27] or to the left point of the excitation [10.26]. Therefore, considering that $x_j < x_0 < x_{j+1}$, it follows:

$$\begin{aligned} \alpha(x_0) = & -\sum_{i=1}^{j} M(x_i) \frac{1}{kGI_0} \frac{\sin k(x_0 - L)}{\sin kL} \sin kx_i \, \Delta_i \\ & - \sum_{i=j+1}^{N} M(x_i) \frac{1}{kGI_0} \frac{\sin kx_0}{\sin kL} \sin k(x_i - L) \, \Delta_i . \end{aligned} \qquad [10.83]$$

The expression [10.81] is sometimes interpreted as passing to the limit of the expression [10.83].

The procedure that we have just applied is completely general and consists of solving a problem using the Green function. The basic tool is the solution of the problem to a unitary localized excitation, which is the Green function of the problem. We then proceed as shown to obtain the response to an excitation distributed by an integral equation. We leave it to the reader to apply this step to longitudinal and bending vibrations as an exercise; the calculations are heavier in the latter case.

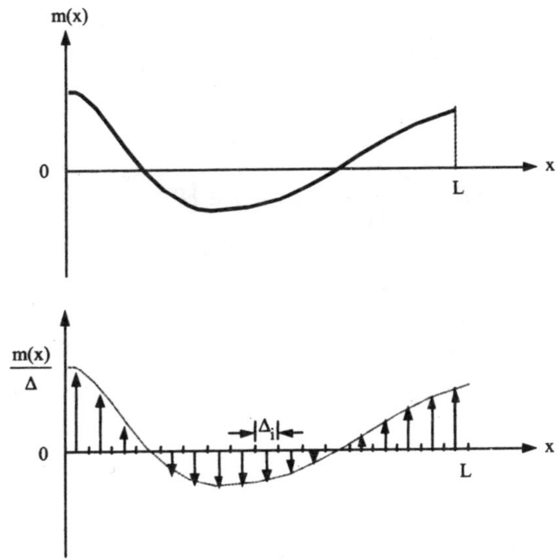

Figure 10.5. *Approximation of the distributed excitation torque by a comb of localized torques*

10.5.2. *Non-harmonic excitations*

The method of decomposition into forced waves is based on the harmonic excitation of structures; we may, however, use it for excitations with unspecified time variation thanks to the Fourier transformation.

To consolidate our ideas let us once again take the example of the beam in torsion from section 10.2.1, but for a non-harmonic excitation.

Equations [10.1] – [10.5] remain unchanged, and the discontinuity of moment of torsion [10.6] becomes:

$$GI_0 \frac{\partial \alpha_1}{\partial x}(x_0, t) - GI_0 \frac{\partial \alpha_2}{\partial x}(x_0, t) = M(t) . \qquad [10.84]$$

Let us take the time Fourier transforms of these equations. Noting the Fourier transform of the angle of torsion in section i as $\tilde{\alpha}_i(x, t)$:

$$\tilde{\alpha}_i(x, \omega) = \int_{-\infty}^{+\infty} \alpha_i(x, t)\; e^{-j\omega t}\, dt , \qquad [10.85]$$

it follows:

$$\rho I_0\, \omega^2\, \tilde{\alpha}_1(x, \omega) + GI_0 \frac{d^2\, \tilde{\alpha}(x, \omega)}{dx} = 0\ ,\ x \in \left]0, x_0\right[, \qquad [10.86]$$

$$\rho I_0\, \omega^2\, \tilde{\alpha}_2(x, \omega) + GI_0 \frac{d^2\, \tilde{\alpha}_2(x, \omega)}{dx} = 0\ ,\ x \in \left]x_0, L\right[, \qquad [10.87]$$

$$\tilde{\alpha}_1(0, \omega) = 0 , \qquad [10.88]$$

$$\tilde{\alpha}_2(0, \omega) = 0 , \qquad [10.89]$$

$$\tilde{\alpha}_1(x_0, \omega) = \tilde{\alpha}_2(x_0, \omega) , \qquad [10.90]$$

$$GI_0 \frac{d\tilde{\alpha}_2}{dx}(x_0, \omega) - GI_0 \frac{d\tilde{\alpha}_1}{dx}(x_0, \omega) = \tilde{M}(\omega) , \qquad [10.91]$$

where $\tilde{M}(\omega)$ is the Fourier transform of the excitation torque:

$$\tilde{M}(\omega) = \int_{-\infty}^{+\infty} M(t)\; e^{-j\omega t}\, dt . \qquad [10.92]$$

Equations [10.86] – [10.91] are formally identical to equations [10.9] – [10.14] and, therefore, lead to the same results for $\tilde{\alpha}_1(x,\omega)$ and $\tilde{\alpha}_2(x,\omega)$ as for

$\overline{\alpha}_1(x)$ and $\overline{\alpha}_2(x)$ obtained for a forced movement harmonic of an angular frequency ω, equations [10.26] and [10.27], that is:

$$\tilde{\alpha}_1(x,\omega) = -\frac{\tilde{M}(\omega)}{kGI_0}\,\frac{\sin k(x_0 - L)}{\sin kL}\,\sin kx \;,\; x \in \left]0,x_0\right[\;,$$

$$\tilde{\alpha}_2(x,\omega) = -\frac{\tilde{M}(\omega)}{kGI_0}\,\frac{\sin kx_0}{\sin kL}\,\sin k(x - L) \;,\; x \in \left]x_0, L\right[\;, \qquad [10.93]$$

with: $k = \omega \Big/ \sqrt{\dfrac{G}{\rho}}$.

It is now possible to obtain the vibrations of torsion taking the inverse Fourier transform, for example, for $\tilde{\alpha}_1(x,t)$:

$$\alpha_1(x,t) = \frac{1}{2\pi}\int_{-\infty}^{+\infty} -\frac{\tilde{M}(\omega)}{kGI_0}\,\frac{\sin k(x_0 - L)}{\sin kL}\,\sin kx\, e^{j\omega t}\,d\omega . \qquad [10.94]$$

Of course, the Fourier integral is not necessarily easy to calculate and a numerical problem may arise here.

10.5.3. *Unspecified homogenous mono-dimensional medium*

The method is applicable to any homogenous mono-dimensional structure whose harmonic movement is governed by a homogenous differential equation of the type:

$$\sum_{n=0}^{2N} \beta_n \frac{d^n U(x)}{dx^n} = 0 . \qquad [10.95]$$

The quantities β_n are constant coefficients, U(x) is the unknown of the problem, and 2N is the order of derivation of the equation of motion. For example, for the equation of bending of beams, we have:

$$N = 2 \;,\; \beta_0 = -\rho S\omega^2 \;,\; \beta_1 = \beta_2 = \beta_3 = 0 \;,\; \beta_4 = EI .$$

The number of boundary conditions associated with this equation is equal to the order of derivation 2N, that is N conditions at each end. They have the general form of N alternatives:

$$j = 0, \ldots, N-1 \quad \begin{cases} \text{either} \quad \dfrac{d^j U}{dx^j}(x) = 0 , \\ \\ \text{or} \quad \beta_{2N-1-j} \dfrac{d^{2N-1-j} U}{dx^{2N-1-j}}(x) = 0 . \end{cases} \qquad [10.96]$$

These conditions are to be verified at both ends x = 0 and x = L. The excitation can occur via N types of forces applied to the beam.

Let us take the case of a source point x_0: the beam is sub-structured into two sections where vibratory displacement is given respectively by $U_1(x)$ and $U_2(x)$. The simultaneous application of N localized forces leads to the discontinuity of the N quantities:

$$j = N, \ldots, 2N-1 \quad \beta_j \frac{d^j U_2}{dx^j}(x_0) - \beta_j \frac{d^j U_1}{dx^j}(x_0) = F_j \qquad [10.97a]$$

and to the continuity of the N quantities:

$$j = 0, \ldots, N-1 \quad \frac{d^j U_2}{dx^j}(x_0) = \frac{d^j U_1}{dx^j}(x_0) . \qquad [10.97b]$$

The two unknown functions $U_1(x)$ and $U_2(x)$ are solutions of equation [10.95], i.e.:

$$\begin{cases} U_1(x) = \sum_{j=1}^{2N} a_1^j \; e^{k_j x} \\ \\ U_2(x) = \sum_{j=1}^{2N} a_2^j \; e^{k_j x} , \end{cases} \qquad [10.98]$$

where k_j are the solutions of the characteristic equation associated with equation [10.95]:

$$\sum_{n=0}^{2N} \beta_n k^n = 0. \quad [10.99]$$

The solutions reveal 2N unknowns per section, that is, 4N unknowns in all, a_1^j and a_2^j. Writing down the 2N boundary conditions [10.96] and the 2N connection conditions [10.97a] and [10.97b] leads to a linear system 4N × 4N, whose solution provides the unknowns of the problem.

The method directly extends to the case of structures homogenous by parts as in section 10.2.2; it is enough to sub-structure it into a sufficient number of sections.

10.6. Forced vibrations of rectangular plates

The method of decomposition into forced waves is adapted to the resolution of the problem of mono-dimensional vibration; we may, however, apply it to rectangular plates. The forced response is no longer obtained by an analytical expression but by a simple series. It is a simplification compared to the classical modal decomposition, which leads to a double series. Let us take the example of a rectangular plate with length a and width b and consider that the opposite edges $x_1 = a$ and $x_1 = 0$ are simply supported. The plate is excited at the point (X_1, X_2) by a harmonic force:

$$Fe^{j\omega t}\,\delta(x_1 - X_1)\;\delta(x_2 - X_2). \quad [10.100]$$

Taking into account the harmonic excitation, the forced response is of the type:

$$W(x_1, x_2, t) = \overline{W}(x_1, x_2)\; e^{j\omega t}. \quad [10.101]$$

The equations to respect in order to solve this problem are the following:

– the equation of motion for $(x_1, x_2) \in \left]0, a\right[\times \left]0, b\right[$:

$$-\omega^2 \rho h \overline{W}(x_1, x_2) + D\left(\frac{\partial^4}{\partial x_1^4} + 2\frac{\partial^4}{\partial x_1^2\,\partial x_2^2} + \frac{\partial^4}{\partial x_2^4}\right)\overline{W}(x_1, x_2)$$
$$= Fe^{j\omega t}\,\delta(x_1 - X_1)\;\delta(x_2 - X_2)\;; \quad [10.102]$$

– the boundary conditions at the supported edges are given by:

$$\overline{W}(0, x_2) = 0 \ , \ \overline{W}(a, x_2) = 0 \tag{10.103}$$

and:

$$\frac{\partial^2}{\partial x_1^2}\overline{W}(0, x_2) = 0 \ , \ \frac{\partial^2}{\partial x_1^2}\overline{W}(a, x_2) = 0 . \tag{10.104}$$

We will seek the solution of the problem in the form of a series whose each term verifies the boundary conditions [10.103] and [10.104] at the edges $x_1 = 0$ and $x_1 = a$:

$$\overline{W}(x_1, x_2) = \sum_{n=1}^{\infty} h_n(x_2) \sin\left(\frac{n\pi}{L} x_1\right). \tag{10.105}$$

Let us introduce equation [10.105] into the equation of motion [10.102]:

$$\sum_{n=1}^{\infty}\left(D\frac{d^4h_n}{dx_2^4}(x_2) - 2D\left(\frac{n\pi}{L}\right)^2\frac{d^2h_n}{dx_2^2}(x_2) + \left(D\left(\frac{n\pi}{L}\right)^4 - \rho h\omega^2\right)h_n(x_2)\right)\sin\left(\frac{n\pi}{L}x_1\right) = F\delta(x_1 - X_1)\ \delta(x_2 - X_2). \tag{10.106}$$

Let us multiply this equation by $\sin\left(\frac{p\pi}{L}x_1\right)$ and integrate the two members between 0 and L; after having introduced the property of orthogonality [10.107], equation [10.108] follows:

$$\text{if } n = p : \quad \int_0^L \sin\left(\frac{n\pi}{L}x_1\right) \sin\left(\frac{p\pi}{L}x_1\right) = \frac{L}{2}, \tag{10.107a}$$

$$\text{if } n \neq p : \quad \int_0^L \sin\left(\frac{n\pi}{L}x_1\right) \sin\left(\frac{p\pi}{L}x_1\right) = 0, \tag{10.107b}$$

$$D\frac{d^4h_n}{dx_2^4}(x_2) - 2D\left(\frac{n\pi}{L}\right)^2\frac{d^2h_n}{dx_2^2}(x_2) + \left(D\left(\frac{n\pi}{L}\right)^4 - \rho S\omega^2\right)h_n(x_2)$$
$$= \frac{2}{L}F\sin\left(\frac{n\pi}{L}X_1\right)\delta(x_2 - X_2). \qquad [10.108]$$

Equation [10.108] is of the type that can be solved by the method of forced waves.

Let us break up the function $h_n(x_2)$ into two parts:

– for $x_2 \in \left]0, X_2\right[$ we introduce the unknown $h_n^1(x_2)$ satisfying the equation:

$$D\frac{d^4h_n^1}{dx_2^4}(x_2) - 2D\left(\frac{n\pi}{L}\right)^2\frac{d^2h_n^1}{dx_2^2}(x_2)$$
$$+\left(D\left(\frac{n\pi}{L}\right)^4 - \rho S\omega^2\right)h_n^1(x_2) = 0\ ; \qquad [10.109]$$

– for $x_2 \in \left]X_2, L\right[$ we introduce the unknown $h_n^2(x_2)$ satisfying the equation:

$$D\frac{d^4h_n^2}{dx_2^4}(x_2) - 2D\left(\frac{n\pi}{L}\right)^2\frac{d^2h_n^2}{dx_2^2}(x_2)$$
$$+\left(D\left(\frac{n\pi}{L}\right)^4 - \rho h\omega^2\right)h_n^2(x_2) = 0. \qquad [10.110]$$

To begin with, let us examine the boundary conditions. It is possible to choose all the configurations: it is necessary to write down the connection conditions at $x_2 = X_2$ and the boundary conditions at $x_2 = 0$ and at $x_2 = L$, support, clamped, free or guided edge.

To consolidate, we choose the conditions of support at both ends. The displacement of the plate must verify equations [10.111] and [10.112]:

$$\overline{W}(x_1, 0) = 0\ ,\ \overline{W}(x_1, b) = 0 \qquad [10.111]$$

and:

$$\frac{\partial^2}{\partial x_2^2}\overline{W}(x_1,0) = 0 \ , \ \frac{\partial^2}{\partial x_2^2}\overline{W}(x_1,b) = 0 \, . \qquad [10.112]$$

Taking into account the form [10.105] of vibratory displacement, the preceding equations lead to boundary conditions directly applicable to the unknown functions $h_n^1(x_2)$ and $h_n^2(x_2)$:

$$h_n^1(0) = 0 \ , \ h_n^2(b) = 0 \ \text{ and } \ \frac{d^2h_n^1}{dx_2^2}(0) = 0 \ , \ \frac{d^2h_n^2}{dx_2^2}(b) = 0 \, . \qquad [10.113]$$

To obtain equations [10.113] we made use of the property of orthogonality [10.107].

Let us examine the connection of the solutions over the $x_2 = X_2$ interface; for plates it is necessary to verify the continuity of displacement, its normal derivative at the line of interface, the bending moment carried by the line of interface and the discontinuity of the shearing force due to the presence of the excitation effort. The expressions of the bending moment and the shearing force have been provided in Chapter 4, equations [4.57] and [4.58].

Another approach consists of reasoning directly using equation [10.108]. We note that the Dirac distribution produces the discontinuity of the third derivative of $h_n(x_2)$. The connection conditions are thus:

$$h_n^1(X_2) = h_n^2(X_2) \, , \qquad [10.114]$$

$$\frac{dh_n^1}{dx_2}(X_2) = \frac{dh_n^2}{dx_2}(X_2) \, , \qquad [10.115]$$

$$\frac{d^2h_n^1}{dx_2^2}(X_2) = \frac{d^2h_n^2}{dx_2^2}(X_2) \, , \qquad [10.116]$$

$$\frac{d^3h_n^1}{dx_2^3}(X_2) - \frac{d^3h_n^2}{dx_2^3}(X_2) = -\frac{2}{LD} F \sin\left(\frac{n\pi}{L} X_1\right). \qquad [10.117]$$

The calculation of the solutions of equations [10.109] and [10.110] is well known; we obtain:

$$h_n^1(x_2) = A_{1n} \sin(k_n x_2) + B_{1n} \cos(k_n x_2) + C_{1n} \, \mathrm{sh}(\gamma_n x_2) + D_{1n} \, \mathrm{ch}(\gamma_n x_2), \qquad [10.118]$$

$$h_n^2(x_2) = A_{2n} \sin\left(k_n (x_2 - L)\right) + B_{2n} \cos\left(k_n (x_2 - L)\right) + C_{2n} \, \mathrm{sh}\left(\gamma_n (x_2 - L)\right) + D_{2n} \, \mathrm{ch}\left(\gamma_n (x_2 - L)\right), \qquad [10.119]$$

with: $$k_n = \sqrt{\sqrt{\frac{\rho h \omega^2}{D}} - \left(\frac{n\pi}{L}\right)^2} \quad \text{and} \quad \gamma_n = \sqrt{\sqrt{\frac{\rho h \omega^2}{D}} + \left(\frac{n\pi}{L}\right)^2}. \qquad [10.120]$$

These solutions consist of vanishing waves with wave numbers γ_n and traveling waves with a wave number k_n; this situation is true as long as $\omega > \left(\frac{n\pi}{L}\right)^2 \sqrt{\frac{D}{\rho h}}$. For weaker pulsations of excitation, there are four vanishing waves, since the wave number k_n is then imaginary.

Applying boundary conditions [10.114] – [10.117] provides the solutions:

$$h_n^1(x_2) = A_{1n} \sin(k_n x_2) + C_{1n} \, \mathrm{sh}(\gamma_n x_2) \qquad [10.121]$$

and:

$$h_n^2(x_2) = A_{2n} \sin\left(k_n (x_2 - L)\right) + C_{2n} \, \mathrm{sh}\left(\gamma_n (x_2 - L)\right). \qquad [10.122]$$

The constants of integration are given by the solution of the linear system [10.123].

$$\begin{pmatrix} \sin k_n L_1 & \mathrm{sh}\gamma_n L_1 & -\sin k_n(L_1-b) & -\mathrm{sh}\gamma_n(L_1-b) \\ k_n \cos k_n L_1 & \gamma_n \mathrm{ch}\gamma_n L_1 & -k_n \cos k_n(L_1-b) & -\gamma_n \mathrm{ch}\gamma_n(L_1-b) \\ -k_n^2 \sin k_n L_1 & \gamma_n^2 \mathrm{sh}\gamma_n L_1 & k_n^2 \sin k_n(L_1-b) & -\gamma_n^2 \mathrm{sh}\gamma_n(L_1-b) \\ -k_n^3 \cos k_n L_1 & \gamma_n^3 \mathrm{ch}\gamma_n L_1 & k_n^3 \cos k_n(L_1-b) & -\gamma_n^3 \mathrm{ch}\gamma_n(L_1-b) \end{pmatrix} \begin{Bmatrix} A_{1n} \\ C_{1n} \\ A_{2n} \\ C_{2n} \end{Bmatrix} = \begin{Bmatrix} 0 \\ 0 \\ 0 \\ -\dfrac{2F}{DL}\sin\left(\dfrac{n\pi}{L}X_1\right) \end{Bmatrix}. \quad [10.123]$$

The vibratory response of the plate is obtained, finally, using [10.121] and [10.122] in [10.105]:

– for $x_2 \in \left]0, X_2\right[$ we have:

$$\overline{W}(x_1, x_2) = \sum_{n=1}^{\infty} \left(A_{1n} \sin(k_n x_2) + C_{1n}\,\mathrm{sh}(\gamma_n x_2)\right) \sin\left(\frac{n\pi}{L} x_1\right); \quad [10.124]$$

– for $x_2 \in \left]X_2, b\right[$ we have:

$$\overline{W}(x_1, x_2) = \sum_{n=1}^{\infty} \left(A_{2n} \sin\left(k_n(x_2 - L)\right) + C_{2n}\,\mathrm{sh}\left(\gamma_n(x_2 - L)\right)\right) \sin\left(\frac{n\pi}{L} x_1\right). \quad [10.125]$$

This method of calculation of the vibratory response extends to plates the decomposition into forced waves. The response is expressed as a semi-modal decomposition: modes in one direction and forced waves in the other. The domain of application of this approach is quite narrow; in fact, it is necessary that the vibration modes $\Phi_Q(x_1, x_2)$ can be expressed by separation of space variables:

$$\Phi_Q(x_1, x_2) = f_{nQ}(x_1)\, f_{mQ}(x_2). \quad [10.126]$$

Under these conditions, the calculation of response by modal decomposition can be expressed in the form:

$$\overline{W}(x_1, x_2) = \sum_{nQ=1}^{\infty} \sum_{mQ=1}^{\infty} a_{nQmQ}\ f_{nQ}(x_1)\ f_{mQ}(x_2). \qquad [10.127]$$

Let us introduce the function $h_{nQ}(x_2)$ given by [10.128]:

$$h_{nQ}(x_2) = \sum_{mQ=1}^{\infty} a_{nQmQ}\ f_{mQ}(x_2). \qquad [10.128]$$

Under these conditions, we may rewrite the modal decomposition [10.127] in the form of a semi-modal decomposition:

$$\overline{W}(x_1, x_2) = \sum_{nQ=1}^{\infty} h_{nQ}(x_2)\ f_{nQ}(x_1). \qquad [10.129]$$

The property [10.126] which the modes must satisfy is very restrictive. In fact, the shape of the plate is already very limited, rectangular or circular; moreover, the boundary conditions cannot be unspecified. For rectangular plates, the method requires that two opposite edges be supported or guided, while the boundary conditions for the other edges are unspecified subject to being the same for a given edge.

When the method functions, it offers an unquestionable advantage by limiting calculations to a mono-dimensional series. For the application to a network of coupled plates, see [REB 97].

10.7. Conclusion

The method of calculation of response by decomposition into forced waves is applicable to mono-dimensional structures homogenous by parts. It is based on a sub-structuring, each section being delimited by two singular points (discontinuity of structure or point of excitation). The solution is thus provided by parts.

The advantage of the method is to give an analytical expression of the solution. This avoids the problems involved in the expression in the form of a series of solutions resulting from modal decomposition.

The case of damped structures is easily considered via complex moduli of elasticity.

The technique of calculation naturally reveals resonances, but also anti-resonances.

The method can be generalized to distributed harmonic excitations thanks to the construction of the Green function by forced waves, which is the solution of the problem with localized unit excitation.

The method is also widely used for non-harmonic excitations, thanks to the use of the Fourier transformation of time signals.

Finally, the method can be applied to all the problems of beams homogenous by parts: it suffices to solve the differential equation of space associated with harmonic movement; technically the method is weighed down but remains identical in its principle.

For plates the approach by forced waves is used in the case of a semi-modal decomposition of the vibratory response. However, in order to be usable, the modes of the plate must be written down in the shape of the product of two modal functions, each depending on only one variable of space. This situation occurs, in particular, for rectangular plates with particular limiting conditions.

Chapter 11

The Rayleigh-Ritz Method based on Reissner's Functional

11.1. Introduction

In the majority of cases of elastic solid media vibrations, obtaining exact analytical solutions is impossible. Therefore, it is necessary to make use of approximation methods. In this light the Rayleigh-Ritz method is an important method, because it constitutes the basis for energy methods, such as, for example, the finite elements method. The goal of this and the following chapters is not to provide a discourse on numerical methods (there is already an excellent selection of literature on this subject), but rather to present the groundwork for the energy method.

As we will see, the Rayleigh-Ritz method uses the variational form of the equations of the vibrations of the continuous mediums. There are two principal alternatives, which we have presented in this course: the formulation in displacements stemming from Hamilton's functional and the formulation in stresses and displacements stemming from Reissner's functional. If the methods are basically identical, their forms are rather specific and we have chosen to cover them both. In this chapter we expose the formulation stemming from the two-fields Reissner's functional.

The discourse is based on a rather simple example of reference: vibrations of flexion of beams. This choice aims at revealing the foundations of the method without weighing down the presentation by abstractions linked to a general case.

11.2. Variational formulation of the vibrations of bending of beams

We consider Bernoulli's model of beam characterized by two unknown functions $W(x,t)$ and $\sigma(x,t)$ that are, respectively, the transverse displacement and longitudinal stress. Reissner's functional for this problem was provided in Chapter 3, equation [3.60] (the notation used here is simplified):

$$R(W,\tau) = \int_{t_0}^{t_1}\int_0^L \left(\frac{\rho S}{2}\left(\frac{\partial W}{\partial t}\right)^2 - I\sigma\frac{\partial^2 W}{\partial x^2} + \frac{I}{2E}\sigma^2 \right) dxdt. \qquad [11.1]$$

In this expression, E is the Young modulus and ρ is the density of the material, while S and I are respectively the cross-section and the inertia of bending of the beam.

Let us note that in equation [11.1], the effect of rotational inertia was not covered in order to simplify matters.

The solution of the problem of vibration consists of determining the particular functions $\overline{W}(x,t)$ and $\overline{\sigma}(x,t)$ rendering the functional stationary [11.1]. Following the functional space, where the calculation of extremum is performed, we obtain solutions corresponding to various boundary conditions.

Let us consider, to begin with, the spaces $\mathcal{W}$ and Σ of the functions $W(x,t)$ and $\sigma(x,t)$, sufficiently regular so that the integral [11.1] exists, and without any restrictive conditions at the 0 and L ends. The calculation of the extremum of the functional, as carried out in Chapter 3, leads to respecting equations [11.2] – [11.7] at any moment t:

$$\rho S\frac{\partial^2 \overline{W}}{\partial t^2} - \frac{\partial^2}{\partial x^2}(I\,\overline{\sigma}) = 0\ ,\ x \in \left]0,L\right[, \qquad [11.2]$$

$$-E\frac{\partial^2 \overline{W}}{\partial x^2} = \overline{\sigma}\ ,\ x \in \left]0,L\right[, \qquad [11.3]$$

$$\overline{\sigma}(0,t) = 0, \qquad [11.4]$$

$$\overline{\sigma}(L,t) = 0, \qquad [11.5]$$

$$\frac{\partial \overline{\sigma}}{\partial x}(0,t) = 0\,, \qquad [11.6]$$

$$\frac{\partial \overline{\sigma}}{\partial x}(L,t) = 0\,. \qquad [11.7]$$

This case is representative of the free-free beam, and we will note the space W by W^{LL}.

Let us modify the functional space of displacements $W(x,t)$ by restricting ourselves to the sub-space of the null functions in 0 and L:

$$W(0,t) = 0\,, \qquad [11.8]$$

$$W(L,t) = 0\,. \qquad [11.9]$$

This new space of functions, noted W^{AA}, because it leads to the boundary conditions of support at both ends, is included in the previous space:

$$W^{AA} \subset W^{LL}\,.$$

The calculation of the extremum leads to equations [11.2], [11.3], [11.4] and [11.5] only, equations [11.6] and [11.7] no longer appear, taking into account the conditions [11.8] and [11.9] imposed on displacement. A very important aspect of the variational method appears here: the boundary stress conditions result from variational calculation and are, therefore, not necessarily imposed *a priori* by the choice of the functional space where we search for the extremum.

On the contrary, the boundary displacement conditions are not produced by the calculation of extremum and need to be imposed *a priori* by the choice of the functional space W^{AA}.

Equations [11.2] to [11.7] are representative of the beam free at both ends. Equations [11.2] - [11.5] and [11.8], [11.9] are those of the beam supported at both ends. Let us give a third example: that of the beam clamped at both ends; the space W^{EE} to be considered is the sub-space of W^{LL} such that:

$$W(0,t) = 0\,, \qquad [11.10]$$

$$W(L, t) = 0\,, \qquad [11.11]$$

$$\frac{\partial W}{\partial x}(0, t) = 0\,, \qquad [11.12]$$

$$\frac{\partial W}{\partial x}(L, t) = 0\,. \qquad [11.13]$$

The calculation of extremum leads only to the respect of the equation of motion [11.2] and of the stress-strain relation [11.3]. The boundary conditions all are imposed *a priori* by the restriction on the functional sub-space W^{EE}.

The variational presentation of the problem of free vibrations of beams will thus be the following (we take the example of a beam on two supports): find the pair $(\overline{W}(x,t)\,,\overline{\sigma}(x,t)) \in W^{AA} \times \Sigma$ that returns the extremum of the functional [11.1]. Using a more compact notation:

$$R(\overline{W}, \overline{\sigma}) = \underset{W^{AA} \times \Sigma}{\operatorname{Ext}} R(W, \sigma)\,. \qquad [11.14]$$

We can propose an alternative of the formulation which uses boundary conditions which the stresses must verify. For example, in the supported case, we can restrict the functional space of working stresses to the sub-space of constraints Σ^{AA}, nil at the ends of the beam.

Equations [11.4] and [11.5] are thus verified *a priori*. The problem of the supported beam is stated in the following way: find the pair $(\overline{W}(x,t)\,,\overline{\sigma}(x,t)) \in W^{AA} \times \Sigma^{AA}$ returning the extremum of the functional [11.1]. That is, in compact notation:

$$R(\overline{W}, \overline{\sigma}) = \underset{W^{AA} \times \Sigma^{AA}}{\operatorname{Ext}} R(W, \sigma)\,. \qquad [11.15]$$

The advantage of [11.15] compared to [11.14] is that it allows improved convergence in the calculation of the approximate solutions which are sought in a more restricted Σ^{AA} space rather than in Σ. The disadvantage lies in the difficulty of construction of the sub-space Σ^{AA}, taking into account larger requirements of the conditions to be verified *a priori*.

In conclusion, the kinematic boundary conditions need to be verified *a priori*, whereas their *a priori* respect is optional for stresses.

11.3. Generation of functional spaces

The Rayleigh-Ritz method relates to a particular technique of construction of functional spaces W and Σ. By decomposition on a functional basis we will write:

$$W(x, t) = \sum_{n=1}^{\infty} a_n(t)\ \phi_n(x), \tag{11.16}$$

$$\sigma(x, t) = \sum_{n=1}^{\infty} b_n(t)\ \psi_n(x). \tag{11.17}$$

The functions $\phi_n(x)$ and $\psi_n(x)$ constitute functional bases defined *a priori*; the amplitudes $a_n(t)$ and $b_n(t)$ are the unknowns of the problem.

Let us take the example of the space W^{AA}; it will be generated as follows:

$$W^{AA} = \left\{ W(x,t) = \sum_{n=1}^{\infty} a_n(t)\ \phi_n(x) /\ \phi_n(0) = 0 \ \text{ and } \ \phi_n(L) = 0 \right\}. \tag{11.18}$$

Let us notice that each basic function must verify the conditions $\phi_n(0) = 0$ and $\phi_n(L) = 0$. This stems from the fact that any function of the [11.16] type must verify the boundary conditions [11.8] and [11.9]. The particular function where all the amplitudes but one, the nth one, are nil leads to:

$$a_n(t)\ \phi_n(0) = 0 \quad \forall\, t \quad \text{and} \quad a_n(t)\ \phi_n(L) = 0 \quad \forall\, t,$$

i.e.:

$$\phi_n(0) = 0 \quad \text{and} \quad \phi_n(L) = 0.$$

The bases of functions that are usually considered are either of the polynomial type or of the Fourier series type. We will have the occasion to consider both of these cases, but in this chapter only the example of polynomial development will be used.

11.4. Approximation of the vibratory response

The approximation of the solution of the problem is obtained simply by carrying out the extremum calculation over the sub-spaces of functional spaces W and Σ. In the Rayleigh-Ritz method, the sub-spaces are built by truncating the series [11.16] and [11.17]. For example, the sub-space of W^{AA} with N elements W_N^{AA} will be obtained by considering the N first terms of the series:

$$W_N^{AA} = \left\{ W(x,t) = \sum_{n=1}^{N} a_n(t)\ \phi_n(x)\ /\ \phi_n(0) = 0 \quad \text{and} \quad \phi_n(L) = 0 \right\}. \qquad [11.19]$$

For the problem of the beam supported at the ends, the search for solutions approximated using the variational technique will, therefore, consist of finding the pair $(\overline{W}(x,t), \overline{\sigma}(x,t)) \in W_N^{AA} \times \Sigma_N$ returning the extremum of Reissner's functional:

$$R(\overline{W}, \overline{\sigma}) = \underset{W_N^{AA} \times \Sigma_M}{\text{Ext}} R(W, \sigma). \qquad [11.20]$$

It is a form identical to [11.14], where functional spaces have a finite dimension. As we will see in examples later on, using this technique, we construct a discrete system with N degrees of freedom approximating the vibratory behavior of the beam.

In [11.20], the functional space Σ_M is the sub-space of Σ such that:

$$\Sigma_M = \left\{ \sigma(x,t) = \sum_{m=1}^{M} b_m(t)\ \psi_m(x) \right\}, \qquad [11.21]$$

where the functions $\psi_m(x)$ constitute a functional base of Σ_M.

Note: we took functional spaces of identical dimensions for W_N^{AA} and Σ_N; this is not obligatory, but has the advantage of leading to manipulating square matrices during computerized processing of the method.

11.5. Formulation of the method

Let us consider the approximations of displacements and stresses, defined in the previous section:

$$W(x,t) = \sum_{n=1}^{N} a_n(t)\ \phi_n(x), \qquad [11.22]$$

$$\sigma(x,t) = \sum_{m=1}^{N} b_m(t)\ \psi_m(x)\,. \qquad [11.23]$$

The calculation of Reissner's functional [11.1] after using [11.22] and [11.23] gives:

$$\begin{aligned} R(a_n, b_m) = \int_{t_0}^{t_1} \Bigg[& \frac{1}{2}\sum_n \sum_p \dot{a}_n(t)\ \dot{a}_p(t) \int_0^L \rho S\, \phi_n(x)\ \phi_p(x)\ dx \\ & - \sum_n \sum_m a_n(t)\ b_m(t) \int_0^L \psi_m(x)\ \frac{d^2\phi_n}{dx^2}(x)\ dx \\ & + \frac{1}{2}\sum_m \sum_q b_m(t)\ b_q(t) \int_0^L \frac{1}{E}\psi_m(x)\ \psi_q(x)\, dx \Bigg] dt\,. \end{aligned} \qquad [11.24]$$

We may propose a matrix expression of the functional:

$$R(\{a\},\{b\}) = \int_{t_0}^{t_1} \frac{1}{2}\{\dot{a}\}^t (A)\{\dot{a}\} - \{a\}^t (B)\{b\} + \frac{1}{2}\{b\}^t (C)\{b\}\ dt \qquad [11.25]$$

where:

$$\{a\}^t = \big(a_1(t)\ a_2(t)\ \ldots\ldots\ a_n(t)\big), \qquad [11.26]$$

$$\{b\}^t = \big(b_1(t)\ b_2(t)\ \ldots\ldots\ b_n(t)\big), \qquad [11.27]$$

$$(A) = (A_{np}) \quad \text{with} \quad A_{np} = \int_0^L \rho S\ \phi_n(x)\ \phi_p(x)\ dx, \qquad [11.28]$$

$$(B) = (B_{np}) \quad \text{with} \quad B_{np} = \int_0^L \psi_m(x)\ \frac{d^2\ \phi_n(x)}{dx^2}\ dx, \qquad [11.29]$$

$$(C) = (C_{mq}) \quad \text{with} \quad C_{mq} = \int_0^L \frac{1}{E} \psi_m(x)\ \psi_q(x)\, dx. \qquad [11.30]$$

The calculation of extremum leads to the matrix relations:

$$(A)\{\ddot{a}\} + (B)\{b\} = \{0\}, \qquad [11.31]$$

$$(B)^T\{a\} = (C)\{b\}. \qquad [11.32]$$

Incorporating the second matrix equation into the first, we obtain:

$$(A)\{\ddot{a}\} + (B)(C^{-1})(B)^t\{a\} = \{0\}. \qquad [11.33]$$

This equation is, in fact, the traditional representation of mass-spring vibrating systems with N degrees of freedom where (A) is the matrix of mass and $(B)(C^{-1})(B)^t$ is the matrix of stiffness. These two matrices are symmetrical.

We may apply the standard results of the discrete vibrating systems: there are N modes of vibrations, each characterized by a normal angular frequency ω_i and a normal vector $\{\bar{a}_i\}$. The general solution is expressed by the sum of modal movements:

$$\{a(t)\} = \sum_{i=1}^{N} (\alpha_i \cos\omega_i t + \beta_i \sin\omega_i t)\{\bar{a}_i\}. \qquad [11.34]$$

The normal angular frequencies are equal to the square roots of the eigenvalues of the matrix:

$$(A^{-1})(B)(C^{-1})(B)^t. \qquad [11.35]$$

The associated normal vectors are orthogonal with respect to the matrices of mass and stiffness:

$$\{\bar{a}_j\}^t(A)\{\bar{a}_i\} = 0 \quad \text{if } i \neq j, \qquad [11.36]$$

$$\{\bar{a}_j\}^t(B)(C^{-1})(B)^t\{\bar{a}_i\} = 0 \quad \text{if } i \neq j; \qquad [11.37]$$

finally:

$$\omega_i^2 = \frac{\{\bar{a}_i\}^t (B)(C^{-1})(B)^t \{\bar{a}_i\}}{\{\bar{a}_i\}^t (A)\{\bar{a}_i\}}. \qquad [11.38]$$

These results stemming from the discrete mass-spring system provide an approximation of the vibratory response of the beam. Using [11.34] in [11.22], we obtain:

$$W(x, t) = \sum_{i=1}^{N} \left(\alpha_i \cos(\omega_i t) + \beta_i \sin(\omega_i t)\right) f_i(x). \qquad [11.39]$$

The functions $f_i(x)$ are the approximated mode shapes. They are given by:

$$f_i(x) = \sum_{n=1}^{N} \bar{a}_{n_i} \, \phi_n(x) \qquad [11.40]$$

where $\bar{a}_{n_i}$ is the nth the component of the ith normal vector.

These mode shapes possess the property of orthogonality [11.41]:

$$\int_0^{\rho} \rho S \, f_i(x) \, f_j(x) \, dx = 0 \quad \text{if } i \neq j. \qquad [11.41]$$

Indeed, substituting $f_i(x)$ and $f_j(x)$ by their respective expressions resulting from [11.40] gives:

$$\int_0^{\rho} \rho S \, f_i(x) \, f_j(x) \, dx = \sum_{n=1}^{N} \sum_{p=1}^{N} \bar{a}_{n_i} \, \bar{a}_{p_j} \int_0^{L} \rho S \, \phi_n(x) \, \phi_p(x) \, dx, \qquad [11.42]$$

that is, with the notation [11.28]:

$$\int_0^{L} \rho S \, f_i(x) \, f_j(x) \, dx = \sum_{n=1}^{N} \bar{a}_{n_i} \sum_{p=1}^{N} A_{np} \, \bar{a}_{p_j}, \qquad [11.43]$$

and in matrix notation:

$$\int_0^{\rho} \rho S\, f_i(x)\ f_j(x)\ dx = \left\{\bar{a}_i\right\}^t (A) \left\{\bar{a}_j\right\}. \qquad [11.44]$$

With [11.36] we then deduce that the approximated mode shapes $f_i(x)$ and $f_j(x)$ are orthogonal in the sense of the integral [11.41].

The calculation of stresses is straightforward thanks to the relation [11.32], we draw the vector $\left\{b\right\}$ knowing the vector $\left\{a\right\}$:

$$\left\{b(t)\right\} = (C^{-1})(B)^t \left\{a(t)\right\}, \qquad [11.45]$$

that is, with [11.34]:

$$\left\{b\right\} = \sum_{i=1}^{N} (\alpha_i \cos \omega_i t + \beta_i \sin \omega_i t) \left\{\bar{b}_i\right\} \qquad [11.46]$$

where:

$$\left\{\bar{b}_i\right\} = (C^{-1})(B)^t \left\{\bar{a}_i\right\}. \qquad [11.47]$$

After an obvious calculation, we draw:

$$\sigma(x, t) = \sum_{i=1}^{N} (\alpha_i \cos \omega_i t + \beta_i \sin \omega_i t)\ h_i(x)\,. \qquad [11.48]$$

The functions $h_i(x)$ are the approximated stresses mode shapes given by:

$$h_i(x) = \sum_{n=1}^{N} \overline{b}_{n_i} \psi_n(x), \qquad [11.49]$$

where the quantity $\overline{b}_{n_i}$ is the nth the component of the ith normal vector $\{\overline{b}_i\}$.

We leave it to the reader, as an exercise, to demonstrate the two other properties of orthogonality:

$$\int_0^L h_i(x) \frac{d^2 f_j(x)}{dx^2} dx = 0 \quad \text{if } i \neq j \qquad [11.50]$$

and:

$$\int_0^L \frac{I}{E} h_i(x)\, h_j(x)\, dx = 0 \quad \text{if } i \neq j. \qquad [11.51]$$

The Rayleigh-Ritz method makes it possible to approximate the vibrations of a continuous medium by a discrete mass-spring system. The vibration modes of the mass-spring system later lead to an approximation of the vibratory response of the continuous medium.

11.6. Application to the vibrations of a clamped-free beam

In this section, we will use a simple case to put the method into practice and we will show some tendencies characteristic thereof on the basis of calculations of systems with a low number of degree of freedom.

11.6.1. *Construction of a polynomial base*

One way of building the functional bases is to use a polynomial decomposition of the functions $W(x, t)$ and $\sigma(x, t)$. The general form is of the type:

$$W(x,t) = \sum_{n=0}^{\infty} a_n(t)\, x^n . \qquad [11.52]$$

Such a decomposition is suggested by a development in Taylor series:

$$W(x,t) = W(0,t) + x\frac{\partial W}{\partial x}(0,t) + \frac{x^2}{2!}\frac{\partial^2 W}{\partial x^2}(0,t) + \dots \qquad [11.53]$$

Let us take the example of the case of a clamped-free beam. Displacement in 0 and its first derivative must be zero to verify the boundary displacement conditions. From this we deduce that:

$$W(x,t) = \frac{x^2}{2!}\frac{\partial^2 W}{\partial x^2}(0,t) + \frac{x^3}{3!}\frac{\partial^3 W}{\partial x^3}(0,t) + \dots \qquad [11.54]$$

that is, introducing the unknowns $a_n(t)$ into [11.54]:

$$W(x,t) = \sum_{n=2}^{\infty} a_n(t)\ x^n . \qquad [11.55]$$

The stress field is left free of any boundary conditions:

$$\sigma(x,t) = \sum_{n=0}^{\infty} b_n(t)\ x^n . \qquad [11.56]$$

Functional spaces W_N^{EL} and Σ_M where the extremum calculations will be carried out are thus defined by:

$$W_N^{EL} = \left\{ W(x,t) = \sum_{n=2}^{N+1} a_n(t)\ x^n \right\}, \qquad [11.57]$$

$$\Sigma_M = \left\{ \sigma(x,t) = \sum_{m=0}^{M-1} b_m(t)\ x^m \right\}. \qquad [11.58]$$

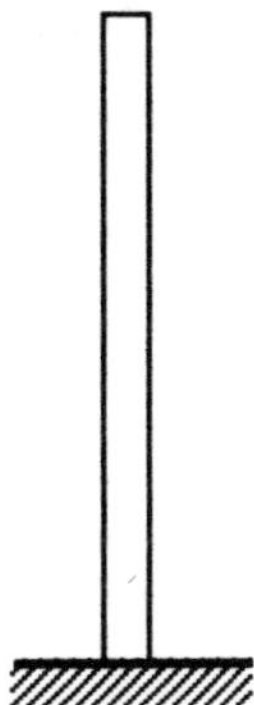

Figure 11.1. *Schematic representation of the case treated in the example of reference*

11.6.2. *Modeling with one degree of freedom*

To have a simple approximation of the technique of calculation we will adopt the maximum restriction of the sub-spaces, i.e. N = M = 1:

$$W(x,t) = a_2(t)\, x^2 \text{ and } \sigma(x,t) = b_{0(t)}. \qquad [11.59]$$

It is, of course, a very rough approximation, since it supposes that bending stress is constant with x.

The calculation of Reissner's functional leads to:

$$R\big(a_2(t), b_0(t)\big) = \int_{t_0}^{t_1} \left[\frac{\rho S}{2} \big(\dot{a}_2(t)\big)^2 \frac{L^5}{5} - 2Ib_0(t)\ a_2(t)\ L + \frac{I}{2E} \big(b_0(t)\big)^2 L \right] dt. \qquad [11.60]$$

It is a functional of the rigid body mechanics type, since the only unknown are functions of time.

The calculation of its extremum is traditional. Noting the integrand of [11.60] as $F\left(a_2(t), b_0(t)\right)$ we obtain:

$$\frac{d}{dt}\frac{\partial F}{\partial \dot{a}_2} - \frac{\partial F}{\partial a_2} = 0\,;$$

$$\frac{\partial F}{\partial b_0} = 0\,,$$

that is, after calculation:

$$\frac{\rho S L^5}{5}\,\ddot{a}_2(t) + 2IL\,b_0(t) = 0\,, \qquad [11.61]$$

$$-\,2IL\,a_2(t) + \frac{IL}{E}\,b_0(t) = 0\,. \qquad [11.62]$$

Substituting the expression of $b_0(t)$ stemming from [11.62] in the expression [11.61], we obtain the equation of a system with one degree of freedom of the unknown $a_2(t)$:

$$\frac{\rho S L^5}{5}\,\ddot{a}_2(t) + 4ILE\,a_2(t) = 0\,. \qquad [11.63]$$

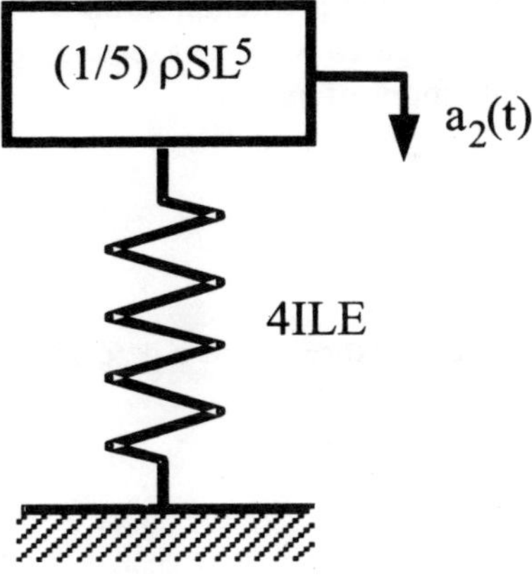

Figure 11.2. *The system with one degree of freedom approximating the first oscillatory mode of the beam*

The solution provides:

$$a_2(t) = A \cos \omega_1 t + B \sin \omega_1 t \,, \quad [11.64]$$

with:

$$\omega_1 = 4.472\sqrt{\frac{EI}{\rho SL^4}}. \quad [11.65]$$

We also draw from it:

$$b_0(t) = 2EI\,(A \cos \omega_1 t + B \sin \omega_1 t)\,, \quad [11.66]$$

that is, finally, using the expressions [11.59]:

$$W(x,t) = (A \cos \omega_1 t + B \sin \omega_1 t)\ x^2\,, \quad [11.67]$$

$$\sigma(x,t) = 2EI\,(A \cos \omega_1 t + B \sin \omega_1 t)\,. \quad [11.68]$$

An indicator of the quality of the prediction is the value of the normal angular frequency, because it can be compared with the exact value ω_1^{ex}, which we gave in Chapter 6 for the bending vibrations of beams.

$$\omega_1^{ex} = 3.52\sqrt{\frac{EI}{\rho SL^4}}. \qquad [11.69]$$

The error committed evaluated expressed as a percentage gives:

$$\frac{\omega_1 - \omega_1^{ex}}{\omega_1} = 27\%. \qquad [11.70]$$

This error is large; a finer approximation can be obtained by pushing the developments further. Let us examine the case of a development with the two terms of displacement and of bending stress.

11.6.3. *Model with two degrees of freedom*

Let us consider the displacements $W(x, t)$ and the stresses $\sigma(x, t)$ in the following form:

$$W(x, t) = a_2(t)\, x^2 + a_3(t)\, x^3,$$

$$\sigma(x, t) = b_0(t) + b_1(t)\, x.$$

The calculation of Reissner's functional leads to equation [11.71]:

$$R(a_2, a_3, b_0, b_1) = \int_{t_0}^{t_1} \left[\frac{\rho S}{2}\left(\dot{a}_2^2 \frac{L^5}{5} + 2\dot{a}_2\, \dot{a}_3 \frac{L^6}{6} + \dot{a}_3^2 \frac{L^7}{7} \right) - I\left(b_0\, a_2\, 2L + b_1\, a_2\, L^2 + b_0\, a_3\, 3L^2 + b_1\, a_3\, 2L^3\right) + \frac{I}{2E}\left(b_0^2 L + b_0\, b_1\, L^2 + b_1^2 \frac{L^3}{3} \right) \right] dt. \qquad [11.71]$$

The four Euler equations resulting from the calculation of extremum are given below:

$$\rho S \frac{L^5}{5} \ddot{a}_2 + \rho S \frac{L^6}{6} \ddot{a}_3 + 2IL\, b_0 + IL^2\, b_1 = 0\,, \tag{11.72}$$

$$\rho S \frac{L^7}{7} \ddot{a}_3 + \rho S \frac{L^6}{6} \ddot{a}_2 + 3IL^2\, b_0 + 2IL^3\, b_1 = 0\,, \tag{11.73}$$

$$-\,2IL\, a_2 - 3IL^2\, a_3 + \frac{IL}{E} b_0 + \frac{IL^2}{2E} b_1 = 0\,, \tag{11.74}$$

$$-\,IL^2\, a_2 - 2IL^3\, a_3 + \frac{IL^2}{2E} b_0 + \frac{IL^3}{3E} b_1 = 0\,. \tag{11.75}$$

From [11.74] and [11.75] we may draw the relations [11.76]:

$$b_0 = 2E\, a_2\ ,\ b_1 = 6E\, a_3\,. \tag{11.76}$$

then, replacing in [11.72] and [11.73], there follows the matrix system [11.77], characteristic of a system with two degrees of freedom:

$$\rho S \begin{bmatrix} \dfrac{L^5}{5} & \dfrac{L^6}{6} \\ \dfrac{L^6}{6} & \dfrac{L^7}{7} \end{bmatrix} \begin{bmatrix} \ddot{a}_2 \\ \ddot{a}_3 \end{bmatrix} + EI \begin{bmatrix} 4L & 6L^2 \\ 6L^2 & 12L^3 \end{bmatrix} \begin{bmatrix} a_2 \\ a_3 \end{bmatrix} = \begin{bmatrix} 0 \\ 0 \end{bmatrix}. \tag{11.77}$$

The solution of this equation is classical and consists of seeking solutions in the following form:

$$\begin{Bmatrix} a_2(t) \\ a_1(t) \end{Bmatrix} = \begin{Bmatrix} \bar{a}_1 \\ \bar{a}_2 \end{Bmatrix} e^{j\omega t}\,.$$

The characteristic equation resulting from it is given by:

$$\omega^4 \frac{1}{1260} - \frac{34}{35} \frac{EI}{\rho SL^4} \omega^2 + 12 \left(\frac{EI}{\rho SL^4} \right) = 0 . \qquad [11.78]$$

We obtain two solutions corresponding to two normal angular frequencies:

$$\omega_1 = 3.53273 \sqrt{\frac{EI}{\rho SL^4}} \qquad [11.79]$$

$$\omega_2 = 34.937 \sqrt{\frac{EI}{\rho SL^4}} . \qquad [11.80]$$

The error for the first normal angular frequency is now much weaker $\varepsilon_1 = 0.3\%$, while that for the second angular frequency remains very large $\varepsilon_2 = 50\%$.

These results are characteristic of the convergence of the Rayleigh-Ritz method:

– the convergence of a normal angular frequency towards exactitude improves when we increase the number of terms of the developments describing displacements and stresses, i.e., on a more physical level, when the number of the degrees of freedom of the associated discrete system grows;

– for a number of terms of fixed developments, the convergence of normal angular frequencies worsens when the order of the mode increases. We clearly note here a much better convergence for mode 1 than for mode 2.

11.6.4. *Model with one degree of freedom verifying the displacement and stress boundary conditions*

As we have emphasized earlier, respecting the boundary stress conditions is optional, since the calculation of the extremum of Reissner's functional leads to their verification.

The functional space Σ_N defined by the expression [11.58] has previously been used with a truncation into one or two terms. It does not *a priori* observe the boundary stress conditions and a significant number of terms may be necessary to approximate them. In a case of this type, the convergence of the method can be rather slow and it may be interesting to restrict the space of stresses to the sub-space of Σ_N *a priori* verifying the boundary stress conditions to accelerate the convergence.

To observe this effect, we will use the functional sub-space Σ_N^{EL} which respects the boundary conditions at the free end in $x = L$, rather than the space Σ_N employed in the preceding examples.

The space Σ_N^{EL} is defined by:

$$\Sigma_N^{EL} = \left\{ \sigma(x,t) = \sum_{m=2}^{N+1} b_m(t)(x-L)^m \right\}. \qquad [11.81]$$

The basic functions $(x - L)^m$ for $m \geq 2$ are clearly nil and have zero first derivatives, at point $x = L$ as required by the boundary conditions of the free end ($\sigma(L, t) = 0$ and $\partial\sigma / \partial x\ (L,t) = 0$).

We restrict ourselves to the first term of each development so as not to weigh down the calculations:

$$W(x,t) = a_2(t)\, x^2 \text{ and } \sigma(x,t) = b_{2(t)}\ (x-L)^2. \qquad [11.82]$$

The calculation of Reissner's functional is straightforward and gives:

$$R(a_2, b_2) = \int_{t_0}^{t_1} \left[\frac{\rho S}{2} (\dot{a}_2)^2 \frac{L^5}{5} - 2I\, b_2\, a_2 \frac{L^3}{3} + \frac{I}{2E} (b_2)^2 \frac{L^5}{5} \right] dt. \qquad [11.83]$$

The calculation of extremum is immediate and leads to the equations:

$$\rho S \frac{L^5}{5} \ddot{a}_2 + \frac{2IL^3}{3} b_2 = 0, \qquad [11.84]$$

$$-\frac{2L^3}{3} a_2 + \frac{IL^5}{5} b_2 = 0, \qquad [11.85]$$

i.e..

$$b_2 = \frac{10}{3} \frac{E}{L^2} a_2, \qquad [11.86]$$

$$\rho S \frac{L^5}{5} \ddot{a}_2 + \frac{20}{9} \text{IEL}\, a_2 = 0\,. \qquad [11.87]$$

Equation [11.87] is that of a system with one degree of freedom, its integration leading to the solution:

$$a_2(t) = A\cos\omega_1 t + B\sin\omega_1 t \qquad [11.88]$$

with ω_1, the normal angular frequency of the system with one degree of freedom, given by:

$$\omega_1 = \sqrt{EI / \rho SL^4}\ \ 0.3333. \qquad [11.89]$$

The relative error for the normal angular frequency ε_1, defined in equation [11.70], gives:

$$\varepsilon_1 = -5.3\% \qquad [11.90]$$

This error is definitely less than 27% obtained with the model with one degree of freedom in section 11.6.2, which shows the clear interest to observe the boundary stress conditions *a priori*.

It as should be noticed as the error with respect to the exact normal angular frequency can be positive or negative, i.e. the method over-estimates or underestimates the exact value according to the case. We will reconsider this point in Chapter 12 where Hamilton's functional with one field is used instead of Reissner's functional.

11.7. Conclusion

The Rayleigh-Ritz method based on Reissner's functional with two fields makes it possible to find a discrete mass-spring system whose vibratory characteristics, normal angular frequencies and vectors, make it possible to give an approximation of the vibration modes of the continuous medium. Our discourse based on the example of reference of the beam in bending vibration has revealed the basic aspects of the method, in particular, related to the choice of functional spaces where the calculation of extremum is carried out.

The generalization to other cases of continuous media is simple; it may be summarized to a modification of the functional, which will be representative of the case considered, for example, the functional [3.10] from Chapter 3 for longitudinal vibrations, or the functional [4.30] from Chapter 4 for transverse vibration of plates

(Mindlin's hypotheses), or the functional [4.44] from Chapter 4 for the Love-Kirchhoff hypotheses. Each unknown function is then developed on a truncated functional basis. For example, in the case of a Love-Kirchoff plate, we will have:

$$W_3^0(x_1, x_2, t) = \sum_{i=1}^{N} a_i(t)\ \phi_i(x_1, x_2),$$

$$\sigma_{11}^3(x_1, x_2, t) = \sum_{i=1}^{N} b_i(t)\ \psi_i(x_1, x_2),$$

$$\sigma_{12}^3(x_1, x_2, t) = \sum_{i=1}^{N} c_i(t)\ \delta_i(x_1, x_2),$$

$$\sigma_{22}^3(x_1, x_2, t) = \sum_{i=1}^{N} d_i(t)\ \gamma_i(x_1, x_2).$$

The calculation of extremum of the functional is then performed with respect to the amplitudes $a_i(t)$, $b_i(t)$, $c_i(t)$ and $d_i(t)$, which, finally, leads to a matrix problem with eigenvalues of the [11.33] type and to the solution exhibited in section 11.6 to obtain the approximate vibration modes of the structure.

We do not develop these very heavy calculations here. In the case of the Rayleigh-Ritz method based on Hamilton's functional with one field covered in the next chapter, we will develop the case of plates.

An important point is the convergence of the approximate solutions; it will be studied in the case of Hamilton's functional in the following chapter. We will see then that the Rayleigh-Ritz method ensures a convergence by a higher value of normal angular frequencies. Nothing stems from it here, as shown by the results of section 11.6, where the approximated normal angular frequencies are either higher or lower than the exact normal angular frequencies towards which they converge. We will reconsider this point in the following chapter during the study of convergence.

Chapter 12

The Rayleigh-Ritz Method based on Hamilton's Functional

12.1. Introduction

In this chapter we present the most common Rayleigh-Ritz method: it is based on Hamilton's variational formulation of the problems of vibrations of elastic solids. Our discourse again follows the broad outline of the steps taken in Chapter 11, in particular, the application of the method to a reference example. The Rayleigh-Ritz method that we present in this chapter is the most used, because, on the one hand, its formulation is simpler, but also because it has convergence properties that the approach of Chapter 11 does not have. These properties of convergence are examined at the end of the chapter and the link with Rayleigh's quotient is established.

12.2. Reference example: bending vibrations of beams

12.2.1 *Hamilton's variational formulation*

The Rayleigh-Ritz method uses a variational formulation for support when calculating the approximated solutions of a vibration problem. In Chapter 11, we presented the method stemming from Reissner's functional. Here we develop the Rayleigh-Ritz method based on Hamilton's functional.

The problem of bending of beams resulting from Bernoulli's hypothesis has been defined in Chapter 3. It is a matter of finding the field of displacement $W(x,t)$, returning the extremum of Hamilton's functional provided by equation [3.71]:

$$H(W(x,t)) = \int_{t_0}^{t_1}\int_0^L \left[\frac{\rho S}{2}\left(\frac{\partial W}{\partial t}\right)^2 - \frac{EI}{2}\left(\frac{\partial^2 W}{\partial x^2}\right)^2 \right] dxdt. \qquad [12.1]$$

The functional space where the calculation of extremum must be carried out depends on the boundary conditions of the beam. The boundary displacement conditions relate to two quantities: transverse displacement $W(x,t)$ and rotation of cross-sections $\frac{\partial W}{\partial x}(x,t)$; these conditions must be *a priori* observed.

The force conditions relate to the bending moment $EI\frac{\partial^2 W}{\partial x^2}(x,t)$ and the shearing force $EI\frac{\partial^3 W}{\partial x^3}(x,t)$. They do not have to be respected *a priori*, but as we saw with Reissner's functional with two fields, their respect *a priori* accelerates convergence.

Let us take the case of a clamped-free beam to consolidate the ideas. The functional space where the calculation of extremum must be carried out is the set of functions that are sufficiently regular for the integral [12.1] to exist and that verify the two boundary displacement conditions imposed in x = 0:

$$W(0,t) = 0 \quad \text{and} \quad \frac{\partial W}{\partial x}(0,t) = 0.$$

We will note this functional space as $W^{E\text{-}L}$. The field of displacement $\widetilde{W}(x,t)$ of this functional space, which returns the extremum of the functional $H(W(x,t))$, is the one verifying the three equations:

$$\rho S\frac{\partial^2 \widetilde{W}}{\partial t^2} + \frac{\partial^2}{\partial x^2}\left(EI\frac{\partial^2 \widetilde{W}}{\partial x^2}\right) = 0 \quad \forall\ x \in \left]0,L\right[\ ,\ \forall \in \left]t_0, t_1\right[, \qquad [12.2]$$

$$EI\frac{\partial^2\widetilde{W}}{\partial x^2}(L,t)=0 \quad \forall\ t \in \left]t_0,t_1\right[, \qquad [12.3]$$

$$\frac{\partial}{\partial x}\left(EI\frac{\partial^2\widetilde{W}}{\partial x^2}\right)(L,t)=0 \quad \forall\ t \in \left]t_0,t_1\right[. \qquad [12.4]$$

12.2.2. *Formulation of the Rayleigh-Ritz method*

In variational terms, the problem of free bending vibrations of clamped-free beams is stated as follows: find the field of displacement $\widetilde{W}(x,t)$ of the functional space $W^{E\text{-}L}$, returning the extremum of Hamilton's functional.

$$H\left(\widetilde{W}(x,t)\right)=\underset{W^{E\text{-}L}}{\mathrm{Ext}}\ H\left(W(x,t)\right).$$

Note: the solution $\widetilde{W}(x,t)$ is a particular $W(x,t)$ displacement; however, for convenience of writing we will note the solution without the tilde, thus confusing at the notation level the solution of the problem and an unspecified displacement.

The Rayleigh-Ritz method is characterized by a particular technique of functional space generation obtained by decomposition over a functional basis $\varphi_n(x)$ verifying the boundary displacement conditions:

$$\varphi_n(0)=0 \ \text{ and } \ \frac{d\varphi_n}{dx}(0)=0\cdot \qquad [12.5]$$

The approximation of the solution comes from the truncation of the functional base with N terms:

$$W(x,t)=\sum_{n=1}^{N} a_n(t)\ \varphi_n(x)\,. \qquad [12.6]$$

The unknowns are the amplitudes $a_n(t)$, which have to be adjusted in order to return the extremum of the functional.

By introducing the approximated expression of displacements [12.6] into the functional [12.1], after calculation we obtain:

$$H\big(a_n(t)\big) = \int_{t_0}^{t_1} \left(\sum_n \sum_p \left(\dot{a}_n \, \dot{a}_p \int_0^L \rho S \, \varphi_n(x) \; \varphi_p(x) \; dx \right.\right.$$
$$\left.\left. - \, a_n \, a_p \int_0^L EI \frac{d^2\varphi_n}{dx^2} \, \frac{d^2\varphi_p}{dx^2} \, dx \right) \right) dt . \qquad [12.7]$$

We may propose a matrix expression of [12.7]:

$$H(\{a\}) = \int_{to}^{t_1} \left(\frac{1}{2} \{\dot{a}\}^t \, (M) \, \{\dot{a}\} - \frac{1}{2} \{a\}^t \, (K) \, \{a\} \right) dt \qquad [12.8]$$

where:

$$\{a\}^t = \big(a_1(t) \,, a_2(t) \,, ..., a_n(t)\big), \qquad [12.9]$$

$$(M) = (M_{np}) \quad \text{with } M_{np} = \int_0^L \rho S \; \varphi_n(x) \;\; \varphi_p(x) \;\; dx, \qquad [12.10]$$

$$(K) = (K_{np}) \quad \text{with } K_{np} = \int_0^L EI \frac{d^2\varphi_n}{dx^2}(x) \; \frac{d^2\varphi_p}{dx^2}(x) \, dx. \qquad [12.11]$$

The calculation of extremum classically leads to the equation:

$$(M) \{\ddot{a}\} + (K) \{a\} = \{0\}. \qquad [12.12]$$

This equation is to be compared to that of a system with N degrees of freedom, where (M) is the matrix of mass and (K) is the matrix of stiffness. The traditional results for the vibrations of discrete systems are, consequently, directly usable.

The solution of [12.12] is provided by joining N modal movements:

$$\{a\} = \sum_{i=1}^{N} (\alpha_i \cos \omega_i t + \beta_i \sin \omega_i t) \{\bar{a}_i\} \qquad [12.13]$$

where ω_i is the ith normal angular frequency and $\{\bar{a}_i\}$ is the associated normal vector. These quantities are calculated with respect to the matrix $(M)^{-1}$ (K) whose eigenvalues are equal to ω_i^2 and the normal vectors are $\{\bar{a}_i\}$. The constants α_i and β_i are fixed by the initial conditions at the origin of the vibratory movement.

There are, moreover, two following properties of orthogonality:

$$\{\bar{a}_i\}^t (M) \{\bar{a}_j\} = 0 \quad \text{if } i \neq j, \qquad [12.14]$$

$$\{\bar{a}_i\}^t (K) \{\bar{a}_j\} = 0 \quad \text{if } i \neq j. \qquad [12.15]$$

Finally, normal angular frequencies verify the relation:

$$\omega_i^2 = \frac{\{\bar{a}_i\}^t (K) \{\bar{a}_i\}}{\{\bar{a}_i\}^t (M) \{\bar{a}_i\}}. \qquad [12.16]$$

Vibratory amplitudes calculated with [12.13] are introduced into the decomposition [12.6] of the vibratory movement of the beam; after grouping of the terms of the expression [12.17] of vibratory displacement, we deduce:

$$W(x, t) = \sum_{k=1}^{N} (\alpha_k \cos \omega_k t + \beta_k \sin \omega_k t) \; f_k(x). \qquad [12.17]$$

In expression [12.17] the function $f_k(x)$ is the mode shape of the mode k. It is given by [12.18] where $\bar{a}_{ki}$ is the ith component of kth normal vector:

$$f_k(x) = \sum_{i=1}^{N} \bar{a}_{ki} \, \varphi_i(x). \qquad [12.18]$$

The properties of orthogonality [12.14] and [12.15] induce properties of orthogonality on the mode shape $f_k(x)$:

$$\int_0^L \rho S\, f_i(x)\ f_j(x)\ dx = 0 \quad \text{if } i \neq j \qquad [12.19]$$

and:

$$\int_0^L EI \frac{d^2 f_i}{dx^2}(x)\ \frac{d^2 f_j}{dx^2}(x)\ dx = 0 \quad \text{if } i \neq j\,. \qquad [12.20]$$

The demonstration is straightforward: we replace the normal strains by their expressions [12.18] in [12.19] and [12.20], and we then use the results [12.14] and [12.15].

12.2.3. *Application: use of a polynomial base for the clamped-free beam*

This polynomial base was presented in the preceding chapter in section 11.6. We adopt it again here without justification; the reader may refer to the previous discussion for more information on this issue.

The functional space where the calculation of extremum takes place is the one described by equation [11.57] from Chapter 11:

$$W_N^{EL} = \left\{ W(x,t) = \sum_{n=2}^{N+1} a_n(t)\ x^n \right\}. \qquad [12.21]$$

Each basic function clearly satisfies the boundary displacement conditions in point 0, but not the stress ones in L.

As an example, we consider the simplest case where N = 1, the displacement of the beam being approximated by:

$$W(x,t) = a_2(t)\, x^2\,. \qquad [12.22]$$

Introducing the expression [12.22] into the functional [12.7], we obtain after all the calculations:

$$H(a_2) = \int_{t_0}^{t_1} \left[\frac{\rho S}{2} \frac{L^5}{5} \cdot (\dot{a}_2)^2 - 2\,EIL(a_2)^2 \right] dt. \qquad [12.23]$$

The calculation of extremum is straightforward and leads to the differential equation [12.24]:

$$\rho S \frac{L^5}{5} \ddot{a}_2 + 4\,EIL\,a_2 = 0. \qquad [12.24]$$

The solution is:

$$a_2(t) = A_1 \cos \omega_1 t + B_1 \sin \omega_1 t \qquad [12.25]$$

with:

$$\omega_1 = \sqrt{\frac{EI}{\rho S L^4}}\, 4.472. \qquad [12.26]$$

This result is identical to the previous chapter (equation [11.65]), which was obtained using the technique with two fields and one degree of freedom. Thus, at this level there is no decisive advantage in using one or other of the variational techniques. The method with one field is, however, definitely easier to implement.

Introducing the *a priori* respect of boundary stress conditions was rather simple using the technique with two fields; that is an advantage since this would be difficult to formulate with the functional with one field using a polynomial base. However, the use of another functional base type which we will encounter later overcomes this difficulty.

12.3. Functional base of the finite elements type: application to longitudinal vibrations of beams

The finite elements method uses a different technique to generate the functions approximating the solution. It is, however, closely linked to the Rayleigh-Ritz method by the use of the variational method to obtain the approximated solutions. We do not pretend to provide a total presentation of the finite elements method in these few lines, but rather to show its connection to the Rayleigh-Ritz method.

The functional to be considered for the case in point is naturally the longitudinal vibrations of beams functional provided in Chapter 3, equation [3.26]:

$$H\big(W_l(x,t)\big) = \int_{t_0}^{t_1}\int_0^L \left[\frac{1}{2}\rho S\left(\frac{\partial W_l^0}{\partial t}\right)^2 - \frac{1}{2}ES\left(\frac{\partial W_l^0}{\partial x}\right)^2\right] dxdt\,. \qquad [12.27]$$

Generating the functions of approximation is a particular process based on functions defined piece by piece. Let us cut up the beam into N equal segments with a length of Δ (these segments are denoted elements) and introduce the N functions $\psi_n(x,t)$ defined in [12.28]:

$$\psi_n(x,t) = \begin{cases} 0 \quad \text{if } x \notin \big[(n-1)\Delta\,,\, n\Delta\big] \\ \dfrac{U_n^{(t)} - U_{n-1}^{(t)}}{\Delta} x + U_n^{(t)} + n\big(U_{n-1}^{(t)} - U_n^{(t)}\big) \\ \qquad\qquad \text{if } x \in \big](n-1)\Delta\,, n\Delta\big[. \end{cases} \qquad [12.28]$$

Figure 12.1 provides the graph of the function $\psi_n(x,t)$. These functions, when reassembled, offer the possibility to approximate the vibratory displacement $W_l^0(x,t)$ by a continuous line.

$$W_l^0(x,t) = \sum_{n=1}^{N} \psi_n(x,t)\,. \qquad [12.29]$$

Figure 12.2 gives an example of approximation resulting from the decomposition of vibratory movement by [12.29] for N = 5. We may note that the functions $U_n(t)$, which represent displacements at point $n\Delta$ of the beam, (or displacement at node n) constitute the new unknowns of the problem.

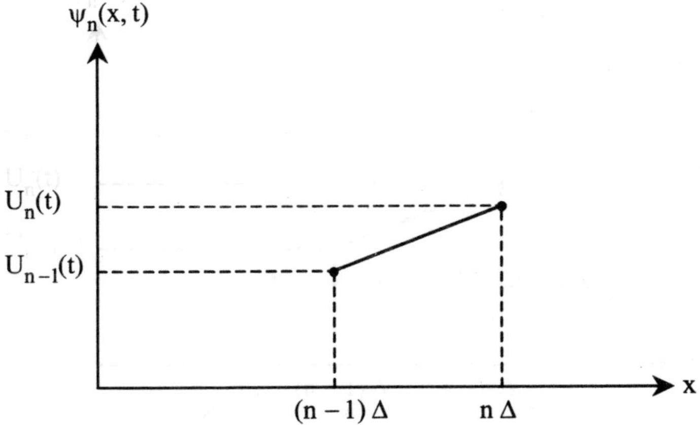

Figure 12.1. *Function* $\psi_n(x, t)$ *at a given moment t*

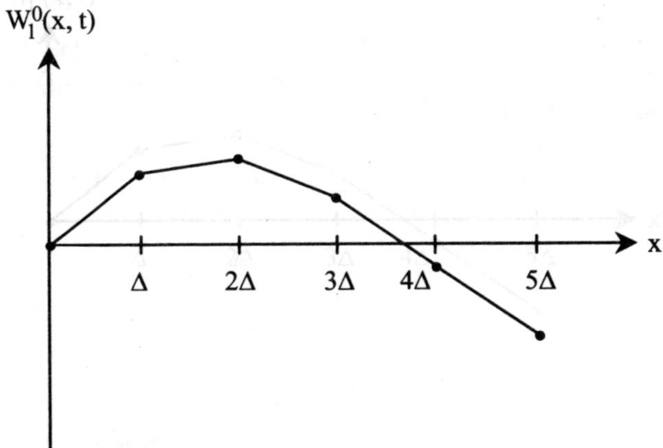

Figure 12.2. *Example of approximated vibratory displacement at a fixed moment t (case where N=5)*

To consolidate our ideas without performing too heavy a calculation, let us take the case where N = 2. The two functions of approximation are:

$$\psi_1(x,t) = \frac{U_1(t) - U_0(t)}{L/2} x + U_0(t) \quad \text{if } x \in \left] 0, L/2 \right[, \qquad [12.30]$$

$$\psi_2(x,t) = \frac{U_2(t) - U_1(t)}{L/2} x - U_2(t) + 2\,U_1(t) \quad \text{if } x \in \left] L/2, L \right[. \qquad [12.31]$$

The calculation of the functional is carried out by introducing the approximated expression of $W_1^0(x,t)$:

$$H\big(U_0(t), U_1(t), U_2(t)\big) = \frac{1}{2}\int_{t_0}^{t_1}\int_0^{L/2}\left[\rho S\left(\frac{d\psi_1}{dt}(x,t)\right)^2 - E\delta\left(\frac{d\psi_1}{dx}(x,t)\right)^2\right] dxdt + \frac{1}{2}\int_{t_0}^{t_1}\int_{L/2}^{L}\left[\rho S\left(\frac{d\psi_2}{dt}(x,t)\right)^2 - ES\left(\frac{d\psi_2}{dx}(x,t)\right)^2\right] dxdt. \qquad [12.32]$$

The functions $\psi_1(x,t)$ and $\psi_2(x,t)$ given in [12.30] and [12.31] correspond to the case of the free beam since displacements $U_0(t)$ and $U_2(t)$ are left free and represent displacements at the ends of the beam.

To treat the case where the beam is clamped in 0, it is sufficient to write $U_0(t) = 0$. The function of approximation $\psi_1(x,t)$ given in [12.30] is then reduced to:

$$\psi_1(x,t) = \frac{U_1(t)}{L/2} x.$$

If the beam is clamped in L, we pose $U_2(t) = 0$. The function of approximation $\psi_2(x,t)$ results from [12.31] in an obvious manner.

Thus, in the case of the beam fixed at its two ends, [12.30] and [12.31] are reduced to:

$$\psi_1(x,t) = \frac{U_1(t)}{L/2} x, \qquad [12.33]$$

$$\psi_2(x, t) = -\frac{U_1(t)}{L/2} x + 2\, U_1(t)\,. \qquad [12.34]$$

The functional [12.32] then takes the expression [12.35]:

$$H\left(U_1(t)\right) = \int_{t_0}^{t_1} \left[\rho S \frac{L}{6} \left(\dot{U}_1(t)\right)^2 - ES \frac{2}{L} \left(U_1(t)\right)^2 \right] dt\,. \qquad [12.35]$$

The calculation of extremum leads to the equation of a system with one degree of freedom:

$$\rho S \frac{L}{6}\, \ddot{U}_1(t) + \frac{2ES}{L}\, U_1(t) = 0\,, \qquad [12.36]$$

that is, with the solution:

$$U_1(t) = (\alpha_1 \cos \omega_1 t + \beta_1 \sin \omega_1 t) \qquad [12.37]$$

with the normal angular frequency equal to:

$$\omega_1 = \sqrt{\frac{E}{\rho}}\, \frac{2\sqrt{3}}{L}\,. \qquad [12.38]$$

We can estimate the quality of this approximation by comparing it with the exact normal angular frequency provided in Chapter 4, ($\omega_1^{ex} = \sqrt{\frac{E}{\rho}} \frac{\pi}{L}$). Calculation yields a relative error of approximately 10%.

Let us notice that the functions $\psi_n(x, t)$ given in [12.28] could not be used in the case of the bending of beams. Indeed, taking into account the discontinuity of their first derivative, these functions are not derivable twice with respect to x as the functional [12.1] requires. In this case, it would be necessary to consider parabolic instead of linear functions to ensure the continuity of the first derivative and, thus, the existence of second derivatives.

With this approximation technique, the sufficient regularity of the used functions of space is the most important issue. This regularity obviously depends on the degree of derivation of the functions appearing in the functional.

The advantage of this method of approximation compared to the traditional development of the Rayleigh-Ritz method, whose example is provided in section 12.2, lies in the shape of the matrices of mass and stiffness. These matrices are generally full in the traditional case, whereas they are band matrices in the case of the finite elements type approximation. This particular property can be detected on the basis of the functions [12.30] and [12.31] and of the functional [12.32]; indeed, it is obvious that the variables $U_0(t)$ and $U_2(t)$ do not have direct coupling since they do not appear in the same element. It follows that the matrix of mass will have the following tri-band form:

$$\begin{pmatrix} M_{00} & M_{01} & 0 \\ M_{01} & M_{11} & M_{12} \\ 0 & M_{12} & M_{22} \end{pmatrix}. \qquad [12.39]$$

This property of matrices remains regardless of the number of sections. Thus, it is possible to use much more powerful adapted numerical algorithms than those applicable to the general case of full matrices.

12.4. Functional base of the modal type: application to plates equipped with heterogenities

In many problems, the vibrating structure consists of a carrier structure equipped with various heterogenities. To consolidate, we will consider the case of a rectangular plate with added mass and distributed springs. This will enable us to give an example of application to continuous 2D mediums and to use a modal functional base, which, in fact, constitutes the principal approach used in the Rayleigh-Ritz method.

Let us take a rectangular plate with the dimensions a by b, supported at the edges, with an added mass M at the point (x_M, y_M) and with a distributed spring K positioned at the line x = xR. The functional considered is that of the plate supporting the mass and the distributed spring; its construction is rather simple, since it uses the property of addition of energies. Let us suppose that the plate has a transverse movement $W(x, y, t)$ governed by the Love-Kirchhoff hypothesis; the functional representative of the transverse movement of the plate was provided in Chapter 4. The functional of the whole system results from joining the energies of the plate, of the mass and of the spring, that is:

$$H(W(x,y,t)) = \int_{t_0}^{t_1} \left(\int_0^a \int_0^b \frac{\mu}{2}\left(\frac{\partial W}{\partial t}\right)^2 - \frac{D}{2}\left[\left(\frac{\partial^2 W}{\partial x^2}\right)^2 + \left(\frac{\partial^2 W}{\partial y^2}\right)^2 \right.\right.$$

$$\left.\left. + 2\nu \frac{\partial^2 W}{\partial x^2}\frac{\partial^2 W}{\partial y^2} + 2(1-\nu)\left(\frac{\partial^2 W}{\partial x \partial y}\right)^2\right] dxdy \right) dt \qquad [12.40]$$

$$+ \int_{t_0}^{t_1} \left[\frac{1}{2} M \left(\frac{\partial W}{\partial t}(x_m, y_m, t)\right)^2 - \frac{1}{2}\int_0^b K \left(W(x_R, y, t)\right)^2 dy \right] dt,$$

where μ is the mass per unit of area of the plate and D is its bending stiffness.

Vibratory displacement must now be approximated by decomposition on a truncated functional basis. This functional base has to verify the kinematic boundary conditions. An interesting way to build this base consists in using the normal modes of the support plate, which of course verify the kinematic boundary conditions by construction. Moreover, they verify the boundary conditions with respect to the forces, which guarantee faster convergence as we saw in the previous chapter.

In the case of the rectangular plate supported at its 4 edges, the normal modes $\varphi_{nm}(x, y)$ have a simple analytical expression, which we provided in Chapter 7:

$$\varphi_{nm}(x,y) = \sin\frac{n\pi}{a}x \, \sin\frac{m\pi}{b}y. \qquad [12.41]$$

The vibratory response is sought in the form:

$$W(x,y,t) = \sum_{n=1}^{N}\sum_{m=1}^{M} a_{nm}(t)\, \varphi_{nm}(x,y). \qquad [12.42]$$

Introducing this decomposition of the response into the functional, after the calculation of the double integrals over the surface of the plate we find:

$$H\left(a_{nm}(t)\right) = \int_{t_0}^{t_1} \left[\frac{1}{2}\left\{\dot{a}_{nm}\right\}^t (M) \left\{\dot{a}_{nm}\right\} - \frac{1}{2}\left\{a_{nm}\right\}^t (K)\left\{a_{nm}\right\} \right] dt \qquad [12.43]$$

where (M) and (K) are the matrices of mass and stiffness whose generic terms have the form:

$$M_{nmpq} = \mu \frac{ab}{4} \delta_{np} \delta_{mq} + M \sin \frac{n\pi}{a} x_m \sin \frac{m\pi}{b} y_m \sin \frac{p\pi}{a} x_m \sin \frac{q\pi}{b} y_m, \qquad [12.44]$$

$$K_{nmpq} = \mu \frac{ab}{4} \omega_{nm}^2 \delta_{np} \delta_{mq} + K \sin \frac{n\pi}{a} x_R \sin \frac{p\pi}{a} x_R \frac{b}{2} \delta_{mq} \qquad [12.45]$$

where δ_{ij} is the Kronecker symbol and $\omega_{nm} = \sqrt{\frac{D}{\mu}} \left(\frac{n^2\pi^2}{a^2} + \frac{m^2\pi^2}{b^2} \right)$.

It should be noted that if M = 0 and K = 0, the matrices of mass and stiffness become those of the bare plate; under these conditions the matrices of mass and stiffness are diagonal. This property relates to the fact that the functions $\varphi_{nm}(x, y)$ are the mode shapes of the bare plate.

The calculation of the extremum of the functional [12.43] is classical; it leads to the matrix system:

$$(M)\{\ddot{a}_{nm}\} + (K)\{a_{nm}\} = \{0\}.$$

Thus, we have built a discrete mass-spring system, approximating the vibratory characteristics of the heterogenous plate.

The advantage of the use of the modes of the carrier structure stems from two aspects which allow a good convergence of the result:

– the functional base verifies the kinematic boundary conditions, as well as the boundary stress conditions, which ensures accelerated convergence as we saw in the previous chapter;

– the modal base $\varphi_{nm}(x, y)$ is the exact solution when heterogenities of mass and stiffness tend towards zero. We may, therefore, consider that for low heterogenities, the modal base of the heterogenous plate will be close to $\varphi_{nm}(x, y)$ and that, consequently, the developments with a small number of terms will be sufficient.

12.5. Elastic boundary conditions

12.5.1. *Introduction*

In order to be able to apply the Rayleigh-Ritz method with the same functional base regardless of the boundary conditions, we can employ the concept of elastic boundary condition. We outline this approach on a very simple case of a beam in longitudinal vibrations, in order to enable the reader to understand the foundation of the method. The more complicated cases are treated in a similar way. In fact, it suffices to adapt the functional to the treated case. On this subject we will provide some results taken from the works given in the bibliography.

12.5.2. *The problem*

We consider the longitudinal vibrations of a beam clamped in 0 and assign a yield stiffness in L.

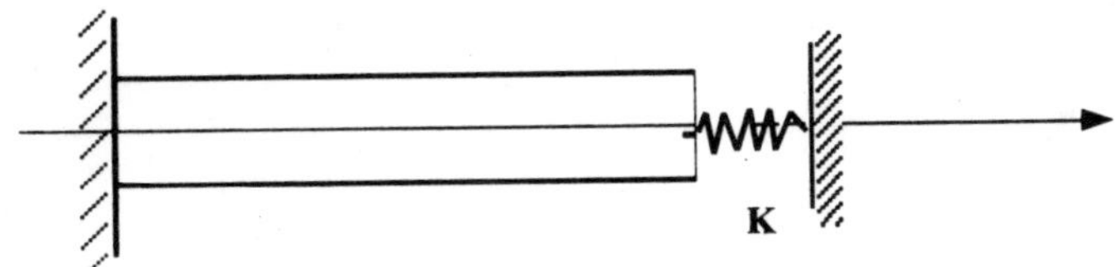

Figure 12.3. *Beam clamped in 0 and with yield stiffness in L*

The equations of free vibrations governing this case are the following:

$$\rho S \frac{\partial^2 \overline{U}(x,t)}{\partial t^2} + ES \frac{\partial^2 \overline{U}(x,t)}{\partial x^2} = 0\,, \qquad [12.46]$$

$$\overline{U}(0,t) = 0\,, \qquad [12.47]$$

$$ES \frac{\partial \overline{U}}{\partial x}(L,t) = K\overline{U}(L,t)\,. \qquad [12.48]$$

$\overline{U}(x,t)$ is the longitudinal displacement solving the problem, ρ is the density, E is the Young modulus and S is the section of the beam. Relation [12.48] translates the elastic boundary condition of stiffness K applied at the end L. It makes it possible to treat the cases starting at the free end for $K = 0$ to the clamped end for $K \to \infty$.

In variational terms, the problem presents itself in the following light: find $\overline{U}(x, t)$ that verifies:

$$H\left(\overline{U}(x,t)\right) = \underset{U^{EL}}{\mathrm{Ext}} \left\{ H\left(U(x,t)\right) \right\}. \qquad [12.49]$$

Hamilton's functional is obtained by joining the beam and spring functionals:

$$H\left(U(x,t)\right) = \int_{t_0}^{t_1} \left(\int_0^L \rho S \left(\frac{\partial U}{\partial t} \right)^2 - \frac{1}{2} \left(\frac{\partial U}{\partial x} \right)^2 dx - \frac{1}{2} K U^2(L,t) \right) dt. \qquad [12.50]$$

The functional base to consider is that of a clamped-free beam, since displacement is not imposed in L. In the example given here, let us choose a polynomial base:

$$U^{EL} = \sum_{n=1}^{\infty} a_n(t)\, x^n. \qquad [12.51]$$

The index $n = 0$ is excluded so that the boundary condition in $x = 0$ is verified.

12.5.3. *Approximation with two terms*

We choose a truncated functional base with two terms, that is:

$$U(x,t) \approx a_1(t)\, x + a_2(t)\, x^2. \qquad [12.52]$$

Introducing this approximation into the functional [12.50] leads to:

$$\begin{aligned} H\left(a_1(t), a_2(t)\right) = \int_{t_0}^{t_1} & \left[\frac{1}{2} \rho S \left(\dot{a}_1^2 \frac{L^3}{3} + 2\, \dot{a}_1\, \dot{a}_2 \frac{L^4}{4} + \dot{a}_2^2 \frac{L^5}{5} \right) \right. \\ & \left. - \frac{1}{2} \left(ES \left(a_1^2 L + 2\, a_1\, a_2 L^2 + a_2^2\, 4 \frac{L^3}{3} \right) \right) - \frac{1}{2} K (a_1 L + a_2 L^2)^2 \right] dt. \end{aligned} \qquad [12.53]$$

The calculation of extremum provides the system with two degrees of freedom [12.54]. To simplify calculations we take the case of unitary length:

$$\begin{bmatrix} \rho S/3 & \rho S/4 \\ \rho S/4 & \rho S/5 \end{bmatrix} \begin{Bmatrix} \ddot{a}_1 \\ \ddot{a}_2 \end{Bmatrix} + \begin{bmatrix} ES+K & ES+K \\ ES+K & \dfrac{4ES}{3}+K \end{bmatrix} \begin{Bmatrix} a_1 \\ a_2 \end{Bmatrix} = \begin{Bmatrix} 0 \\ 0 \end{Bmatrix}. \qquad [12.54]$$

The two normal angular frequencies associated to [12.54] are calculated in a classical fashion. We come to equation [12.55]:

$$\omega^4 \frac{(\rho S)2}{240} - \omega^2 \left[\rho S\,(ES+K)\,1/30 + \frac{\rho SES}{3} \right] + (ES+K)\left(\frac{ES}{3}\right) = 0. \qquad [12.55]$$

If K is null, the model corresponds to a clamped-free beam, the calculation then leads to the two normal angular frequencies:

$$\omega_1 = \sqrt{\frac{E}{\rho}}\,1.579 \quad \text{and} \quad \omega_2 = \sqrt{\frac{E}{\rho}}\,5.67,$$

These two values are to be compared with the exact angular frequencies:

$$\omega_1 = \sqrt{\frac{E}{\rho}}\frac{\pi}{2} \quad \text{and} \quad \omega_2 = \sqrt{\frac{E}{\rho}}\frac{3\pi}{2}.$$

In the case where K tends towards infinity, the model corresponds to a clamped-clamped beam. The calculation of the two roots of [12.55] leads to an infinite normal angular frequency and to:

$$\omega_1 = \sqrt{\frac{E}{\rho}}3.16227.$$

This angular frequency is to be approximated to the exact normal pulsation:

$$\omega_1 = \pi\sqrt{\frac{E}{\rho}}.$$

When the stiffness of the boundary varies from 0 to infinity, the first normal angular frequency of the model varies from $\sqrt{\frac{E}{\rho}}\,1.579$ to $\sqrt{\frac{E}{\rho}}\,3.162$, and the second pulsation varies from $\sqrt{\frac{E}{\rho}}\,5.67$ *ad infinitum*.

This approach makes it possible to sweep boundary stiffness to infinity over all the boundary conditions between the free and the clamped ends without modifying the functional base, which remains that of the clamped-free beam. The approximation over the first mode is completely correct; over the second mode it is worse for the free end and is completely degraded for the clamped end. It should also be noted that, at the numerical level, using very large rigidities K renders the matrices ill conditioned and poses numerical problems. Thus, despite an apparent simplicity, this technique comes up against the choice of the value K that needs to be chosen to describe a clamped end correctly. Too low a value does not model a clamped end, while too strong a value creates numerical problems; in fact the value of K depends on the structure and the eigenfrequency considered, and it requires a numerical study of the solutions in each considered case.

12.6. Convergence of the Rayleigh-Ritz method

12.6.1. *Introduction*

The property of convergence of the Rayleigh-Ritz method is important because it largely explains the success of the method in this field. As we demonstrate, normal angular frequencies converge by higher values when the functional space, where the calculation of extremum is performed, grows. On a physical plane, the reduction in the normal angular frequency when additional movements are allowed amounts to releasing the system, making it more flexible. Let us note that this property of convergence is specific to the Rayleigh-Ritz method based on Hamilton's functional and is not true if the variational formulation used is Reissner's functional.

12.6.2. *The Rayleigh quotient*

Let us consider Hamilton's functional [12.1] describing the bending of beams. To consolidate, let us adopt boundary conditions of clamped type at both ends. The extremum of the functional is attained for the solution field of displacement $\overline{W}(x,t) \in W^{EE}$ that verifies:

$$H\left(\overline{W}(x,t)\right) = \underset{W^{EE}}{\mathrm{Ext}}\left\{ H\left(W(x,t)\right) \right\}.$$

Let us formulate equation [12.56] where λ is a real number:

$$W(x,t) = \overline{W}(x,t) + \lambda\, v(x,t)\,. \qquad [12.56]$$

We observe with [12.56] that, if $\lambda = 0$, $W(x,t) = \overline{W}(x,t)$ and thus that the calculation of the extremum is provided by the condition:

$$\frac{d}{d\lambda}\left(H\left(\overline{W} + \lambda v(x,t)\right)\right)(\lambda = 0) = 0 \quad \forall\, v(x,t)\,. \qquad [12.57]$$

This amounts to saying that the directional derivative of H is nil in $\overline{W}(x,t)$.

The calculation of the extremum of the functional [12.1] using [12.57] leads to the result:

$$\int_{t_0}^{t_1}\int_0^L \left[\rho S \frac{\partial \overline{W}}{\partial t}\frac{\partial v}{\partial t} - EI \frac{\partial^2 \overline{W}}{\partial x^2}\frac{\partial^2 v}{\partial x^2} \right] dx\, dt = 0 \quad \forall\, v(x,t)\,. \qquad [12.58]$$

By restricting the functions $v(x,t)$ to verify the two conditions $v(x,t_0) = v(x,t_1) = 0$, by integration by parts over time, the integral [12.58] becomes:

$$\int_{t_0}^{t_1}\int_0^L \left[-\rho S \frac{\partial^2 \overline{W}}{\partial x^2} v - EI \frac{\partial^2 \overline{W}}{\partial x^2}\frac{\partial^2 v}{\partial x^2} \right] dx\, dt = 0 \quad \forall\, v(x,t)\,. \qquad [12.59]$$

We know that the solutions of the problem of free vibrations have the form:

$$\overline{W}(x,t) = (A \cos\omega_n t + B \sin\omega_n t)\; f_n(x) \qquad [12.60]$$

where ω_n is the normal angular frequency of mode n and $f_n(x)$ is the mode shape.

Let us further restrict the test functions $v(x,t)$ to take the form [12.61], with $g(t_0) = g(t_1) = 0$:

$$v(x,t) = g(t)\;\psi(x)\,. \qquad [12.61]$$

Taking into account [12.60] and [12.61], the integral [12.59] becomes:

$$\left(\omega_n^2 \int_0^L \rho S\, f_n(x)\ \psi(x)\, dx - \int_0^L EI \frac{d^2 f_n}{dx^2} \frac{d^2 \psi}{dx^2} dx \right) K_n = 0 \tag{12.62}$$

where:

$$K_n = \int_{t_0}^{t_1} (A\ \sin \omega_n t + B\ \sin \omega_n t)\ g(t)\, dt\,,$$

that is, finally:

$$\omega_n^2 = \frac{\displaystyle\int_0^L EI \frac{d^2 f_n}{dx^2} \frac{d^2 \psi}{dx^2} dx}{\displaystyle\int_0^L \rho S\, f_n(x)\ \psi(x)\, dx} \quad \forall\, \psi(x)\,. \tag{12.63}$$

In the particular case where we take $\psi(x) = f_n(x)$, we obtain the Rayleigh quotient, which provides the normal angular frequency according to the mode shape:

$$\omega_n^2 = \frac{\displaystyle\int_0^L EI \left(\frac{d^2 f_n}{dx^2} \right)^2 dx}{\displaystyle\int_0^L \rho S\, f_n^2\, dx}\,. \tag{12.64}$$

12.6.3. *Introduction to the modal system as an extremum of the Rayleigh quotient*

In this section, we present the method of calculation of the modal system stemming from that of extremum of the Rayleigh quotient. Let us introduce the functional $\Omega(y(x))$, called the Rayleigh quotient associated to the shape function $y(x)$.

$$\Omega\left(y(x)\right) = \frac{\int_0^L EI\left(\frac{d^2 y}{dx^2}\right)^2 dx}{\int_0^L \rho S\, y(x)^2\, dx}. \qquad [12.65]$$

Let us calculate the directional derivative of $\Omega(y)$ when $y(x) = f_n(x)$. We obtain:

$$\frac{d\Omega\left(f_n + \lambda\psi\right)}{d\lambda}(\lambda = 0) = \qquad [12.66]$$

$$\frac{\int_0^L EI \frac{d^2 f_n}{dx^2}\ \frac{d^2\psi}{dx^2} dx\ \int_0^L \rho S\, f_n^2(x)\, dx - \int_0^L \rho S\, f_n(x)\ \psi(x)\, dx\ \int_0^L EI\left(\frac{d^2 f_n}{dx^2}\right) dx}{\left(\int_0^L \rho S\, f_n^2(x)\, dx\right)^2}.$$

Using the relation [12.63] we can give a simpler form of [12.66]:

$$\frac{d\Omega\left(f_n + \lambda\psi\right)}{d\lambda}(\lambda = 0) = \qquad [12.67]$$

$$\left(\omega_n^2 \int_0^L \rho S\, f_n^2(x)\, dx - \int_0^L EI\left(\frac{d^2 f_n}{dx^2}\right) dx\right) \frac{\int_0^L \rho S\, f_n(x)\, \psi(x)\, dx}{\int_0^L \rho S\, f_n^2(x)\, dx}.$$

Finally, with [12.64], we note that the directional derivative of Ω is nil in $f_n(x)$:

$$\frac{d\,\Omega\,(f_n + \lambda\psi)}{d\lambda}(\lambda = 0) = 0 \quad \forall\, \psi(x). \qquad [12.68]$$

The Rayleigh quotient is stationary for each mode shape of the problem and its value is equal to the corresponding normal angular frequency.

$$\omega_n^2 = \underset{Y}{\text{Ext}}\, \Omega\big(y(x)\big) = \Omega\big(f_n(x)\big) \tag{12.69}$$

where Y is the functional space of the admissible y(x) functions.

There arises the additional question of the nature of the extremum. To clear this point up, it is necessary to calculate the second directional derivative leading to the following result:

$$\frac{d^2\,\Omega\,(f_n + \lambda\psi)}{d\lambda^2}(\lambda = 0) = \frac{\displaystyle\int_0^L EI\left(\frac{d^2\,\psi}{dx^2}\right)^2 dx - \omega_n^2 \int_0^L \rho S\,\psi^2\,dx}{\displaystyle\int_0^L f_n^2\,dx}. \tag{12.70}$$

Let us take the case of the first vibration mode $\left(f_1(x), \omega_1\right)$:

$$\frac{d^2\,\Omega\,(f_1 + \lambda\psi)}{d\lambda^2}(\lambda = 0) = \frac{\displaystyle\int_0^L EI\left(\frac{d^2\,\psi}{dx^2}\right)^2 dx - \omega_1^2 \int_0^L \rho S\,\psi^2\,dx}{\displaystyle\int_0^L \rho S\, f_1^2\,dx}. \tag{12.71}$$

The normal angular frequency considered is the one with the smallest value; consequently, the function $f_1(x)$ is the one returning the smallest possible Rayleigh quotient and, thus, for any function $\psi(x)$ we will have:

$$\omega_1^2 \leq \frac{\displaystyle\int_0^L EI\left(\frac{d^2\,\psi}{dx^2}\right)^2 dx}{\displaystyle\int_0^L \rho S\,\psi^2\,dx}. \tag{12.72}$$

The consequence of this inequality for equation [12.71] is:

$$\frac{d^2\,\Omega\,(f_1 + \lambda\psi)}{d\lambda^2}(\lambda = 0) \geq 0 \quad \forall\,\psi\,. \tag{12.73}$$

The extremum is thus a minimum for the first mode shape:

$$\omega_1^2 = \underset{Y}{\text{Min}}\ \Omega\left(y(x)\right) = \Omega\left(f_1(x)\right). \qquad [12.74]$$

For higher order modes, this property does not apply directly, since the relation [12.72] which characterizes ω_1 is not true for other modes. It remains true if functional space where the calculation of extremum is performed excludes the mode shapes of lower order mode than the considered mode. For example, for mode 2, we will calculate the extremum in the functional space Y^1, excluding mode 1: $Y^1 = Y - \{a\ f_1(x)\}$ where a is a real number.

We then have:

$$\omega_2^2 = \underset{Y^1}{\text{Min}}\ \Omega\left(y(x)\right) = \Omega\left(f_2(x)\right). \qquad [12.75]$$

In general:

$$\omega_n^2 = \underset{Y^{n-1}}{\text{Min}}\ \Omega\left(y(x)\right) = \Omega\left(f_n(x)\right) \text{ with : } Y^{n-1} = Y^{n-2} - \left\{a\ f_{n-1}(x)\right\}. \qquad [12.76]$$

Normal angular frequencies thus appear as minima of the Rayleigh quotient taken in increasingly restricted functional spaces.

12.6.4. *Approximation of the normal angular frequencies by the Rayleigh quotient or the Rayleigh-Ritz method*

Let us consider a sub-space of the size N of the space Y defined in section 12.6.3 and write it down as Y_N. This sub-space is constructed, as in the Rayleigh-Ritz method, by the linear combination of N basic kinematically admissible functions. The calculation of the minimum of the Rayleigh quotient in the sub-space Y_N leads to an approximation ω_1^N of the first mode of vibration:

$$\omega_1^N = \underset{Y_N}{\text{Min}}\ \Omega\left(y(x)\right) = \Omega\left(f_1^N(x)\right) \qquad [12.77]$$

where $f_1^N(x)$ corresponds to an approximation of the mode shape $f_1(x)$.

Let us consider the sub-space of dimension N+1 of the functional space Y built by adding one base function to the space Y_N ; we thus have:

$$Y_N \subset Y_{N+1}.$$

Under these conditions, the approximation ω_1^{N+1} obtained by minimizing the Rayleigh quotient over Y_{N+1} is necessarily smaller than ω_1^N, since the minimum is sought over a larger space containing Y_N. We thus have:

$$\omega_1^N \geq \omega_1^{N+1} \quad [12.78]$$

The convergence of the Rayleigh quotient is thus carried out by a higher value when we increase the size of the sub-space where the extremum is calculated. Taking into account the identity of the solutions obtained using the Rayleigh quotient and using the Rayleigh-Ritz method, the same applies to the latter.

The property is repeated for the higher order modes insofar as $Y_i^N \subset Y_i^{N+1}$; we will have $\omega_i^N > \omega_i^{N+1}$.

We may thus conclude that the normal angular frequency obtained by the Rayleigh-Ritz method converge by higher values. On the physical plane, this tendency indicates an increased flexibility of the system when we increase the number of basic functions. We may also state that limiting the possible movements of the vibrating continuous medium by restricting the functional spaces where the solution is sought amounts to blocking the possible movements through an added stiffness, which leads to normal angular frequencies that are higher the more we limit the possible movements.

12.7. Conclusion

In this chapter we have presented the most widespread Rayleigh-Ritz method based on Hamilton's functional. Compared to the method presented in the preceding chapter it retains the same basic idea, which consists of building an equivalent discrete system, although this approach has the advantage of leading to normal angular frequencies converging by higher value, which is not the case when we use the Reissner's functional.

The procedure consists of expressing vibratory displacements as a linear combination of functions constituting a subspace of finite dimension of the admissible functional space. This approach is at the origin of the finite elements method, which, in fact, amounts to approximating the solutions with particular basic functions.

We have also provided several simple examples to illustrate the important aspects of the method and established the link with the Rayleigh quotient.

Bibliography and Further Reading

[AKG 93] AKGÜN M.A., "A new family of mode superposition methods for response calculation", *J. Sound and Vib.*, 167(2), p. 289-302, 1993, Chapter 9.

[BAT 82] BATHE K.J., *Finite Element Procedures in Engineering Analysis,* Prentice Hall, New Jersey, 1982, Chapters 11, 12.

[BHA 84] BHAT R.B., "Obtaining natural frequencies of elastic systems by using an improved strain energy formulation in Rayleigh-Ritz method", *J. Sound and Vib.*, 93 (2), p. 314-320, 1984, Chapters 11, 12.

[BLE 84] BLEVINS R.D., *Formulas for Natural Frequency and Mode Shape,* R.E. Kriegen Publishing, 1984, Chapters 5, 6, 7.

[BUH 84] BUHARIWALA K.J. and HANIEN J.S., *Dynamics of Viscoelastic Structures,* Institute for Aerospace Studies, University of Toronto, Toronto, Canada, 1984, Chapter 8.

[CAL 68] CALLOTE L.R., *Introduction to Continuum Mechanics,* Di Van Nostrand Co Princeton, New Jersey, 1968, Chapter 1.

[COU 53] COURANT R. and HILBERT D., *Method of Mathematical Physics,* Intescience Publishing, 1953, Chapter 2.

[COW 66] COWPER G.R., "The shear coefficient in Timoshenko's Beam Theory", *J. Applied. Mechanics*, Transaction of the ASME, Vol. 33, p. 335, 340, 1966, Chapter 3.

[DHA 73] DHARMARAJAN S. and MCCUTCHEN H.JR., "Shear coefficients for orthotropic beams", *J. Composite Materials,* Vol. 7, p. 530-535, 1973, Chapter 3.

[DIT 86] DITARANTO R.A. and Mc GRAW J.R., "Vibratory bending of damped laminated plates", *J. Engineering for Industry,* Trans-ASME, p. 1081-1090, 1986, Chapter 4.

[DUV 90] DUVAUT G., *Mécanique des milieux continus*, Collection Mathématiques Appliquées pour la Maîtrise, Masson, Paris, 1990, Chapter 1.

[DYM 73] DYM C.L. and SHAMES I.H., *Solid Mechanics: A Variational Approach,* McGraw-Hill, New York, 1973, Chapter 2.

[ERI 71] ERINGER A.C., *Continuum Physics,* Academic Press, 1971, Chapter 1.

[GAR 96] GARIBALDI L. and ONAH H.N., *Viscoelastic Material Damping Technology,* Becchis Osiride, Torino, 1996, Chapter 8.

[GER 62] GERMAIN P., *Mécanique des milieux continus,* Masson, Paris, 1962, Chapter 1.

[GER 73] GERMAIN P., *Cours de Mécanique des milieux continus,* Masson, Paris, 1973, Chapter 1.

[GER 80] GERMAIN P. and MULLER P., *Introduction à la mécanique des milieux continus,* Masson, Paris, 1980, Chapters 1, 2.

[GIB 88] GIBERT R.J., *Vibrations des structures,* Eyrolles, Paris, 1988, Chapters 5-9.

[GOR 82] GORMAN D.J., *Free Vibration of Rectangular Plates,* Elsevier, North Holland, 1982, Chapters 4, 7.

[GOU 71] GOULD S.H., *Variational Methods for Eigenvalue Problems,* Oxford University Press, Oxford, 1971, Chapter 2.

[GUY 86] GUYADER J.L., "Méthode d'étude des vibrations de milieux continus élastiques imparfaitement caractérisés. Application à la justification de l'hypothèse de plaque de Love-Kirchhoff", *J. Mec. Th. Appl.*, 6(2), p. 231-252, 1986, Chapter 2.

[JUL 71] JULLIEN Y., *Vibration des Systèmes Élastiques,* Naturalia et Biologia, 1971, Chapters 3-7.

[KEL 00] KELLY S.G., *Fundamental of Mechanical Vibrations,* 2nd edn., McGraw-Hill, New York, 2000, Chapters 5, 6, 9.

[KER 59] KERWIN E.M., "Damping of flexural waves by constrained viscoelastic layer", *J. Acoustical Soc. America,* 31(7), p. 952-962, 1959, Chapter 3.

[KIN 74] KING W.W. and CHIEN-CHANG LIN, "Application of Bolotin's method to vibrations of plates", *AIAA Journal,* p. 399-401, 1974, Chapter 7.

[LAN 77] LANCZOS C., *The Variational Principles of Mechanics,* University of Toronto Press, 1977, Chapters 2, 11, 12.

[LAN 67] LANDAU L. and LIFCHTZ E., *Théorie de l'élasticité,* MIR Editions, 1967, Chapter 2.

[LEI 93] LEISSA A., *Vibration of Plates,* ASA edition, 1993, Chapters 4, 7.

[LOV 64] LOVE A.E.H., *A Treatise of the Mathematical Theory of Elasticity,* Dover Publication, 1964 (4th edn.), Chapter 1.

[MAL 88] MALIKARJUNA and KANT T., "Dynamics of laminated composite plates with higher-order theory and finite element discretization", *J. Sound and Vib.,* 126(3), p. 463-475, 1988, Chapter 4.

[MAN 66] MANDEL J., *Cours de mécanique des milieux continus,* Gauthier-Villars, Paris, Vol. I and II, 1966, Chapter 1.

[MEA 69] MEAD D.J. and MARKUS S., "The forced vibration of three-layer damped sandwich beam with arbitrary boundary conditions", *J. Sound and Vib.,* 10(2), p. 163-175, 1969, Chapter 3.

[MEA 82] MEAD D.J., "A comparison of some equations for the flexural vibration of damped sandwich beams", *J. Sound and Vib.,* 83(3), p. 363-377, 1982, Chapter 3.

[MEI 01] MEIROVITCH L., *Fundamentals of Vibration,* McGraw-Hill, New York, 2001, Chapters 5-9.

[NAD 25] NADAI A., *Die Elastiche Platten,* Julius Springer (Berlin), p. 164-173, 1925, Chapter 7.

[NAS 85] NASHIF A.D., JONES D.I.G. and HENDERSON J.P., *Vibration Damping,* John Wiley and Sons, 1985, Chapter 8.

[REB 97] REBILLARD E. and GUYADER J.-L., "Vibrational behaviour of lattices of plates: basic behaviour and hypersensitivity phenomena", *J. Sound and Vib.*, 205(3), p. 337-354, 1997.

[REI 50] REISSNER E., "On a variational theorem in elasticity", *J. of Math. and Phys.,* Vol. 29, p. 90-95, 1950, Chapter 2.

[REI 75] REISSNER E., "On transverse bending of plates including the effect of transverse shear deformation", *Int. J. Solids and Structures,* 11(5), p. 569-573, 1975, Chapter 4.

[RIV 73] RIVAUD J., *Cours de mécanique de l'Ecole Polytechnique,* Hermann, Paris, 1973, Chapter 1.

[ROS 84] ROSEAU M., *Vibration des systèmes mécaniques*, Masson, Paris, 1984, Chapter 9.

[ROY 96] ROYER D. and DIEULESAINT E., *Ondes élastiques dans les solides,* Masson, Paris, Vol. I and II, 1996, Chapters 5, 6, 10.

[SAA 74] SAADA A.S., *Elasticity Theory and Applications,* Pergamon Press, 1974, Chapter 1.

[SAD 74] SADASIWA RAO Y.V.K. and NAKRA B.C., "Vibration of unsymmetrical sandwich beams and plates with viscoelastic cores", *J. Sound and Vib.,* 34(3), p. 309-326, 1974, Chapter 3.

[SAG 61] SAGAN H., *Boundary and Eigenvalue Problems in Mathematical Physics,* John Wiley, 1961, Chapter 2.

[SED 75] SEDOV L., *Mécanique des milieux continus,* MIR Editions, 1975, Chapters 1, 2.

[SNO 68] SNOWDON J.C., *Vibration and Shock in Damped Mechanical Systems,* John Wiley and Sons, 1968, Chapters 8, 9, 10.

[SOE 81] SOEDEL W., *Vibrations of Shells and Plates,* Marcel Dekker Inc, New York, 1981, Chapter 7.

[SOI 01] SOIZE C., *Dynamique des structures,* Ellipses, Paris, 2001, Chapters 5, 6, 9.

[SOL 68] SOLOMON L., *Elasticité linéaire,* Masson, Paris, 1968, Chapter 1.

[TRU 74] TRUESDELL C., *Introduction à la mécanique rationnelle des milieux continus,* Masson, Paris, 1974, Chapter 1.

[VAL 77] VALID R., *La mécanique des milieux continus et le calcul des structures,* Eyrolles, 1977, Chapter 1.

[WIL 45] WILLIAMS D., Dynamic loads in aeroplanes under given impulsive loads with particular reference to lauding and gust load on a large flying boat, RAE reports SMU 3309 and 3316, 1945, Chapter 9.

[ZIE 77] ZIENKIEWICZ O.C., *Finite Element Method,* McGraw-Hill, New York, 1977, Chapters 11, 12.

Index

E

F

G

H, I

L

M

N, O

P

Q, R

S

T, U, V

W